O9-AIF-516

# THE CIRCUIT RIDERS

# THE CIRCUIT RIDERS

## Rockefeller Money and the Rise of Modern Science

GERALD JONAS

W. W. NORTON & COMPANY
NEW YORK LONDON

Published simultaneously in Canada by Penguin Books Canada Ltd., 2801 John Street, Markham, Ontario L3R 1B4.
Printed in the United States of America.

The text of this book is composed in Baskerville,
with display type set in Corvinus Skyline.
Composition and manufacturing by The Maple-Vail Book Manufacturing Group.
Book design by Jo Anne Metsch.

First Edition

Library of Congress Cataloging-in-Publication Data

Jonas, Gerald, 1935–
The circuit riders : Rockefeller money and the rise of modern science / by Gerald Jonas.
p. cm.
Bibliography: p.
Includes index.
1. Endowment of research. I. Title.
Q180.55.G7J66 1989 88–14179
509—dc19

ISBN 0-393-02640-X

W. W. Norton & Company, Inc., 500 Fifth Avenue, New York, N.Y. 10110
W. W. Norton & Company Ltd., 37 Great Russell Street, London WC1B 3NU

1 2 3 4 5 6 7 8 9 0

*This is for Barbara*

# Contents

# Acknowledgments

To Henry Romney for encouragement and patience
beyond the call of duty.
To the staff members of the Rockefeller Foundation's
Public Information Office, current and former,
for their able assistance and counsel.
To Joseph Ernst, William Hess, Darwin Stapleton,
Thomas Rosenbaum, Melissa Smith, and the
entire staff at the Rockefeller Archive Center
for their gracious and efficient advice and assistance.
To Muriel Regan and Jane Allen for research.
To Ed Barber for patience, for honesty, for support.

The following members of the international scientific community gave generously of their time to talk to me about the role of the Rockefeller Foundation in the support of modern science:

Guy Gamus, Jean Coulomb, Antonio Brito de Cunha, Honor B. Fell, Mario Guimaraes Ferri, Newton Freire-Maia, Eloy Julius Garcia, Scott B. Halstead, Luiz Carlos Uchoa Junqueira, Warwick Kerr, Hans A. Krebs, Andre M. Lwoff, Szolem Mandelbrojt, Peter B. Medawar, Harry M. Miller, Manoel da Frota Moreira, Crodowaldo Pavan, Mauricio Matos Peixoto, Max F. Perutz, Gerard A. Pomerat, Isaias Raw, Oscar Sala, Francisco Mauro Salzano, Helena Lopes de Souza Santos, Persio de Souza Santos, Simon Schwartzman, Marina Silveira Santos, Piotr Slonimski, John Z. Young.

This book was written with the support and cooperation of the Rockefeller Foundation. However, all views expressed and accounts of facts and conclusions contained in this book are those of the author only and do not necessarily reflect the views of that Foundation, which is in no way responsible for the contents.

# FOUNDATIONS

# 1

# From Charity to Philanthropy

History is never kind to historians. Any history that aims at being more than a mere compilation of names, dates, and places tries to tell a story. The best stories have a beginning, a middle, and an end—or at least a sense of movement that carries the reader from one event to the next, with the promise that the entire narrative will add up to something more that the sum of its parts.

The conscientious historian has trouble making good on this promise. Behind every event looms a legion of antecedents. When examined closely, strict genealogies of cause and effect tend to unravel; lines of demarcation blur; what appeared at first glance to be a clear signpost to an unequivocal lesson sinks into swampy uncertainties of might-haves, maybes, and if-thens. After a time one cries out for firm ground, an unmistakable landmark, a place where one can stand, with some assurance of remaining upright, to survey the entire field.

Historians of modern philanthropy are luckier than most. With only a little fudging it can be asserted that modern philanthropy and the twentieth century began together. The key event was the creation of the large nonprofit foundation that treated its endowment as venture capital for the promotion of the common good. Although many of the early projects of these foundations were directed toward education—that ancient and honorable target of charity—the most distinctive gifts broke new ground by supporting scientific research on a large scale.

This was something new not only in the annals of philanthropy but also in the history of science. In fact, big philanthropy and big science came of age at the same time and in a complementary fashion. For both these developments the year 1900 was a watershed. It was in the years just prior to the turn of the century that the two most influential philanthropists of the era—Andrew Carnegie and John D. Rockefeller—began to dream of gift giving on a hitherto unprecedented scale. And it was in the first decade of the new century that these dreams came to fruition, with an impact that even their far-seeing founders could not have anticipated.

The "venture capital" of the Carnegie and Rockefeller benefactions came, of course, from the enormous personal fortunes that the two men had accumulated during the last three decades of the nineteenth century. This was the era before personal or corporate income taxes, before substantial inheritance taxes, an era when the "partnership" between government and business was even more openly rigged in favor of business than it is today. Behind all the truly great fortunes made during this period lay the extraordinary resources of a continent whose surface had barely been scratched (literally as well as figuratively) by previous generations.

Delving into the ground for iron ore, rock oil (petroleum), copper, and coal, laying down a network of railroad tracks to carry goods and passengers more efficiently (and more lucratively) than ever before, manipulating instruments of trade and trust that had been devised for much simpler commercial transactions, and, when necessary, inventing their own, the Carnegies, Rockefellers, and J. P. Morgans of the second half of the nineteenth century played the game of laissez-faire capitalism so well that they almost succeeded in bringing it to an end.

The goal of competition is to outdo your competitors. Rockefeller was so successful an industrialist that by 1880 his Standard Oil was refining 95 percent of all the oil drilled in the United States. J. P. Morgan was so successful a financier that during the Panic of 1893 the federal government solicited his aid in shoring up the national currency; a banking syndicate headed by Morgan helped raise the country's gold reserves above the $100 million mark—at an estimated profit to the syndicate of $10 million. As one historian put it, until the creation of the Federal Reserve System in 1913, Morgan, in effect, "was the central bank."

Perhaps Morgan's greatest achievement was the creation in 1900 of the United States Steel Corporation, a holding company that dominated America's premier industry from raw material to finished product. The largest corporate enterprise the world had ever seen, it accounted for half of the country's steel output, employed 168,000 men, and was capitalized at the then-unheard-of figure of $1.4 billion. The largest chunk in this gigantic trust-to-end-all-trusts was Andrew Carnegie's coke and steel combine, which the canny Scot sold to U.S. Steel for nearly half a billion dollars.

On September 6, 1901, President William McKinley was shot by an anarchist named Leon Czolgosz; he died eight days later and was succeeded by Theodore Roosevelt. McKinley had been proud to be known as a friend of big business. His election to a first term as president in 1896 had been masterminded by Mark Hanna, a wealthy Cleveland businessman and former school friend of John D. Rockefeller's. McKinley's election to a second term in 1900, amid general prosperity and the victorious completion of the Spanish-American War, symbolized the commanding position of the American businessman in the new century—not only within the continental boundaries but throughout the world. Standard Oil had already opened up new markets in Europe and Asia; by 1890 it was selling more oil overseas than in the United States.

The death of McKinley changed the symbolism. Roosevelt was known to sympathize with those forces in American political life that sought to disband the business cartels known as trusts, or at least to cut them down to size. The Sherman Antitrust Act of 1890 had barely slowed the trend toward monopoly. But Teddy Roosevelt's antitrust rhetoric, emanating from his "bully pulpit" in the White House, was a warning that the day of unabashed cartelization was coming to a close.

If the nineteenth century saw the development of financial instruments that permitted accumulations of capital far beyond anything dreamed of in Adam Smith's philosophy, the twentieth century witnessed a parallel development in philanthropy. New ways had to be found to give away really large sums of money. As Andrew Carnegie wrote, "It requires the exercise of not less ability than that which acquires it, to use wealth so as to be really beneficial to the community."

Certainly, there was no dearth of conventional charity in America. The Puritan ethic, both in its pure form in the Massachusetts Bay colony and in the much softened Quaker version seen in Pennsylvania, stressed the Christian doctrine of "stewardship of wealth." God may reward the righteous with riches on Earth, but woe unto the rich man who does not treat his wealth as a kind of loan, to be used to succor and uplift the less fortunate members of God's flock. "It shall be easier for a camel to pass through the eye of a needle than for a rich man to enter into heaven" is the relevant biblical text.

The cynical may view charity under such auspices as just one more shrewd investment on the part of the rich—the financial equivalent of Pascal's wager. Whatever the motive, the well-to-do in America, both before and after the Revolution, generously supported local charitable organizations. Writing in the *North American Review* in 1833, Samuel Gridley Howe noted, "Every infirmity, every misfortune, every vice even, has a phalanx of philanthropists to oppose its effects."

But Andrew Carnegie had not been talking about this kind of "ameliorative" or "palliative" charity. He repeatedly quoted an estimate that nineteen out of twenty dollars given to such benevolent projects are wasted: "Neither the individual nor the race is improved by almsgiving." What he had in mind was something much more ambitious than charity wards for the ill or soup kitchens for the indigent.

In the June 1889 issue of the *North American Review,* Carnegie—a voracious reader and self-taught social philosopher—set forth what came to be called his "gospel of wealth." Like many of his contemporaries, rich and poor, he viewed the concentration of wealth in the hands of the few as inevitable. A disciple of the philosopher Herbert Spencer, he believed that there was no arguing with success; the fittest always rose to the top of the heap, because of the inexorable operation of what Carnegie, following Spencer, called the "laws upon which civilization is founded." This comforting pastiche of Darwin's theory of evolution—known as Social Darwinism—was perhaps the most influential popular philosophy of the day. Many in positions of power and affluence used Spencer's Social Darwinism to justify a hands-off attitude toward all social problems; it was argued that any attempt to tinker with the "laws" of nature that favored some men over others could only produce disaster.

But Carnegie combined his Social Darwinism with the Puritan-Quaker doctrine of "stewardship," which enjoined a rich man to treat his fortune as a sacred trust. After reserving sufficient funds for a "modest, unostentatious" way of life, and making "moderate" provisions for his dependents, the man of wealth should earmark "all surplus revenues" for good works. Furthermore, he should begin giving money away during his own lifetime. Purely posthumous philanthropy showed both a grasping nature and poor management. In Carnegie's words, "The man who dies thus rich dies disgraced."

To help his fellow millionaires avoid disgrace, Carnegie drew up a list of causes to which a man of means might comfortably disburse his "surplus." The list began with universities and libraries, went on to "hospitals, medical colleges, laboratories and other institutions connected with the alleviation of human suffering, and especially with the prevention rather than the cure of human ills," and concluded with institutions and activities that bring "sweetness and light" into men's lives: parks, gardens, music halls, "swimming baths," and churches.

In carrying out this program, Carnegie had no qualms about exploiting those benefits of scale that gave the big businessman such an advantage over his lesser competitors. The small businessman was ever at the mercy of periodic fluctuations in supply and demand, not to mention the more vertiginous swings in the economy that the nineteenth century knew as booms and panics. By contrast, the big businessman could control his sources of supply; he could eliminate competitors by buying them out or driving them to the wall with price wars and other pressure tactics; he could ensure favorable treatment from the courts and legislatures through bribes; and he could not only ride out recessions and depressions but actually profit from them by picking up bargains in temporarily undervalued goods and stocks.

Like the small businessman, the person who gives a few dollars—or even a few hundred or a few thousand dollars—to charitable organizations like churches, hospitals, and schools is pretty much a prisoner of the status quo. If he does not like a minister's sermons, or a school's admissions policy, or a hospital's accounting practices, he can, of course, reduce or withdraw his contributions, but that is all he can do. His voice, unless amplified by some occupational, cultural, or communal distinction, will not carry much weight with those who set the policies of charitable organizations. But the big philanthropist—the man who gives hundreds of thousands, or millions of dollars—can set conditions that recipients must meet before they qualify for a gift; and if he finds no suitable institutions already in existence, he can create a new one in his own image.

As an avowed Social Darwinist, Carnegie felt fully justified in promoting his private vision of the public good. Wasn't the self-made rich man—confident in his "superior wisdom, experience, and ability to administer"—likely to do better for the poor "than they would or could do for themselves"?

Carnegie's gifts to his hometown of Dunfermline, Scotland, began with the construction of public baths in 1873. Eventually, he provided a $4 million trust fund that paid for everything from communal flower beds to regular medical examinations for the local children. The result was a kind of private welfare state that prompted some residents to complain that their "Andy" was even trying to tell them how to eat.

There had been few public libraries when Andy Carnegie was young, but in Allegheny, Pennsylvania, he had enjoyed access to a 400-volume private library that a local benefactor opened to the working boys of the town on Saturday afternoons. In 1881 Carnegie offered to build and endow free public libraries, which he referred to as "the people's universities," in Dunfermline and Pittsburgh. Pittsburgh at first rejected Carnegie's offer of $250,000 for a library building because he had made the gift contingent on a municipal pledge to spend $15,000 a year on upkeep. When the city fathers repented a few years later and came to Carnegie hat in hand, he not only forgave their orginal ingratitude but upped the gift to $1 million to provide an art museum and music hall as well. Carnegie went on to offer a similar deal to communities throughout the country and the world.

In both its scale and the intensity of his personal involvement, the library-construction program was a true Carnegie benefaction. When the Carnegie Corporation terminated the program, in 1919, funds totaling $60 million had been used to erect some two thousand libraries in the United States and another eight hundred overseas.

By the time of his death, Carnegie had given away more than $350 million—and one way or another he left his personal stamp on all the philanthropic institutions he created. The one-time penniless "bobbin boy" was especially proud of the Carnegie Hero Funds, which he set up (with gifts of $10 million) in the United States, England, and several European countries. These endowments were designed to honor "heroes of peace" the way governments have traditionally honored wartime heroes. Medals and cash awards were given to people who saved others from fires, drownings, and other natural and manmade disasters—and if the hero lost his life while performing the deed, the Carnegie funds came to the financial rescue of his widow and heirs. Carnegie claimed that this was the only one of his benefactions that

*John D. Rockefeller, Sr., ca. 1880s*

was not suggested to him by someone else; he called it fondly his "ain bairn."

One of the most enthusiastic admirers of Carnegie's "gospel of wealth" was John D. Rockefeller, whose hold on the nation's oil refineries was virtually complete by 1889. In a fan letter to Carnegie, he wrote, "I would that more men of wealth were doing as you are doing with your money, but, be assured, your example will bear fruits, and the time will come when men of wealth will more generally be willing to use it for the good of others."

The grandeur of Carnegie's vision obviously excited Rockefeller's imagination: the benefits of civilization were to be purchased for the masses wholesale by the privileged few. But Rockefeller did not need to be reminded of the Christian obligation to give alms. Ever since childhood he had been tithing—reserving 10 percent of his income for charity, as the Old Testament enjoins in no fewer than three places.

Ledger A, a daily accounts book that he kept after getting his first job, at the age of sixteen, shows gifts of five and ten cents for the poor and for various charities of the Baptist church. As his income increased, so did his giving. For many years the principal beneficiaries of Rockefeller's charity were Baptist and other Christian groups engaged in education, missionary work, and care of the poor and sick. Indeed, Rockefeller never abandoned these causes in his personal giving. By 1889, however, he had taken his first big step toward the role of public benefactor that Carnegie so vigorously espoused.

# 2

# Scientific Giving

Toward the end of the 1880s, Baptist leaders began talking to Rockefeller about their plans to build a major educational institution under denominational auspices. At first he dragged his feet; the project seemed too large, support from the Baptist community was by no means assured, and the leaders themselves were divided on crucial matters of policy, including the question of where to put the new institution. In 1889, however, Rockefeller subscribed $600,000 to a campaign to help launch the new University of Chicago (or, to be more precise, to raise from the dead a moribund Baptist institution by that name).

The scope of the project kept growing, and Rockefeller kept giving, long after the original idea of a small sectarian school had given way to a vision of a midwestern citadel of higher education to rival the great universities of the eastern seaboard. By the end of his life, his gifts to the University of Chicago had reached $35 million; and his total benefactions exceeded Andrew Carnegie's by a hundred million dollars.*

"The best philanthropy," Rockefeller once said, "is to give the

*Rockefeller's gifts to the University of Chicago came to a grand total of $34,702,375.28. The Rockefeller Foundation, the Rockefeller Institute, and the General Education Board received $466,719,371.22—calculated in terms of the market value of stocks and bonds conveyed on the days the gifts were made.

opportunity for progress and healthful jobs where it did not exist before." The fact that he was talking about opportunities created not by one of his philanthropic "boards" but by the Standard Oil Corporation is very much to the point. Like Carnegie, Rockefeller saw no conflict between his philosophy of getting and his philosophy of giving; he felt that mere charity, which he defined as money obtained "without effort on the recipient's part," only reinforced the weakness of character that led to penury in the first place. But unlike Carnegie, Rockefeller wrote no tracts, endowed no quirky do-good schemes of his own devising. Rockefeller's philosophy of giving was more bound up with principles of management than with moral exhortation or theories of social organization.

From his earliest days in business, he recognized that his genius lay in two areas: he had a bookkeeper's eye for the little details that other men either missed or felt it beneath their dignity to notice, and he had a knack for picking brilliant associates and ensuring their loyalty by offering them broad responsibilities with commensurate financial rewards.

Throughout his career he put his passion for minutiae to work in ferreting out all forms of waste. It is part of the Standard Oil legend that he once observed a workman sealing a five-gallon tin of kerosene with forty drops of solder. Why forty drops? he asked. When the workman replied that it had always been done that way, Rockefeller had tests conducted; these proved that thirty-nine drops were sufficient. From then on, all five-gallon tins of Standard Oil were sealed with thirty-nine drops of solder. The savings, Rockefeller later claimed, added up to "a fortune."

As a philanthropist Rockefeller was similarly concerned with getting top value for his dollar: "I am so constituted as to be unable to give away money with any satisfaction until I have made the most careful inquiry as to the worthiness of the cause." The quote comes from a conversation he had in March 1891 with Frederick Taylor Gates, an outspoken Baptist minister who had come to his attention during the University of Chicago campaign. It was Gates's carefully documented presentation that finally won Rockefeller over to that project. And in 1891, when Rockefeller decided he no longer had time to judge the worthiness of all the charitable causes proposed to him, he asked Gates to come to New York to head what he called his "department of benevolence."

In hiring Gates, Rockefeller once again demonstrated a tolerance for strong-willed subordinates that was all but unique among the strong-willed entrepreneurs of his day. His chief lieutenants at Standard Oil formed a "brains trust" whose reputation for acumen (and ruthlessness) was based on such public testimony as this comment by William Vanderbilt, son of the rapacious "Commodore" and a formidable financial power on his own: "I never came in contact with any class of men as smart and able as they are in their business."

Frederick Gates turned out to be quite a businessman in his own right. At his employer's request he was soon appraising investments and managing some of Rockefeller's vast business interests; in these dealings he demonstrated a nose for profits that might have earned him a name as one of the era's premier entrepreneurs if he had devoted himself to it full-time. But Gates's true calling lay elsewhere: he was determined to bring order to Rockefeller's personal benefactions through a concept he called "scientific giving."

The phrase was a felicitous one. What he meant by it, originally, was that charity should be directed as much by the head as by the heart. On his arrival in New York, Gates had been astonished to discover that Rockefeller was "constantly hunted, stalked, and hounded almost like a wild animal" by alms seekers of all kinds. "Neither in the privacy of his home nor at his table, nor in the aisles of his church, nor on his trips to and from his office, nor during his business hours, nor anywhere else . . . was Mr. Rockefeller secure from insistent appeal." Gates found it particularly annoying that many of these supplicants were Baptist ministers and missionaries.

Gates set out to do what Rockefeller, for all his protestations, had never succeeded in doing: determine by careful investigation which of the many competing claims were likely to produce more benevolence per dollar. Inevitably, he eliminated the smaller benefactions altogether, referring individual alms seekers to denominational agencies that Rockefeller was already funding. "In no long time," Gates later wrote in his autobiography, Rockefeller found himself "laying aside retail giving almost wholly, and entering safely and pleasurably into the field of wholesale philanthropy."

Once Rockefeller's benefactions were being conducted "wholesale," the phrase "scientific giving" took on a new meaning. In looking for worthy activities that could absorb large sums of money and give a good philanthropic "return," Gates found himself drawn more and

*Frederick T. Gates*

more to the science—especially the medical science—of the day. The Rockefeller Institute for Medical Research was his idea. He described it as "a sort of Theological Seminary," adding, "In these sacred rooms He is whispering His secrets."

By this time (1911) Gates was speaking not only as a former Baptist minister but also as a former Baptist. What was called the higher criticism—the close scholarly analysis of biblical texts that flourished in the nineteenth century, especially at German universities—had convinced him "that Christ had neither founded nor intended to found the Baptist Church, nor any church."

On coming east to work for Rockefeller, Gates had settled his family in Montclair, New Jersey. After becoming dissatisfied with the Baptist church of Montclair, he shifted his allegiance to the local Congregational church, which demanded no doctrinal commitment except a desire to "seek to exercise the Spirit of Jesus." This must have seemed a pale creed to a man who once characterized himself as having an "eager, ardent, nervous temperament," and whom a colleague remembered this way: "He was an intense man, deepset eyes close enough together to suggest the zealot. . . . His wisdom was of the fiery, mordant kind, not mellow and tolerant." From his own writings and from his actions as John D. Rockefeller's lieutenant in charge of benevolence, it is clear that his evangelical fervor had been redirected not toward Congregationalism but to the glories of modern science.

Rarely has the aspect of science as an intellectual faith been so clearly stated as in Gates's quick gloss of his conversion: "I came fully to accept the methods and the results of modern Biblical criticism and also the spirit and the results of modern scientific research." And at the tenth anniversary of the founding of the Rockefeller Institute, he went even further in associating science with the former object of his devotion:

> Has it not often occurred to us that, after all, science is about the only thing that is destined to live forever in this world? . . . For what is human progress? Ultimately, it is this, just this, and nothing else—an ever closer approach to the facts, the laws, the forces of Nature, considered of course in its largest meaning. Nothing else is progress and nothing else will prove to be permanent among men.

He went on to say, "Theology is already being reconstructed in the light of science, and the reconstruction is one of the most important of the services which scientific research is performing for humanity."

This was the vision that lay behind Frederick Gates's notion of "scientific giving." Among those who fell under its spell was John D. Rockefeller's son and heir, John D., Jr., who graduated Phi Beta Kappa from Brown University in 1897 and came to New York to work in his father's office. There, according to Gates, he received "a post-graduate course in business and benevolence." The young Rockefeller found benevolence more to his liking (a preference his father approved of) and was soon deeply involved in the vast philanthropic empire that Gates was building.

As for John D., Sr., it seems clear that he was attracted more by the businesslike efficiency of "scientific giving" than by Gates's attempt to substitute scientific truth for the failed revelations of nineteenth-century American Protestantism. Not that Rockefeller was unaware of the value of scientific research in business. The petroleum industy—and Standard Oil in particular—had made good use of advanced chemical know-how from the earliest Pennsylvania boom days. But the particular object of Gates's veneration—modern medical science—held no personal appeal for his employer.

*John D. Rockefeller, Jr., in college*

Long after he had endowed the Rockefeller Institute for Medical Research and given millions of dollars to raise the standards of medical schools throughout the world, Rockefeller retained as his personal physician a man devoted to the practice of "homeopathy"—a discipline based on ideas about the cause and cure of illness that were, according to the scandalized Mr. Gates, at least fifty years behind the times.

In living to the age of ninety-seven, John D., Sr., may have earned the right to a few mild I-told-you-sos. But he insisted only on his freedom of action. When his family forced him to receive a more conventional medical doctor, this is how he described their meeting: "The doctor came to see me today. He wouldn't give me the medicine I wanted, and I wouldn't take the medicine he presented, but we had a lovely talk."

Inevitably, a speculation arises: If Rockefeller had been as enamored of his own ideas as Andrew Carnegie was of his, would there have been a multimillion-dollar Rockefeller Institute for Homeopathic Research instead of (or in addition to) the medical institute he allowed Gates to create? The answer is almost surely no.

Throughout his life Rockefeller continued to give a few thousand dollars a year to his favorite personal charities. But in matters of big philanthropy, as in matters of big business, he relied on the best brains he could buy. And in hiring Frederick Gates, he was buying more than sound judgment and management skills; he was purchasing a package of up-to-date ideas about society and the future, a ready-made philosophy of philanthropy oriented toward modern science.

Whether or not he approved of the entire package, he clearly felt that he had made a good deal. The elder Rockefeller retired from active participation in Standard Oil in 1897. If he had had the slightest urge to play a Carnegie-like role of philosopher-philanthropist, this was surely the time. His Standard Oil holdings were worth some $200 million. In the first years of the new century, with the help of Gates and his son, he endowed a series of giant philanthropic trusts—and then removed himself from their direction. Typically, although he lived for forty years after the creation of the Rockefeller Foundation, he never attended a Trustees' meeting. Nor is there any evidence that he tried to influence Foundation policy through public or private pronouncements on major decisions.

That Rockefeller could act decisively to safeguard the *financial* soundness of his benefactions was demonstrated during a budgetary crisis at the University of Chicago in 1903. A long series of annual deficits had caused Rockefeller to lose confidence in the administrative acumen of the man he himself had persuaded fourteen years earlier to accept the presidency, Dr. William R. Harper. When Harper, a former professor of Hebrew at Yale, showed no inclination or ability to curtail expenditures, he was brought into line at a trustees' meeting held in Rockefeller's private office in New York. There was nothing subtle about the pressure that Rockefeller brought to bear. Through his son he let it be known that he would make no more gifts for endowment until the university put its finances in order. Within a few years Harper's successor had balanced the budget, and Rockefeller money started flowing again.

This intervention is the rule-proving exception to Rockefeller's hands-off attitude toward the "benevolent trusts" he funded. His rejection of the philosopher-philanthropist role undoubtedly had something to do with the times. Carnegie's cocky moralizing belonged to the heyday of the Gilded Age. In the decade following the promulgation of his gospel of wealth, the methods and motives of big businessmen had

*John D. Rockefeller, Sr., 1905*

come under ever-closer scrutiny by muckraking journalists, the courts, and congressional investigators—and the picture that emerged did little to justify any Spencerian claims to superior character or moral insight.

The resulting shift in public opinion was unmistakable. When Rockefeller donated $100,000 to the Congregational Board of Foreign Missions in 1905, many ministers publicly urged the board to reject this "tainted money"—a moral stance that proved somewhat embarrassing when it turned out that the Congregationalists had solicited the donation in the first place.

Between 1910 and 1913 Congress repeatedly refused to charter Rockefeller's proposed $50 million foundation. (The Carnegie Institution of Washington had received a federal charter in 1902, but only after a public outcry had forced President Theodore Roosevelt to turn down Carnegie's offer to finance a semipublic agency along the lines of the Smithsonian Institution.) The Rockefeller Foundation finally settled for a charter from New York State, as did Carnegie's last and largest philanthropy, the Carnegie Corporation of New York.

Even if circumstances had been more favorable, however, it is hard to imagine Rockefeller trying to pass himself off as a sage simply because he had amassed the largest personal fortune in history.* John D., Sr., had remarkable powers of concentration which enabled him to pursue goals with a single-minded tenacity that panicked his competitors and awed even his most hardened colleagues. But intellectual pretension was as alien to his makeup as irresolution.

In 1909 he published a small book with the modest title *Random Reminiscences of Men and Events*. This was the only extended piece of writing ever to appear under his name, and the contrast to Carnegie's effusive opinion mongering could not be greater. No doubt there was an element of calculation in this. Two years earlier Standard Oil had hired a former journalist as its first publicity agent, and it is likely that John, Sr., made use of his services. But the tone of the book, which seems to spring from an unassailably clear conscience, could only have been Rockefeller's.

In an era when party favors at a fancy ball might include cigarettes wrapped in hundred-dollar bills, who but Rockefeller would have dared to pen the line "The very rich are just like all the rest of us"? He maintained that his own fortune had come to him partly through luck and partly through his steadfast adherence to something he called "the sound natural laws of business" and "the established laws of high-class dealing." He never elaborated on what these laws were. The only specific advice that a young entrepeneur-on-the-make might have gleaned from Rockefeller's reminiscences was this admonition: "Be sure you are not deceiving yourself at any time about actual conditions."

The evidence shows that John D., Sr., never did deceive himself, about either his abilities or his limitations. Indeed, his clear-minded distinction between his personal beliefs and the policies of the charitable institutions he established may have been among his most important legacies to modern philanthropy. It was Rockefeller's essentially impersonal gospel of wealth, rather than Carnegie's self-aggrandizing model, that became the norm in the new century.

*The government won its long antitrust suit against Standard Oil in 1911, when the Supreme Court ordered the combine broken into its constituent companies. Ironically, the decision set off a stock-market boom, adding some $56 million in value to Rockefeller's personal holdings.

# 3

# Prudent Gamblers

Nothing is easier than giving away money. Nothing is harder than giving it away well. Tossing dollar bills into Times Square on New Year's Eve is not philanthropy. Nor is giving cake to people who are crying for bread. What makes philanthropy so difficult is the inherent conflict between doing something well, which implies a measure of control, and giving something away, which requires an attitude of detachment. How the philanthropist resolves this conflict determines his ultimate success, in his own eyes and in the eyes of society.

True innovations in philanthropy are rare, since the major problems of mankind—hunger, disease, ignorance, war—change hardly at all through the ages. Only when new methods for dealing with old problems come along do opportunities for innovative philanthropy arise, and few individuals will have the resources or courage to seize such opportunities. With enormous sums at their disposal, with competitive temperaments emboldened by their success in laissez-faire capitalism, Carnegie and Rockefeller seemed ideally equipped to gamble on new and untested approaches to curing social ills. Yet the very size of their philanthropic bankrolls argued for caution.

The larger the gift, the more likely that the giver will be held accountable for its impact on the community—whether or not the giver retains any real control over its use. The man who buys a dollar raffle ticket from a nonexistent "charity" may feel cheated, but no one else has to know about his embarrassment, and no one else is likely to suffer from his lapse of judgment. But the man who sets out to do good on a grand scale runs the risk of looking foolish on a grand scale, especially if he seeks to be innovative in his philanthropy. If the recipients of his benefactions turn out to be charlatans or fools, the philanthropist who hopes to "buy" public acclaim while shaping society in his own image may end up being pilloried as a dupe, an incompetent, or worse.

This being so, the giver of large gifts is obliged to take some precautions to prevent their misappropriation. No one would expect a pacifist to sit by idly while his charitable gifts are diverted to subsidize gunrunning. Even gifts that are simply squandered through mismanagement represent significant lost opportunities. But while the prudent philanthropist tries to ensure that his benefactions are not used in ways that violate his deepest beliefs, he must resist the temptation to equate his own tastes and principles with those of society at large.

Getting the balance right is no easy task. Too much detachment can lead to accusations of coldness and inhumanity. Long before Frederick Gates invoked the formula of "scientific giving," Charles Dickens sardonically defined organized charity as a "device for channeling our sympathies, making them impersonal and unavailable in human terms."

Too much control can be equally damaging. Anticipating Andrew Carnegie's gospel of wealth by two decades, Mark Twain paid his respects to the sanctimonious almsgiver in these biting words: "I wish to become rich so that I can instruct the people and glorify honest poverty a little, like those kind-hearted, fat benevolent people do."

One way for a philanthropist to minimize personal risk is to restrict his or her benefactions to areas where the social dividends are so obvious and the existing institutions so strong that the consequences of even the largest gifts are virtually guaranteed to win community approval. The cautious giver will not venture outside the traditional "safe" areas of education, established religion, and medical care. As we have seen, Rockefeller supported Baptist and other charities throughout his life, Andrew Carnegie's shopping list for newly rich philanthropists began with universities and libraries, and educational institutions were the major beneficiaries of Carnegie's and Rockefeller's earliest ventures in big philanthropy.

In placing aid to higher education first on his list, Carnegie solemnly noted, "It is reserved for very few to found universities." This was undoubtedly true even in an expanding laissez-faire economy whose most characteristic product seemed to be millionaires (more than four thousand go-getters had achieved this status by 1900). Yet Carnegie was able to call a roll of nineteenth-century university founders that included Peter Cooper (of New York's Cooper Union), Charles Pratt (of Brooklyn's Pratt Institute), Ezra Cornell, and Leland Stanford. And he did not even mention Johns Hopkins, the Baltimore whiskey merchant whose $7 million bequest made it possible in 1876 for yet another eponymous institution of learning to open its doors. (If nothing else, Rockefeller's contributions to the University of Chicago broke new ground in that the institution that acknowledged him as founder was *not* named after him.)

Between them Carnegie and Rockefeller gave away some $850 million during their lifetimes. This scale of giving dwarfed anything that had come before. If Rockefeller had so chosen, he could have founded ten universities the size of Chicago and still had money left over for other major projects.* Obviously, it would have been difficult to find suitable sites—not to mention qualified teachers and students—for ten new major universities in early-twentieth-century America.

One alternative was to bolster existing institutions, an honorable tradition in philanthropy stretching back through Sir Thomas Bodley, a collector of medieval manuscripts who, in the early 1600s, restored and expanded the library of Oxford University that had been endowed by earlier benefactors in the fourteenth century (Thomas de Cogham) and the fifteenth century (Humphry, duke of Gloucester). Characteristically, Carnegie found a way to put his personal stamp on this venerable form of giving; instead of aiding major universities, he gave some $15 million to what he referred to as the small "freshwater" colleges of the American hinterlands. Meanwhile, Rockefeller's General Education Board (founded in 1903) took as its mandate the improvement of education throughout the American South.

The programs that Carnegie and Rockefeller set in motion ranged from a broad campaign to raise standards in American medical schools to a generous pension plan for American college teachers. But such initiatives, vast as they were by the standards of previous generations, simply could not absorb the enormous sums of money that Carnegie and Rockefeller had to give away. Something else was needed, a major new target for philanthropy that would be just as acceptable to public

*The undergraduates who greeted him in 1896 with the chant "John D. Rockefeller, wonderful man is he/Gives all his spare change to the U. of C." were not far off the mark.

opinion as a traditional hospital or university and, if possible, even more costly.

Neither Carnegie nor Rockefeller had any reserves of public sympathy to draw upon. Whatever they proposed to do was bound to draw the fire of critics who argued that "malefactors of great wealth" (Teddy Roosevelt's term) should not be allowed to enjoy their ill-gotten money, even to the extent of presiding over its redistribution. For any "malefactor" who determined to practice philanthropy on a grand scale, the challenge was to strike out in a new direction without *seeming* to depart too radically from the more traditional forms of generosity. The goal, in short, was innovation within the safest possible context.

As it happened, this was the policy that both Carnegie and Rockefeller had followed in amassing their fortunes. Although neither was afraid of breaking new ground when absolutely necessary, they much preferred to buy innovations that others had already shown to be profitable. As Carnegie liked to put it, "Pioneering don't pay!" Yet once convinced that there were no alternatives, each was ready to gamble for high stakes rather than pass up a lucrative opportunity.

When the new technology of the Bessemer-Kelly steel-making process was introduced in England, Carnegie was slow to see its value; yet once he became convinced of its superiority, on a trip abroad in 1874, he rushed back to America to start building a new steel plant in the middle of the worst economic slump of the century. Similarly, Rockefeller initially resisted involvement in the oil boom that followed Edwin Drake's first gusher at Titusville, Pennsylvania, in 1859. But four years later he and two partners organized a new firm "to refine and deal in oil." He invested $4,000 of his own savings in what he later characterized as a "most hazardous undertaking." If profits were high, so were development costs; and there was the ever-present risk of losing everything in a fire or a string of dry wells. Significantly, Rockefeller was soon pushing his partners to expand the business on borrowed money. When one partner balked, Rockefeller and a third partner bought him out, for $72,500. The year was 1865. Five years later the Standard Oil Company was capitalized at $1 million; by 1883 it was worth more than seventy times that.

Yet the future was by no means assured. Already, the Pennsylvania fields that had fueled the boom were giving out. A little farther west, near Lima, Ohio, even larger deposits of petroleum had been discovered. But there was a problem with the Lima oil; it was so laden with sulfur that it smelled like rotten eggs. Advised that such "sour" or "skunk" oil could never be satisfactorily refined, most Standard executives wanted no part of it. But Rockefeller kept insisting that a way would be found to "sweeten" the oil.

When his optimistic forecasts fell on deaf ears, he offered to put up $3 million of his own money to buy the Lima fields. If the venture proved profitable, he told his colleagues, the company could reimburse him. "And if it goes wrong," he said, "I'll stand the loss." There is no way of knowing whether this was a serious offer or a clever bluff, because his fellow directors decided to purchase the Lima fields for the company after all. As Rockefeller remembered it, one of them said to him after the vote, "I guess I can take the risk if you can." The gamble paid off handsomely. After two years of intensive—and expensive—research, Standard chemists perfected the Frasch process for desulfurizing oil. By 1888 petroleum from the Lima fields, which had been selling for fifteen cents a barrel, was worth a dollar a barrel.

As they turned their attention to philanthropy, both Carnegie and Rockefeller showed a similar blend of caution and risk taking. Only gradually did they move away from the relatively noncontroversial areas of education and cultural uplift. It was not until both men had retired from active participation in business—Rockefeller in 1897, Carnegie in 1901—that the real innovations in giving away large sums of money began.

In 1901 the Rockefeller Institute was founded with a start-up pledge of $200,000. The next year the Carnegie Institution of Washington began operation with $10 million in the till (equal to the entire endowment of Harvard University). That same year Rockefeller pledged another $1 million to the Institute in New York and an additional $1 million to the nascent General Education Board. In 1909 the Rockefeller Sanitary Commission (for the eradication of hookworm) started its work with $1 million. The next year saw the founding of the Carnegie Endowment for International Peace with a grant of $10 million. By this time John D., Sr., had given nearly $6.5 million to the Rockefeller Institute and more than $32 million to the General Education Board; and Carnegie's aid to the Institution in Washington had topped $30 million.

There were some observers who portrayed the rush by Carnegie and Rockefeller to give away money as a personal competition, a kind of gargantuan potlatch on an unprecedented scale. In 1910 one newspaper even published a box score comparing the benefactions and showing Carnegie in the lead, $179,000,000 to $134,271,000. And this was before the *really* big giveaways: $125 million to set up the Carnegie Corporation of New York in 1911, and $182 million to the Rockefeller Foundation between 1913 and 1919.

Except for the size of their fortunes, Carnegie and Rockefeller were, as we have seen, opposites in almost every way. What is most remarkable about their philanthropic "competition" is that their search for suitable recipients for large sums of money led them both further and further from the more traditional channels of charity and deeper and deeper into the unfathomed waters of modern science—which Frederick Gates had divined from afar and prophetically pronounced to be "good."

# 4

# Putting the Science into Scientific Giving

The founding of the Rockefeller Institute for Medical Science and the Carnegie Institution of Washington signified a new relationship between philanthropy and science in America. The very existence of these institutions suggested that modern science was fundamentally different from other scholarly pursuits (like history or philosophy or the classics) and as such deserved special attention from would-be patrons.

In 1901 President Charles William Eliot, of Harvard, had declared, "The endowment of research is becoming an attractive object for private benevolence." Eliot, however, was referring to research in a university context, where science could be treated as one scholarly pursuit among many. If he was right, the time-honored mechanisms for endowing and administering universities could simply be adopted to the support of science. This conservative strategy had obvious appeal to those givers of large gifts who had reason to fear the public's reaction to their benefactions.

The genesis of the Rockefeller Institute began with a memo that Frederick Gates wrote to John D. Rockefeller in the summer of 1897. In it he urged his employer to establish "an institution for scientific medical research" similar to the renowned institutes that had been founded in Paris (1888) and Berlin (1891) under the auspices of the two most famous scientific researchers of the day, Louis Pasteur and Robert Koch.

Although the European models were not affiliated with universities, Gates at first assumed that the Rockefeller institute would make its home at the University of Chicago. The university's decision, in 1898, to associate itself instead with Rush Medical College forced a change in plans. When initial discussions with Harvard and Columbia proved disappointing, and Rockefeller's advisers recommended the creation of a fully independent institute, John D., Sr., responded favorably, perhaps because the new entity would have no history of opposition to the homeopathic remedies that he and his personal physician believed in.

The first order of business for the new Rockefeller Institute was to build a laboratory. But the seven prominent physicians who had been named to the Board of Directors were eager to make a mark in the world of research as quickly as possible. While the Institute's own facilities were under construction, they decided to award small sums, ranging from $250 to $1,500, to promising young investigators around the country. During the Institute's first fiscal year (ending June 30, 1902) these "grants-in-aid" totaled $13,200.

It is fair to say that the grant-in-aid program was never anything but an awkward stopgap. Frederick Gates was vehemently opposed to what he called this "utterly futile . . . system of scattered subventions." Abraham Flexner, a powerful figure in the Rockefeller philanthropies for over two decades, believed that handing out small grants-in-aid was inherently demeaning: "Their main effect is to keep the recipients on their knees, holding out their hats from year to year." Administration of the grants was left to William H. Welch, dean of the Johns Hopkins University Medical School and president of the Institute's Board of Directors, whose famous "filing system" consisted of piles of correspondence balanced on chairs throughout his bachelor apartment in Baltimore.

In April 1906 the Institute moved into new quarters on the East Side of Manhattan, and the grants-in-aid program was severely curtailed. A decade later it was abolished.

Once the brief fling with grants-in-aid was over, the Institute devoted itself to "in-house" research exclusively. In organization it came to resemble a university without undergraduates. There were university-like departments, divided along more or less traditional disciplinary lines and headed by eminent researchers under contract. After much discussion of who should control this curious "university," bylaws were drawn up in 1910 giving oversight of scientific matters to a board of scientific directors and creating a separate board of trustees to look after financial matters.

But final approval of the budget was vested in an interboard committee made up of two trustees and three scientific directors. This amounted to a vote of confidence in the ability of scientists to manage their own affairs, a significant concession by a philanthropist who only a few years earlier had acted to curb what he saw as the profligacy of academics at the University of Chicago. In the words of Simon Flexner, who headed the Institute in its formative years, "Rockefeller and his advisers agreed to a delegation of power such as may never before have existed in an American philanthropic institution. The trustees agreed merely to take care of the funds, leaving the decision of how they were to be spent to a joint committee containing a *majority* of scientists" (emphasis added).

Nevertheless, the administrative apparatus of the Rockefeller Institute (on whose boards and committees sat many of the closest advisers of John D., Sr., including his son and Frederick Gates) must have seemed reassuringly familiar compared with a helter-skelter grants-in-aid program.

Significantly, the Carnegie Institution conducted an almost identical experiment with grants-in-aid, with the same outcome. By its charter the Institution was committed "to encourage, in the broadest and most liberal manner, investigation, research, and discovery, and the application of knowledge to the improvement of mankind."* Many scientists urged the trustees to support individual investigators directly; and in its first few years a large portion of the Institution's budget was earmarked for fellowships and grants-in-aid to scholars in fields ranging from physics and mathematics to history and literature.

But this was no more than a temporary expedient; the trustees had already decided on a very different policy, emphasizing research "under strong effective direction from a central office." By 1912 the Carnegie Institution had settled on a university-like departmental structure, with its own laboratories staffed entirely by its own salaried investigators.

It is not just coincidence that the first ventures in science patronage launched by Carnegie and Rockefeller experimented with, and then

*The only restriction that the Carnegie Institution admitted came as a result of a meeting between Andrew Carnegie and John D. Rockefeller, Jr., in the spring of 1902, when Rockefeller sought and received Carnegie's personal assurance that his Institution would leave the field of medical research to the Rockefeller Institute.

drew back from, the support of outside investigators. There were few precedents for a grants-in-aid program built around awards to independent researchers. It was not at all clear how a science patron could choose rationally among competing requests for aid or how he could ensure that his funds were not put to purposes seriously at odds with his intentions.

The real pioneer in this area was the British government. In 1849, after more than a decade of agitation by the British Association for the Advancement of Science (see chapters 5–6), Parliament began appropriating £1,000 a year to be used by the Royal Society for the promotion of scientific research. The annual appropriation was increased to £4,000 in 1876, and for the first time individuals were invited to apply for stipends.

But the British example was difficult to emulate in the United States because this country lacked any national scientific organization like the Royal Society (whose charter went back to 1662) to lend prestige to the enterprise. The one body that might have undertaken such a task, the National Academy of Sciences (created by an act of Congress in 1863), was little more than an impecunious honor society in the early 1900s. Without a widely respected institutional buffer, a science patron who chose to give money piecemeal to independent investigators would be leaving himself open to second-guessing (or worse) by a public skeptical of "philanthropic" motives.

Both Carnegie and Rockefeller were personally acquainted with the benefits of "in-house" research. Their fortunes, like most of those created in the nineteenth century, were based on the timely application of new technology to problems of production and distribution. Carnegie's rapid conversion to the Bessemer-Kelly steel-making process in the 1870s probably saved his industrial empire from extinction. And as we have seen, Rockefeller's early business success was closely tied to the Frasch process for desulfurizing oil, perfected by chemists on the Standard Oil payroll. Indeed, a professor of chemistry at Yale College, Benjamin Silliman, had laid the foundations of the oil industry in the 1860s when he showed how the crude petroleum recently found in Pennsylvania could be refined into a number of useful products, including kerosene for lighting.

But industrial laboratories (which tended to focus on narrowly defined problems) were hardly suitable models for foundations created to spend vast sums of money on projects that were supposed to be independent of the founders' personal interests. As for the federal agencies that hired scientific investigators—like the U.S. Weather Office (established in 1870), the U.S. Geological Survey (1879), and the National Bureau of Standards (1901)—these labored under congressionally imposed restrictions even tighter than the bottom-line considerations governing the industrial labs.

John D. Rockefeller was fond of saying that his interest in the University of Chicago had been "enhanced" by the fact that the university gave "so much attention to Research." But turn-of-the-century academe had surprisingly little to offer a philanthropist looking for models of efficient, large-scale science patronage. University departments of science could absorb only so much money; in the 1890s there were no more than two hundred practicing physicists in the United States, and only a fifth of these published original research papers. At the beginning of the new century, total enrollment in *all* American graduate schools in *all* subjects did not exceed six thousand.

In the absence of safe institutional channels for science patronage, the most prudent course of action for a philanthropist interested in science is the awarding of prizes for past achievement. Not coincidently, this form of retrospective philanthropy found its definitive expression at the turn of the century, in the will of Alfred Nobel (1833–1896).

The first Nobel Prizes, in physics, chemistry, physiology and medicine, literature, and peace, were announced in 1901. The categories closely paralleled the interests of the donor, a Swedish chemist, engineer, and industrialist who invented dynamite and other explosives, and who was at the same time a pacifist and a man of strong literary opinions. As a maker of explosives, Nobel had learned to live with risk while doing everything in his power to minimize it. As a philanthropist, he did the same. If his principal goal was to have posterity identify his name with man's highest ideals rather than with instruments of destruction, he certainly succeeded. That this success has been due more to the science awards than to the prizes for literature and peace may be viewed as irony by some.

Not all the Nobel awards in science have been immune from second-guessing; in 1949 the prize for medicine went to a Portuguese neurologist who devised the radical brain surgery known as frontal lobotomy, an operation that turned out to have serious side effects and probably caused more suffering than it alleviated. Nevertheless, the roll of Nobel laureates does include most of the major figures in the history of modern science, and very few clinkers. Nobel's gift would seem to be as close as one can get to a sure thing in philanthropy.

The trouble with a sure thing is that you get what you pay for and little else. By stirring worldwide interest in science and by appealing to national and local pride, the Nobel Prize announcements and ceremonies may loosen public and private purse strings. Who could deny a promotion, salary increase, or research grant to a recently anointed Nobel laureate? But in terms of long-range impact on science, the value of such awards is questionable.

In many cases the recipient's best work has already been completed by the time he or she wins the prize. And even when a recipient continues to do important research, it would be hard to prove that the prize made the further contributions possible. Indeed, the effect can be just the opposite. Cynics have observed that many Nobel laureates are too busy giving speeches and interviews to do any serious research for at least a year after their awards are announced.

It is possible that the publicity generated by such awards attracts bright young people to scientific careers or motivates them to switch from one area of research to another. But there is no hard evidence that this is so or that the result, even if it exists, is beneficial to science. The story of the elucidation of the structure of the DNA molecule, told by James Watson in his book *The Double Helix,* is sometimes seen as a race for the Nobel Prize by rival teams of researchers in England and America. A race there undoubtedly was. Whenever researchers tackle a problem of major importance, they are aware that if someone else beats them to the solution, they will get scant reward for their effort. In science, priority is all. Except in rare cases the first person to announce a discovery gets all the recognition—including whatever prizes are available. But it makes no more sense to say that Watson, Francis Crick, and Linus Pauling were competing for a Nobel Prize than it would to say that the candidates in a presidential election are competing for a key to 1600 Pennsylvania Avenue or a seat on *Air Force One.*

Endowing prizes for successful senior scientists is the philanthropic equivalent of buying blue-chip stocks and bonds. Both Carnegie and Rockefeller leaned toward more speculative ventures. In organization and policy the Rockefeller Institute and the Carnegie Institution borrowed different features from the investigator-centered medical institutes that had been created for Koch and Pasteur, from the goal-oriented research labs that paid such dividends for American industry in the latter half of the nineteenth century, and from the departments of science at universities like Johns Hopkins and Yale whose stated purpose was a disinterested search for knowledge for its own sake.

The mix was innovative, and the rewards were commensurate with the risks. But it would take another two decades to bring to fruition (under the banner of the Rockefeller Foundation) an even more innovative and risky kind of science patronage, which aimed at supporting individual researchers of promise wherever they worked. What held back this development was a general lack of understanding of the nature of science itself.

# 5

# What's in a Name?

Most people think of Dr. Victor Frankenstein, the protagonist of Mary Wollstonecraft Shelley's novel *Frankenstein; or, A Modern Prometheus,* as the original "mad scientist." But when Mrs. Shelley's novel first appeared, in 1818, no one, including the author, referred to Dr. Frankenstein as a "scientist," mad or otherwise, because the word had not yet been invented.

Mrs. Shelley's novel was a popular success, and in 1831 she oversaw the publication of a revised edition. She had been just twenty years old when the book first came out; now she went over the manuscript with the eye of an experienced writer, paying close attention to niceties of usage. In just one of many such changes, she replaced the adjective "scientifical" in the 1818 edition with the more modern-sounding "scientific." But whenever she wanted to identify Dr. Frankenstein by his scholarly predilections, she was forced to make do with roundabout phrases like "the most learned philosopher" and "a man of great research in natural philosophy." In the very awkwardness of these circumlocutions, the reader may hear not just the author but the language itself straining for a new way to express the concerns of a new era.

Certainly, Dr. Frankenstein's creator was not alone in sensing the inadequacy of the available English vocabulary. In the same year that the revised edition of Mrs. Shelley's novel came out, the British Association for the Advancement of Science (BAAS) was formed. Membership was open to anyone who shared the group's announced purpose: "to give a stronger impulse and more systematic direction to scientific inquiry" and to remove "those disadvantages which impede its progress."

From the beginning, according to a contemporary account in a London magazine, the members "felt very oppressively at their meetings the want of any name [to describe] students of the knowledge of the material world collectively." Without such a label, it was thought, spokesmen for the BAAS would have difficulty explaining to the public precisely in whose name they spoke. Even the use of the word "science" in the title of the Association was an attempt to carve out a definable area of expertise—an attempt that struck some critics as laughably immodest. Charles Dickens satirized what he saw as the pretentious "quackery" of the BAAS in a series of articles about "the Mudfrog Association for the Advancement of Everything in General, and Nothing in Particular."

How important was this battle over words? One historian states flatly, "Ancient science failed to develop . . . because those who did scientific work did not see themselves, nor were they seen by others, as scientists but primarily as philosophers, medical practitioners or astrologers."

From the vantage point of the late twentieth century, when the power of science—as a method of inquiry, as a collection of observable facts and verifiable laws, as an instrument of control—goes virtually unchallenged, it is hard to believe that the notion of science as a distinct intellectual enterprise in the English-speaking world took hold only within the last hundred years. Most history books (even works of intellectual history) obscure this fact by anachronistically referring to all contributors to the development of modern science (from Copernicus on) as "scientists." But in matters like this the language does not lie.

According to the *Oxford English Dictionary,* the authority on historical usage, the word "science" (derived from the Latin *scire,* "to know") was originally used to refer to any kind of specific learning or knowledge, as in "the science of goode werkes" (Chaucer, 1364) or "the science of metre" (Gower, 1390). Standing alone, the same word stood for knowledge in the most general sense, as in this quotation from a fourteenth-century Psalter: "For God of sciens is lord."

The "seven liberal arts" that the scholar of the Middle Ages was supposed to master—grammar, logic, rhetoric, arithmetic, music, geometry, and astronomy—were also known as "the seven sciences," and theology was often called "the queen of the sciences" because all other species of learning were assumed to be contained in the knowledge of God. As Samuel Taylor Coleridge put it in 1830, "The science of theology was the root and trunk of the knowledge that civilized man." Somewhere between the revelations of theology and the specific skills of the more mundane "sciences" came the systematic knowledge of man and his world known as philosophy (from the Greek for "love of wisdom").

Late medieval European philosophy, as practiced under the watchful eye of the church, was little more than a series of footnotes to the works of Aristotle. In the early seventeenth century there was an awakening of interest in the nature of "the material world." Such worldly inquiries were conducted under the name of natural philosophy or natural science. The more independent-minded students of nature, who called for a complete break with the past, spoke of a "New Philosophy" in which fresh observations and calculations would count for more than cleverly marshaled quotations from Aristotelian texts. The New Philosophers were happy to leave rhetoric and the like to the academic philosophers, while they concentrated on "Physics, Anatomy, Geometry, Astronomy, Navigation, Staticks, Magnetics, Chymicks, Mechanics and Natural Experiments."

As the importance of carefully controlled observations came to be recognized, the innovators began talking about "experimental" philosophy or "experimental" science. But in general, until the beginning of the nineteenth century, "the notion now usually expressed by *science* was commonly expressed by *philosophy*." The giants of natural science in the seventeenth and eighteenth centuries, men like Isaac Newton and John Dalton, were content to be known as natural (or "mathematical," or "experimental") philosophers. Dalton presented most of his early papers on chemistry to the Manchester Literary and Philosophical Society, and his first comprehensive exposition of his atomic theory was published in 1808 under the title *A New System of Chemical Philosophy.*

Critics like Charles Dickens, who made fun of the British Association for the Advancement of Science for claiming too much in its title, missed the point. Far from taking all knowledge as their province, the founders of the BAAS were seeking a label for a division of intellectual labor that could not be adequately expressed in the older vocabulary. For centuries both astronomy and rhetoric had been "liberal sciences." But the very mission of the BAAS was to emphasize the fundamental difference between disciplines like astronomy, mechanics, optics, and chemistry, on the one hand, and disciplines like rhetoric, ethics, poetics, and metaphysics, on the other. The phrase "natural philosophy" was no help, since it implied a close connection to disciplines like moral philosophy that had nothing to do with scientific inquiry.

"In a direct and literal sense," historians tell us, "the BAAS made science visible." It did this most emphatically in its name, which helped transform the old and neutral word "science" into "the party label of a particular mode of understanding, possessed—so it was said—of superior power." Having declared its independence of the older scholarship, science was now ready to present itself as *the* source of authoritative knowledge on all matters short of the supernatural.

The first order of business for the BAAS was to convince the public that this newly redefined enterprise of science deserved its support. The support that the leaders of the BAAS had in mind went beyond mere lip service; what they sought was hard cash to sustain the intellectual descendants of Newton. No sooner had they begun promoting science to the British public than they realized how useful it would be to have a concise, memorable, and, if possible, euphonious name to identify the recipients of the hoped-for aid.

There were, of course, names for people who practiced many of the disciplines that fell under the rubric of natural science. Some of these titles, like chemist, astronomer, and mathematician, had long and honorable lineages; others, like biologist, had been invented but a few years before. What was lacking was an umbrella term, a professional title that could be used to refer to all practitioners of experimental natural science, the way "musician" referred to all practitioners of music, whether they played the lute or the pianoforte or composed for orchestra or sang or fiddled.

At the 1834 meeting of the BAAS, according to an eyewitness report in the *Quarterly Review* of London, "some ingenious gentleman proposed that by analogy with artist, they might form *scientist* . . . but this was not generally palatable." The ingenious gentleman was most likely the polymath William Whewell, who continued to argue for the new word in his extensive writings.

Another coinage of Whewell's that initially met with a cold reception was "physicist." In 1843 a writer for *Blackwood's Magazine* ridiculed the ugly sound of the plural *physicists,* "whose four sibilant consonants fizz like a squib"—a squib being a firecracker that fails to go off. A more substantive objection was that "physicist," like "scientist," had a presumptuous ring; it implied a command of nature's secrets that even the most confident investigators were reluctant to claim.

It was not until the beginning of the twentieth century that "scientist" and "physicist" gained general acceptance. The preferred terms during most in the nineteenth century remained roundabout locutions of the kind that Mary Shelley had struggled with: "scientific investigator," "scientific worker," "science worker," "cultivator of science," "man of science," and the traditional "natural philosopher." (Attempts to import the French term *savant*—literally, one who knows—were unsuccessful, no doubt because "savant" was every bit as presumptuous as "scientist.") As for "researcher," that was coined, as we shall see, in the 1870s, with the purpose of ridiculing a group of British academics who had begun to agitate for increased state aid to science.

At first the British Association for the Advancement of Science made no pitch for government funds, preferring to solicit money from private citizens. Surprisingly, despite indifference and even opposition from segments of the intellectual establishment, the BAAS was a resounding financial success from the start. Its organizational model was a German association of university professors, the Gesellschaft Deutscher Naturforscher und Ärzte, founded in 1822. More than a thousand people attended a meeting of this association in Berlin in 1828; among them was a maverick English mathematician named Charles Babbage who had been trying to break down the insularity of British science. Babbage had worked with Whewell, the mathematician George Peacock, and the astronomer John Herschel to modernize the teaching of mathematics at Cambridge University; more recently he had taken the Royal Society to task for not paying what he considered sufficient attention to a visionary proposal (his own) to build a mechanical calculating machine.

The gathering in Berlin inspired Babbage, who called it "a real association of men in search of the truth." On his return to England, he redoubled his efforts, together with like-minded fellows of the Royal Society, to bring new life to that august if somewhat moribund institution. When the powerful post of president of the Society fell vacant in 1830, the Society's controlling council nominated the duke of Sussex, one of George III's sons. Although the duke himself was a man of liberal views, the reformers would settle for nothing less than a real man of science, so they put forward a candidate of their own, Sir John Herschel.

To rally the forces of reform, Babbage published a book entitled *Reflections on the Decline of Science in England.* In this broadside he denounced the Royal Society for electing to its fellowship so many dilettantes and dull university dons who had done no research in natural science. (The unprofessional attitude common among these fellows was epitomized by one of their own number a few years later, when he wrote, "It is not necessary that the President of the Royal Society should be what is termed a philosopher, that is, eminently distinguished in any particular branch of science. . . . I myself should prefer a gentleman of literary and scientific tastes and habits.") If the word had already existed, Babbage might have summarized his case in this manner: of the 687 fellows of the Royal Society in 1830, the majority could not be considered, by any stretch of the imagination, scientists.

Damning as this statistic sounds to modern ears, the entire argument would have puzzled the founders of the Royal Society, who, back in 1660, had conceived of their new fraternity as a means of bringing together natural philosophers and their well-placed and well-to-do patrons—with emphasis on members of the royal family, up to and including the newly restored King Charles II. Indeed, a roster of fellows better known for their literary than for their scientific "tastes and habits" would run from Dryden and Addison in the seventeenth century to Lord Byron in the nineteenth.

Nevertheless, Babbage, Whewell, Dalton, and others believed that the time had come to shake off the "amateur" status that had so long hindered the development of British science. The French had found ways to support science through governmental subsidies of one kind or another, and so had many of the German states. German industries, especially the big chemical firms, were also becoming active patrons of science. Only England still expected its scientific investigators to be independently wealthy or to conduct their inquiries on a part-time basis while they supported themselves by other means. As Babbage wrote, "Science in England is not a profession, its cultivators are scarcely recognized as a class."

The reformers' decision to openly oppose a member of the royal family for the presidency of the Royal Society seemed little short of lèse-majesté to many Englishmen. Yet the election was close. With just over a third of the membership voting, the duke beat out the astronomer by only eight votes.

Meanwhile, strong liberalizing currents were running in the society at large. In 1830 a Whig government dedicated to political and social reform came to power. Whig leaders spoke in the name of an increasingly affluent middle class, and it was to this new constituency that the reformers eager to shake up the British scientific establishment now turned. Under the leadership of a clergyman turned impresario, the Reverend Vernon Harcourt, the British Association for the Advancement of Science held its first public meeting in York in 1831.

Despite the ridicule of unsympathetic journalists and the generally unsettled conditions throughout the country, this "Festival of Science" attracted large and appreciative crowds; soon, enthusiasts in cities like Bristol, Liverpool, and Manchester were competing to attract the three- or four-day annual gatherings of the BAAS—the way local boosters today fight for the right to host the Olympic Games or the nominating conventions of major political parties.

Brisk sales of "annual" and "life" subscriptions, and of special "ladies tickets" that admitted the bearer to general lectures and to other social events, put the BAAS in the black from the beginning. In 1833 Harcourt proposed using the Association's surplus, which already stood at more than £1,000, to establish a fund for the support of scientific investigation. Over the next decade the BAAS paid out an average of £1,500 a year to investigators chosen by various committees.

These awards were the first research grants by any British (or American) scientific society. Even more significantly, they represented the first fruits of a long campaign to "professionalize" science. Just four years earlier William Hyde Wollaston, a chemist who had invented a lucrative process to render platinum malleable, had left a considerable sum of money to the Royal Society to "promote experimental research." But the leaders of the Society had dragged their feet in fulfilling the terms of Wollaston's bequest, since they were uncomfortable with the idea of giving money to investigators for their "personal use."

The founders of the BAAS never intended to supplant the Royal Society as the official guardian of natural science in Great Britain.

Men like Babbage, Whewell, and Harcourt were all fellows themselves; while working to change the Society from within, they used the free-wheeling Association to keep the pressure on the Society's slow-moving leadership.

# 6

# Who Pays the Piper?

By midcentury the scientific reformers had won most of their original demands. The power of the Royal Society's president to act without the consent of the fellows was curtailed; stiffer requirements for election were imposed; the number of fellows named each year was reduced by more than half, and the quality of new members (as measured by achievements in scientific research) improved dramatically. In less than two decades of agitation, the Royal Society had been transformed from a stodgy gentleman's club into the model of a modern scientific society.

In the closely linked areas of intellectual prestige and public opinion, the BAAS also scored impressive victories. By the late 1840s the use of the word "science" to refer to a body of knowledge that rested on more secure grounds than "ordinary" knowledge had spread across the Atlantic; the American Association for the Advancement of Science was founded in 1848 in frank emulation of the British model. As early as 1839 the implications of the new nomenclature had been correctly foreseen, and bitterly deplored, by a spokesman for John Henry Newman's Christian revivalist Oxford Movement, who wrote, "Physical science, the science in other words of matter and material things, now arrogates in effect the name of 'science' exclusively to itself."

More than anything else, of course, it was the success of scientific ideas (ideas like John Dalton's atomic theory) that secured the linguistic victory. But the promotional efforts of William Whewell undoubtedly played a part. Whewell, who was born in 1794 and educated at Trinity College, Cambridge, made a name for himself as a mathematician, architect, crystallographer, educator, and theologian (he was an ordained minister of the Church of England). His best-known works were *The History of the Inductive Sciences* and *The Philosophy of the Inductive Sciences* (published in various editions between 1837 and 1860).

Whewell knew exactly what he was doing when he coined new words like "scientist" and "physicist," and argued for other neologisms like "biology" and "biologist." "Terms record discoveries," he wrote in 1837. His goal was nothing less than to create "some new word or phrase which becomes part of the common language of the philosophical world"—not unlike an explorer or conqueror altering the geopolitical landscape by planting a flag of sovereignty on newly claimed territory.

Not all of Whewell's coinages were equally successful. The names he proposed for the new sciences of heat and atmosphere—"thermotics" and "atmology"—never caught on. But he was relentless in his campaign to stamp out the old use of "science" to refer to *any* kind of knowledge. He was always ready to explain, to anyone who would listen, "why some portions of knowledge may properly be selected from the general mass and termed SCIENCE."

For Whewell, it was a simple question of maturity. Only "the Material Sciences" had so far created "bodies of exact and enduring knowledge." While conceding that sometime in the future it might again be appropriate to talk about "one" science, for now, he argued, the term should be reserved for those disciplines "in which, by observation of facts and the use of reason, systems of doctrines have been established which are universally recognized as truths among thoughtful men."

Whewell had no qualms about rating the sciences according to their level of maturity. Mechanics, hydrostatics, and physical astronomy were rapidly approaching their "complete and final form." Chemistry and meteorology were still in their infancy. Biology was barely embryonic. And so on. Nor did Whewell and his colleagues at the BAAS have any doubts about which disciplines should be excluded from their annual "Festivals of Science." No presentations were permitted by scholars working in philology, literature, metaphysics, music, or phrenology. (The phrenologists petitioned repeatedly, without success, to have their status reconsidered.)

Whewell's linguistic strategy was to divide and conquer. He knew that once people started talking about "biology," they had already taken the first step, however unwittingly, toward accepting the argument that the "science of life" is more closely related to mathematics, mechanics, and astronomy than to, say, moral philosophy. While he joined his religious antagonists in publicly deploring the fragmentation of knowledge in the modern era, Whewell was interested in unification only on his own terms. He could not force anyone to use the word "science" in the new restricted sense. But there were other ways to nudge public opinion toward the conclusion that one path of inquiry, and one path only, led to reliable knowledge: "We need very much a name to designate a cultivator of science in general. I should incline to call him a Scientist."

The fact that Whewell's coinage did not come into common use until the end of the century is hardly surprising. In a society that had long identified truth with religion, it was the height of audacity to claim that no investigators but those who applied the methods of modern science had a right to be recognized as "knowers." (The term "scientific method" was first use by Thomas Huxley in 1854.) What *is* surprising is that by midcentury—before most people were even aware of the revolutionary nature of the reformers' enterprise—Whewell and his colleagues had obtained a government-financed franchise for seeking knowledge in the scientific manner.

In 1849, thanks to behind-the-scenes exertions by the leaders of the BAAS, Parliament voted to provide £1,000 a year for the next five years to the Royal Society; the Society undertook to pass along the money, in the form of small grants, to eminent scientific investigators who were finding it difficult to purchase equipment and pay assistants out of their own pockets. Since there were no precedents for this sort of assistance, the money was actually taken from a parliamentary account earmarked for "charity." When, after five years, Parliament tried to cut the appropriation as part of a general reduction in charitable funds, the Royal Society successfully fought for continuation of the £1,000 grant on a separate budget line. It was a significant victory, but hardly a conclusive one. The size of the annual appropriation remained unchanged for the next twenty years, while scientific research became more and more costly.

The final battle in the campaign for the professionalization of science was to be fought in a setting that the founders of the Royal Society had once explicitly rejected as a home for their "New Philosophy"—the British universities.

Francis Bacon, whose writings greatly influenced the founders of the Royal Society, professed a deep distrust for the universities of his day, especially Oxford, which he saw as a stronghold of the sterile Aristotelian philosophy that stressed "disputation" instead of close examination of "nature in action." The antiquarian bias that Bacon found among academics was not just an English problem. From the seventeenth century to the nineteenth century, according to one historian, "the major contributions to science (except in medicine) were made outside the universities."

To bring about what he called "a true and lawful marriage between the empirical and the rational faculties," Bacon urged the creation of entirely new institutions unconnected to the universities—a task that he conceded would require large amounts of public funds. The model that Bacon probably had in mind was the Accademia dei Lincei (the Academy of Lynxes or of the Lynx-eyed, depending on the translation), founded in Rome in 1603 by Federico Cesi, an eighteen-year-old nobleman. Among its first members was an independent-minded professor of mathematics named Galileo Galilei.

But the Academia dei Lincei did not long survive the untimely death of its young patron, in 1630; it seems that Cesi's father deplored the squandering of the family fortune on such fruitless ventures. Despite their stuffy traditionalism, the old universities had one important advantage over de novo institutions; they could, as fixtures of the establishment, hold out the promise of regular employment. Even Bacon, when he ran short of money, was not above pulling a few strings in an unsuccessful effort to get himself named head of some college at Cambridge or Oxford.

Like their intellectual godfather, the founders of the Royal Society maintained an ambivalent relationship with the established universities. Under Oliver Cromwell, Oxford and Cambridge were forced to open their doors to exponents of the new learning; in the late 1640s a group of avowed Baconians (including the brilliant young mathematician and architect Christopher Wren) began meeting regularly at Oxford to practice what they called "a free way of reasoning" and to conduct experimental "trials in Chymistry or Mechanicals."

In the last, unsettled years of the Commonwealth, many of these scholars found refuge in London, where they continued to meet for discussion and mutual support. There was general agreement that the old universities could not change quickly enough to accommodate the revolution in thought that was already under way. Wren himself held the chair of astronomy at Gresham College, an upstart institution that had been founded in 1596 to promote practical knowledge of the type that sailors, merchants, and farmers could use.

When the coronation of Charles II signaled the end of the civil war, the Baconians were ready; the spark that led to the founding of the Royal Society was a lecture by Wren in November 1660, calling for the creation of a new "Colledge for the Promotion of Physico-Mathematicall Experimentall Learning."

At least half of the original forty-one members of the Royal Society had connections to either Oxford or Cambridge. Nevertheless, academic authorities looked askance at the new institution. When it petitioned for the right to confer degrees in natural science, the Society was vilified as a papist plot to undermine the English universities, and its members were ridiculed (together with the professors of Gresham College) for spending all their time in such useless activities as "weighing of ayre."

Apologists for the new Society were quick to answer these charges. John Dryden wrote, "Nothing spreads more fast than Science, when rightly and generally cultivated." But the rancorous controversy made it difficult to raise money, which was the founders' most pressing concern. Along with his royal blessings, the king had promised the Society some income-producing lands in Ireland, but these were not forthcoming. In 1667 Thomas Sprat wrote an authorized "history" of the Society that was frankly intended to encourage donations; although he conceded that "the first Benevolence should come from the Experimenters themselves," Sprat noted that the Society's "mighty Design" would require a "mighty Revenue" for its realization.

The original plan was to have the members do their own experiments and pay dues of one shilling a week to defray expenses, but neither expectation proved realistic. In 1662 Robert Hooke was engaged to perform "3 or 4 considerable experiments" at the weekly meetings; his salary was originally set at £80 a year, but the Society was always behind in payments because the majority of members were behind in their dues. By 1676, arrears totaled £2,000.

In 1684 the astronomer Edmund Halley persuaded Isaac Newton, who held a prestigious chair of mathematics at Cambridge, to make public all he knew about planetary motion; the result was the three-volume work *Philosophiae Naturalis Principia Mathematica* (Mathematical Principles of Natural Philosophy), which has been called the single most important scientific book ever written. Newton's colleagues in the Royal Society immediately recognized its importance, and on May 19, 1686, they voted to underwrite the cost of publication. Two weeks later this decision was rescinded; the members had discovered that there was no money in the treasury. The day was saved when Halley agreed to meet the costs of publication out of his own pocket.

A pattern had been set that lasted for over a hundred years. By and large, the pursuit of natural science in England was left to what one historian has called "the amateur activities of the leisured classes." During the eighteenth century the British government supported some scientific activities (surveys, expeditions, the invention and development of measuring instruments) that might prove useful to the Royal Navy or to agriculture or to other props of the state. But the British universities played at best a minor role in the advance of natural science.

Academic interest in the sciences fell to a new low during the first half of the nineteenth century; at Oxford enrollment in astronomy courses declined by more than four-fifths between 1819 and 1838. Not until the 1850s did Oxford and Cambridge start teaching natural science to undergraduates on a regular basis—a change that closely coincided with the easing of rules requiring that candidates for degrees be members of the Church of England.

Once the universities had accepted a responsibility to teach science, the focus of efforts to raise money for scientific research shifted to the academic arena. The idea of the university as a center of both teaching *and* research is so common today that we may find it hard to believe how controversial it was a century ago. Once again, the leaders of the BAAS were in the thick of the controversy.

The poor showing of British industry at the Paris Exhibition of 1867 prompted a new wave of interest in science patronage. In 1868 Mark Pattison, the rector of Lincoln College, Oxford, wrote a book calling for the allocation of funds to support research at the university. "In order to make Oxford a seat of education," he stated, "it must first be made a seat of science and learning." In November of that year the BAAS formed a committee to lay the groundwork for what came to be known as "the endowment of research" movement. Among the committee members were Thomas Henry Huxley, who would soon become known as the foremost apologist for Darwin's theory of evolution, and Norman Lockyer, who was in the process of founding *Nature,* the international journal of science.

The committee's first act was to send a letter to some eminent British men of science, asking whether they believed that the interests of science could be best served by creating new institutions for experimental research, or by strengthening institutions already in existence, or by the "enlargement of grants to individuals for apparatus and materials."

A few respondents declared themselves opposed to any form of state aid to science, on the grounds that the recipients' independence would inevitably be compromised. But most agreed with William Crookes (who later gave his name to the Crookes vacuum tube). Although he himself had inherited a fortune, Crookes complained that laboratory supplies and instruments had lately become so expensive that it was almost prohibitively difficult "for a poor man, unaided, to win his way to eminence" in science. The parliamentary grant of £1,000 a year had not been increased since its inception, twenty years before. In 1869 the BAAS asked the government to look into the entire issue of science patronage, arguing that if the country wanted to enjoy the benefits of scientific research, it must be prepared to support those individuals willing to "pursue a full-time vocation in science."

Early the next year a royal commission was appointed to study the state of science in Britain. The commission took its task seriously: it met eighty-five times over the next six years, heard testimony from 150 witnesses, and published eight reports and four massive volumes of evidence. While it was deliberating, the lines became ever more sharply drawn between the pro- and antiscience parties in the universities. Charles E. Appleton, a classical philologist and fellow of St. John's College, Oxford, became the unlikely leader of the proscience forces, abetted by Pattison and Alexander Strange, a retired Indian Army surveyor who knew the value, and the cost, of reliable instruments.

Strange's instrument of choice when it came to public debate was irony. Why, he wondered, should society insist on treating science as an exception among the professions? "Soldiers, lawyers, physicians and divines" all received fair pay for their services, yet it was commonly expected that "the man of science should work for love and die . . . in poverty." Appleton meanwhile got down to specifics in his "Draft Scheme for the Endowment of Research." He proposed that "a few citizens of the Republic of Science," whose work had won the approval of their peers, be given annual incomes "equal to but not greater than the average income they would receive in a legal, medical or diplomatic career." Appleton was careful not to ask for too much. He characterized the stipends he had in mind as "enough to live upon—perhaps not enough to marry on."

But Appleton's appeal to reason did not save him or his cause from ridicule at the hands of traditionalists. Noting that many people who were arguing for state aid to science would be among the first to benefit from it, *Punch* suggested that the "Endowment of Research Movement" might be more truthfully renamed the "Research for Endowment Movement." To further stigmatize the holders of Appleton's "unpalatable" doctrine, the pejorative term "researcher" was invented.

But by 1875, when the royal commission issued its eighth and last report, strong support for the proscience movement had developed in Parliament. Legislators were impressed by statistics showing that German investigators were outpublishing their British counterparts four, five, or six papers to one; even more humiliating, half the research papers appearing in British chemistry journals were written by Germans. An Oxford reform bill, committing that university to work for the advance, as well as the dissemination, of knowledge (and mandating a number of new chairs in the natural and medical sciences) was introduced in the House of Lords in February 1876. A few months later not only did Parliament increase the annual grant to the Royal Society from £1,000 to £4,000; for the first time, the lawmakers accepted the principle that this money could be used by scientific investigators to meet ordinary living expenses as well as to pay specific bills incurred in the course of experiments.

The very success of the Endowment of Research Movement prompted renewed opposition to the campaign to professionalize British science. Traditionalists who believed that the job of an Oxford or Cambridge don was to teach, not to do research, resented the state's intrusion into academe. Investigators working in state-supported scientific institutions like the British Museum and the Royal Observatory were concerned that their own funds would be cut to pay for what they referred to scornfully as "cottage industry" research.

There were bureaucrats who balked at accepting what one historian has called "a fundamental change in the way in which the government and the universities viewed science." In the past only investigators who developed a patentable device or process (like Wollaston's method for refining platinum) could expect to profit in a big way from their labors. But from now on, scientific investigators were to be paid (at least at the beginning of their careers) on the basis of promise, not results. Behind this shift in policy lay a faith in the social value of science—and in the ability of researchers to deliver society's goods—

that many keepers of the public purse not unreasonably questioned.

Some critics argued that the "humiliating necessity of personal application in each case" would "lower the dignity of the recipients"; others smelled a plot to subvert the vaunted independence of British men of science and turn them into "imitation Germans." Only with reluctance did the Royal Society agree to serve as executor of the government's expanded support of scientific investigators; the treasurer of the Society even went on record with the opinion that it was a disservice to encourage men "not yet of independent income" to interrupt "the business of their life merely for the sake of science."

Meanwhile, the royal astronomer lent his name to a campaign to rescind the parliamentary grant entirely. Humanists like John Ruskin worried out loud about the establishment of a "priestcraft of science." Popular opposition to science, fed by the acrimonious debate over Darwin's theory of evolution, swelled the membership of national leagues against vivisection, smallpox vaccination, rabies inoculation—even the official notification of cases of infectious diseases.

In December 1880 the *Times* of London criticized a tendency among some men of science to worship an "idol, which it is technically fashionable to call 'research,' and to ignore the far higher mental effort which is required for successful ratiocination." Parliament was just then considering a new proposal to spend £2,000 a year for state fellowships in the sciences. Attacked by critics as a scheme to put impecunious investigators on the "dole," and backed only halfheartedly by the Royal Society, the proposal went down to defeat. Not until 1914 would the British government act to increase its annual subsidy to scientific researchers above the £4,000 level set in 1876.

There is little doubt that such a modest level of support was in keeping with the tenor of public opinion. It is true that the work of the British Association for the Advancement of Science and the Endowment of Research Movement had established the principle—now enshrined on a separate line in His Majesty's budget—that science was in some sense "a national resource, like labor and capital, a productive factor to be cultivated."

And yet in 1919, in the wake of a worldwide flu epidemic that took some twenty million lives, a government committee looking into the state of medical research in Britain noted that most people still considered scientific research of any kind essentially "a private hobby."

# 7

# Beggars and Choosers

In the United States, a country less inclined to expect productive labor from its leisured class (which was, in any case, much smaller than England's), the "professionalization" of science in the latter half of the nineteenth century met with less resistance. The main issues were who would pay for the labors of the professional researchers and who would choose the fortunate recipients.

In 1877 Edward C. Pickering, who had been appointed professor of astronomy and director of the observatory at Harvard the year before, took up the "endowment of research" cause. He gave a speech calling for the establishment of a research institute for mathematics, physics, and chemistry—a large, expensive research institute. But Pickering did not ask for a dime of government funds. It was his conviction that government aid was inherently wasteful. Instead, he invited contributions from private individuals, who could expect to share in the glory when the research they had made possible led to important discoveries. But Pickering was a man ahead of his time. None of the wealthy people he approached was eager to buy reflected glory on his terms.

After a decade of frustration Pickering lowered his sights and narrowed his focus. Launching a new campaign for a capital fund of $100,000, he proposed to devote the annual income exclusively to the advancement of astronomy, in the form of small grants that would be awarded solely at Pickering's discretion. Astronomy, perhaps the most ancient of the sciences and a source of indisputably useful information (as any sailor could attest), had always been a favorite of wealthy donors. In 1890 Catherine Wolfe Bruce, who had already bought Harvard its 24-inch telescope, agreed to contribute $6,000 toward Pickering's endowment fund as an "experiment in giving." When Pickering sent a letter to his colleagues asking what they would do with a portion of Miss Bruce's money, he got eighty-four replies, from which he chose the fifteen grant recipients.

Two years later Pickering's great rival among fund-raising astronomers, George Ellery Hale, talked a Chicago streetcar magnate named Charles Tyson Yerkes into putting up the money for what became the 40-inch refracting telescope at the University of Chicago's Yerkes Observatory. Hale, himself the son of a wealthy patron of science, went on to build 60-inch and 100-inch reflecting telescopes at the Carnegie Institution's Mount Wilson Observatory, near Pasadena, California. Hale's motto was "Make no small plans." The 100-inch instrument, known as the Hooker telescope, was paid for by a Los Angeles hardware tycoon named Joseph D. Hooker.

Meanwhile, Pickering, who moved with ease in the most elevated social circles, seized any opportunity that came his way to sell the rich on the idea of astronomical philanthropy. At a breakfast in honor of a Prussian prince in February 1902, he sat next to Henry A. Rogers, a member of John D. Rockefeller's Standard Oil "brains trust." Pickering must have been especially eloquent that morning about the good uses to which American stargazers could put some extra funds; the next day he received a check from Rogers for $20,000.

Obviously, this kind of free-lance cadging at the tables of the rich could support only a limited amount of scientific research. Yet little money was available from institutional sources before 1900. The National Academy of Sciences, established by an act of Congress in 1863 to provide expert advice to the federal government, had less than $100,000 in its endowment funds; other sums were disbursed by learned societies like the American Academy of Arts and Sciences and the American Association for the Advancement of Science and by museums like the Smithsonian Institution. The total came to no more than $3 million. Of this total, Harvard and the Smithsonian controlled over three-quarters; and more than half was earmarked for astronomy alone.

The tide, however, was about to turn. On hearing of the establishment of the Carnegie Institution, Pickering tried to persuade Carnegie to support American astronomy through a large trust fund to be administered at Harvard. But other astronomers protested Pickering's attempt to have himself named trustee; in any case (as we saw in chapter 4), the Carnegie Institution was not about to surrender control of its research funds to outsiders.

Not one to admit defeat easily, Pickering continued to approach Carnegie and his Institution and, a little later in the century, John D. Rockefeller and his Foundation, for research funds. As we will see, Pickering himself never succeeded in making the leap from the personal fund-raising typical of the nineteenth century to the institutional philanthropy in the name of science that would mark the twentieth century. That transition would be left for others to negotiate and administer. But all the pieces were now in place.

The big philanthropists were looking for a "safe" gamble on which they could bet enormous sums of money. Science was just gaining public recognition as a profession with a special knack for uncovering useful truths. But it is impossible to have a profession (at least a financially sound profession) without a convenient title for the professionals who practice it. People, after all, give money to other people, and they want to know with whom they are dealing. Credentials, whether based on the whim of an elite or on the voice of the people, are a social necessity.

Behind the enthusiasm of the editorialist for London's *Daily Telegraph* who declared in 1901, "Nothing is better worth buying than scientific fact," lay the increasingly widespread acceptance of the word "scientist" and of the scientist's mission as purveyor of expertise, keeper of the consensual reality, and socially sanctioned seeker after knowledge. This heady mixture is what the philanthropists bought.

The importance of what has been called "the professionalization of science" cannot be overemphasized. As one historian put it, "The greatest advances in a science come *after* the development of a professional consciousness." In a fully developed profession, only the professionals themselves are deemed competent to judge their own affairs and to dispense funds and favors among their colleagues.

The birth of modern science as a profession was neither accidental nor spontaneous. It was the outcome of nearly a century of propaganda and agitation by a band of apologists who fought for their cause with the confidence and tenacity of true believers. It is remarkable that the men who helped bring about this intellectual and social revolution—Babbage, Harcourt, Lockyer, Whewell, Appleton, Hale, Pickering—should be largely unknown to the general public. If history books deal at all with the popular image of science during the crucial hundred years from 1850 to 1950, they usually focus on the confrontation between defenders of Darwin's theory of evolution like Thomas Huxley and religious traditionalists like Bishop Samuel Wilberforce.

In 1860, a year after the publication of Darwin's *On the Origin of Species by Means of Natural Selection,* Huxley and Wilberforce engaged in public debate at a meeting of the British Association for the Advancement of Science. Before a standing-room-only crowd of seven hundred, Wilberforce asked sarcastically whether Huxley traced his own descent from the apes through his father's or his mother's side. Huxley replied that if he had to choose as an ancestor either the kind of man who would introduce such a remark into a serious scientific discussion or an ape, he would gladly choose the ape.

Huxley's efforts on behalf of Darwinism were certainly memorable and undoubtedly did much to discredit the know-nothing wing of the antiscience opposition. But it is important to see what Huxley did as part of a larger battle. Huxley himself was a member of the X Club, a group of nine BAAS types, mostly self-taught or educated on the Continent, who worked tirelessly both in public and behind the scenes for the professionalization of science.

The last stage in this process—the placing of the new profession on a sound financial basis—was accomplished by another band of men, even less well known today, who sometimes referred to themselves as "philanthropoids." The word was apparently coined by Edwin R. Embree, who served as secretary of the Rockefeller Foundation in the early 1920s. According to Embree, a philanthropoid is someone who makes a living by giving away other people's money.

As we will see, Embree's own attempt to establish an ambitious program of science patronage with Rockefeller money ended in failure. But his contemporaries and successors in the Rockefeller philanthropies succeeded in working out a formula for financing the new breed of scientists, most of whom had academic degrees and were associated with large universities in the United States, Britain, Germany, France, and the Scandinavian countries.

Because of their almost constant travel throughout the international scientific community, the Rockefeller philanthropoids who pioneered in the support of science became known as "circuit riders."

The name refers to the early disciples of John Wesley who traveled ceaselessly on horseback to bring the word of God to out-of-the-way places. Circuit riders were especially active on the American frontier during the nineteenth century; one Methodist preacher is said to have ridden five thousand miles a year to serve his far-flung congregants.

It is not mere metaphor to talk of the religious fervor that the Rockefeller circuit riders brought to their labors. In their official duties they combined the functions of seeker and missionary. Like the Babbages, Whewells, and Appletons of the preceding century, they often spoke as if the future of humanity hung on the outcome of a race between the civilizing light of science and the dark forces of ignorance and superstition. Animated by this belief, they played an influential role in the shaping of the modern scientific community. During the decades between the two world wars, the Rockefeller Foundation set a new standard for science patronage—becoming, in the words of the historian Daniel J. Kevles, "a private precursor to the National Science Foundation and the National Institutes of Health combined."

# OPERATIONS

# 8

# Missionaries and Professionals

On the night of March 19, 1924, the recently installed president of the International Education Board (IEB), sixty-year-old Wickliffe Rose, found himself "sitting in a comfortable Pullman berth looking out over great fields of snow-covered ice" in the middle of the Baltic Sea halfway between Copenhagen and Stockholm. He was making a grand tour of postwar Europe to scout out opportunities for a new scheme of supporting research in the natural sciences that he had conceived and persuaded John D. Rockefeller, Jr., to back. Before he was done, he would visit two hundred scientists in more than fifty institutions in nineteen countries, to see for himself what the need was.

To get from Denmark to Sweden, he had boarded a train that was shunted onto a ferry—an ice-breaking ferry. This is how he described the night-long voyage to Wallace Buttrick, a fellow board member of the IEB, who was also his uncle and godfather and a longtime fishing companion:

> The boat seemed like a living thing, tireless in energy, undismayed, never discouraged by the failure to break through at any particularly stubborn point; but turning, shaking itself like a dog just out of water, it would plunge forward to buck the line again at some more promising place.

The image is pure Rose, and can serve as an introduction to the man who adapted the circuit-riding tradition to the science-support programs of the Rockefeller philanthropies. A slight, unprepossessing figure, whose large eyes peered lucidly at the world from behind glasses that kept slipping down his nose, Rose was a good listener and a great persuader. He brought to science patronage not the routine devotion of the acolyte but the impassioned dedication of the recent convert. Like Frederick Gates, he came to a belief in the saving power of science late in life. But once committed, he gave his all to the cause. "This is an age of science," he wrote in his personal notebook. "All important fields of activity, from the breeding of bees to the administration of an empire, call for an understanding of modern science."

Rose was aware that his gospel was not universally appreciated by his contemporaries. "Here and there throughout the centuries," he said, "occasional individuals have had the passion for scientific investigation, but the organization and endowment of this passion on a large scale are relatively recent. . . ." Far from being disheartened by this state of affairs, Rose found it bracing. Since the truth about science had as yet been revealed only to "a small minority of the peoples of the world . . . it should be feasible to extend the field for the cultivation and service of science almost indefinitely."

Just as Rose was embarking on his five-month tour of the scientific capitals of Europe, another representative of the Rockefeller philanthropies was beginning a very different kind of experiment in science patronage. On December 5, 1923, the trustees of the Foundation had created the Division of Studies to identify new targets for support outside the well-established areas of education, medicine, and public health. Edwin Embree, the forty-year old secretary of the Foundation, was put in charge.

In their curiously interwoven careers Rose and Embree faced many of the same issues as they sought to define policy and set in motion large-scale, long-range programs of support for scientific researchers. Their successes, and failures, had a great impact on the development of science patronage in this century, not only at the Rockefeller philanthropies but also at the other institutions, both private and public, in this country and abroad.

From the beginning of the century, the Rockefeller philanthropies had struggled to reconcile two distinct (some might say conflicting) approaches toward doing good. At one extreme, the Rockefeller

Institute for Medical Research, which provided laboratories for eminent investigators to pursue fundamental problems of their own choosing, epitomized the newly emergent professionalism of modern science; at the other extreme, the General Education Board (GEB), founded in 1903 to help raise educational standards in the American South, adopted the more traditional stance of missionaries bringing light to those still in darkness.

This difference between the philanthropic styles of the two institutions had nothing to do with the post-Darwinian antagonism between science and religion that so exercised Victorian intellectuals. Missionary work is, after all, only incidentally religious. Whether a missionary is spreading the Gospel of Christ to the heathen or the gospel of free public education to the illiterate, his fundamental approach remains the same. He begins with a sense of superiority. He already possesses (or at least understands the value of) something that the target population needs. To accomplish his purpose, the missionary must master what the business man calls marketing skills: he must devise efficient methods of distribution, overcome resistance in the target population, try to cushion the impact of change.

By contrast, the philanthropic patron of science has no reason to feel superior to the recipients of his largess. Quite the contrary. To the extent that he supports scientists who are recognized leaders in their fields, he must buy expertise on the experts' terms, taking on faith that what he buys will eventually result in some good that he can understand.

Both the Rockefeller Institute and the General Education Board were headquartered in New York, the capital of American capitalism (and therefore of American philanthropy) in its most dynamic phase. But the Institute looked to the Old World for inspiration—to those European centers of science where men like Pasteur and Koch had shown how major advances in medical care could grow from new knowledge about fundamental natural processes. By contrast, the General Education Board drew its inspiration from the American South—not because the South had anything to teach the North about education (or farming or medicine) but precisely because, for various historical reasons, the South needed to learn what the North already knew.

Of course, the GEB took pains to avoid any hint of intellectual or moral carpet-bagging in its operations; virtually all its staff people and field representatives were bona fide southerners. But there was no disguising the missionary style in the GEB's approach. Views commonly accepted elsewhere were still news in the South, and the GEB's self-imposed task was to close that gap.

In taking on this task, the General Education Board was following a trail blazed in 1869 by a Massachusetts banker named George Peabody, who set up a $2 million fund to assist education "in the suffering south." (This Peabody Education Fund was later hailed as "the beginning of the foundation as we know it," by a Rockefeller philanthropoid.) Other wealthy northerners followed Peabody's example during the next three decades. But at the end of the century, education in the South was still suffering—a victim of political neglect, racial polarization, and the general poverty of the region.

To coordinate an all-out attack on these evils, a coalition of southern and northern educators and philanthropists began meeting annually in 1898. The young John D. Rockefeller, Jr., only a few years out of Brown University, was drawn into this coalition in 1901, when he and fifty other interested parties toured the South on a chartered train that local newspapers dubbed "the millionaires' special." The train stopped at all the famous but perennially broke Negro schools like Hampton Institute, in Virginia, and Tuskegee Institute, in Alabama.

John D., Jr., never one to use words loosely, later called this trip "one of the outstanding events" in his life. On his return to New York, he conferred with a number of people, including his father, Frederick Gates, and Dr. Wallace Buttrick, secretary of the Baptist Home Mission Society. From these discussions emerged the idea of the General Education Board, which quickly became the centerpiece of a closely interlocking group of philanthropic organizations (including the Peabody Fund) that assumed responsibility for improving southern education for both whites and blacks.

The General Education Board was explicitly chartered (by an act of Congress) to work at all educational levels, from the one-room rural schoolhouse to the "highest university culture," and with "both races."* A method of operations was soon devised that combined the missionary zeal of American Protestantism with the respect for hard facts and nuts-and-bolts detail that characterized every venture, commercial or

*To achieve its goals, the GEB was forced to cooperate with representatives of the dominant white power structure and to acquiesce in Jim Crow rules and customs. Whether these actions inadvertently contributed to the development of a segregated school system in the South is an open question.

philanthropic, in which the Rockefellers became involved. Fundamental to these early projects of educational uplift was a corps of dedicated circuit riders whose job was to gather information and to "evangelize" for innovation.

For example, to help the eleven southern states build a system of high schools comparable to that found in the Northeast, the General Education Board persuaded the state universities to take on a new kind of faculty member—a "professor" whose principal responsibility was not to teach but to travel throughout the state, surveying educational needs, suggesting ways to improve existing secondary schools, and drumming up local support for new high schools. The GEB paid the salaries of these circuit riders, but money alone could not have sustained them on their rounds, where the obstacles they had to overcome ranged from bad food and bad roads to illiterate county officials and a widespread belief that free public schools were somehow "socialistic."

Circuit riding for the GEB was considered a "calling." A constant stream of correspondence offering moral support and inspiration for flagging spirits emanated from GEB headquarters in New York. This was more than mere rhetoric. Before authorizing any programs, Gates and Buttrick (whom Gates had chosen as the first secretary of the GEB) had themselves crisscrossed the South on fact-finding expeditions, traveling by train, river launch, buggy, and horseback, often for weeks and months at a time, getting a feel for the region and its people. One of Buttrick's assistants summed up a day in the life of the circuit rider in a quatrain that found its way into the GEB files;

> This road is not passable,
> Not even jackass-able,
> So when you travel
> Bring your own gravel.

The insights into local conditions gained on such arduous treks were invaluable in shaping the GEB's early programs. For example, to those who knew the territory, it was obvious that tax money to improve the South's public schools could come only from an improved southern economy. Since agriculture was the South's main industry, efforts were undertaken to boost the efficiency of southern farms. In one imaginative program (conducted in collaboration with the U.S. Department of Agriculture), the GEB helped create a network of trained consultants whose services were available, free of charge, to local farmers. Results were immediate and dramatic: markedly higher yields per acre for many staple crops and, perhaps as important, a more positive attitude toward innovation in this most conservative of American regions. (The modern county extension-agent system evolved from the GEB's farm demonstration network.)

The innovations of the GEB worked so well that it was decided in 1908 to create a parallel corps of circuit riders, under the auspices of the Peabody Fund, to do for elementary schools what was already being done for high schools. Operating under the title of "supervisors of rural schools," they were to focus on the smallest, poorest southern counties, where, according to one survey, a quarter of the white children of school age did not "even enter a school." The new circuit riders would be attached to the department of education in each state, although their salaries would be paid by the Peabody fund and they would have permanent tenure. The oversight of the program fell to the general agent of the Peabody Fund, Wickliffe Rose.

Like Frederick Gates, Rose was the son of a clergyman. Born in 1862 in Salisbury, Tennessee, he was educated entirely in his home state. After receiving a bachelor's and a master's degree from the Uni-

*Wickliffe Rose*

versity of Nashville, he started teaching philosophy there and at the affiliated Peabody College for Teachers. Although he had a passion for his subject (later in his life, he told a friend that he still read Plato and Aristotle "with the thrill of those early years"), his gift for administration led to his appointment as dean at both institutions in 1904. Three years later he left the campus to join the Peabody Fund.

Aware that the success of the new program would be determined by the caliber of the circuit riders, Rose personally examined the qualifications of each candidate. Describing the ideal rural school circuit rider, he wrote, "Our man must do pioneer work. He must not hesitate to get off the railroad out into the remote places." By the end of 1912, with "supervisors of rural schools" at work in twelve states, the first signs of progress were visible. Hundreds of "demonstration schools" had been organized, thousands of children were being bused or "horse and wagonned" to centrally located schools (which made more efficient use of limited resources than many scattered one-room schoolhouses), and, as an unexpected by-product of these efforts, an adult-education drive was sparked that not only taught thousands of unlettered men and women to read and write but also created a wider constituency for educational uplift in the most backward counties of the South.

As for Rose himself, his drive and acumen as general agent of the Peabody Fund so impressed the guiding figures of the Rockefeller philanthropies—especially Gates, Buttrick, and John D., Jr.—that in 1910 he was tapped to head a far larger enterprise: the Rockefeller Sanitary Commission for Eradication of Hookworm Disease.

The hookworm project had won the enthusiastic approval of John D., Sr., because it promised the kind of logical, step-by-step progress toward an unmistakably worthwhile goal that the Rockefellers and their aides were always on the lookout for.

The facts seemed clear. Hookworm, a debilitating parasite, was endemic in the southern states. Its life cycle was well known. The eggs hatched in the ground and could survive only in a frost-free climate. The larvae entered a person's body through the soles of the feet. The first symptom was a mild discomfort known as "ground itch," and in many cases that was the worst of it. But once a worm found its way to the lining of the intestines, it could live there for as long as twelve years; and its eggs (excreted in the host's fecal matter) became a source of new infections. With repeated contact, a victim could be host to as many as five thousand worms at one time. Such an infestation led to anemia, which drained the victim's strength and may have been responsible, in part, for the stereotyping of the barefoot southern sharecropper, white or black, as "lazy" and "shiftless." The most severe cases were marked by emaciation, stunted growth, swollen joints, twisted posture, and an uncontrollable urge to eat dirt.

In 1909 all this information about hookworm was brought to the attention of the advisers of John D., Sr., by a dynamic public health officer named Charles Wardell Stiles. Dr. Stiles was a frustrated man. Here was a disease that affected some two million Americans. Not only was the cause known, but a cheap and effective cure had recently been developed in Italy. With about fifty cents worth of an easily available pharmaceutical (thymol) mixed with Epsom salts, a victim could rid himself of worms in eighteen hours. Preventive measures (to break the cycle of reinfection) were also simple; individuals had to wear shoes, and communities had to close or upgrade primitive latrines that allowed the eggs passed in fecal matter to seep into nearby soil. Why, then, was hookworm still a major health problem?

Stiles had the answer to that question too. Official apathy and lack of funds had up to now stymied all his efforts to mount a regional antihookworm crusade. But with the proper backing, the problem could be virtually eliminated in the American South in a short time. After listening to Stiles make his case, Buttrick and Gates became converts. Dr. Simon Flexner of the Rockefeller Institute, who had heard Stiles lecture at Johns Hopkins, vouched for his scientific reliability. And when Gates laid the matter before John D., Sr., the result was a pledge of $1 million to launch a campaign to eradicate the disease once and for all.

Although the Rockefeller Sanitary Commission was set up as an independent organization, with its own budget and a separate office in Washington, D.C., Rose adopted the methods that had already proved effective in the school and farm-demonstration campaigns of the GEB. A survey was conducted to pinpoint centers of infestation; each state was induced to hire a director of sanitation, whose job was to mobilize public officials and private citizens and to get the word about hookworm (along with free medication) to every county afflicted with the disease.

Eventually, the campaign against hookworm reached over a thousand counties in eleven states; more than two million people attended some twenty-five thousand public meetings; millions of informative pamphlets were distributed, and a quarter of a million rural homes were inspected by commission circuit riders, who explained the importance of wearing shoes and maintaining modern sanitary facilities. By 1913 some five hundred thousand people had been treated for hookworm; and the gospel of public health—the good news that the ravages of disease can be halted and prevented by carefully planned community action—was winning adherents throughout the area.

Not surprisingly, both Gates and Rose were eager to repeat the success of the Sanitary Commission on an even wider scale. The chartering of the Rockefeller Foundation by New York State, in the spring of 1913, provided the machinery for internationalizing the work that Rose had begun in the American South. On May 22, 1913, at the first meeting of the RF's Board of Trustees, Rose was asked to draw up a plan of action. A month later he was named director of the International Health Commission of the Rockefeller Foundation. A resolution drafted by Gates empowered Rose to "extend to other countries and people the work of eradicating Hookworm Disease as opportunity offers, and so far as practicable to follow up the treatment and cure of this disease with the establishment of agencies for the promotion of public sanitation and the spread of the knowledge of scientific medicine."

In taking over the work of Rose's Sanitary Commission as its first order of business, the Rockefeller Foundation inherited the traditional missionary outlook characteristic of the General Education Board as well as the faith in the new profession of science that had prompted the founding of the Rockefeller Institute for Medical Research more than a decade earlier. The new Foundation would have to learn how to reconcile these two approaches before it could distinguish itself as the leading patron of science in the first half of the twentieth century. The man who worked out the compromise was Wickliffe Rose.

# 9

# A Missionary for Science

A tug-of-war between the missionary and the scientific approaches to philanthropy is evident in the earliest discussions about the corporate undertaking that eventually became the Rockefeller Foundation. On June 3, 1905, Frederick Gates wrote one of his frank letters to his employer, John D. Rockefeller, Sr., about the "ultimate uses" of the family fortune. The General Education Board was two years old and the Rockefeller Institute less than four; the large gifts to the University of Chicago continued. But Gates knew that these benefactions had barely scratched the surface of Rockefeller's wealth, which was "rolling up like an avalanche." He warned, "You must keep up with it! You must distribute it faster than it grows! If you do not, it will crush you and your children and your children's children."

Gates not only defined the problem; he also offered a definitive solution. The time had come, he declared, for Rockefeller to establish "several permanent corporate philanthropies . . . dedicated to and legally secured for the service of mankind." To Gates, it was primarily a question of "moral responsibility." Those men who accumulated great wealth were, as Andrew Carnegie had argued, naturally the best fitted to give it away. If Rockefeller did nothing, control of the money would eventually "pass into the unknown, like some other great fortunes, with unmeasured and perhaps sinister possibilities." To ensure proper control of the "legally incorporated endowment funds" that Gates had in mind, it would be necessary to delegate authority to carefully chosen men who would "administer these trusts now and in successive generations."

Gates had already introduced John D., Jr., to the world of corporate benevolence, and he took care to include the young heir (whom he referred to as "Mr. John Junior") in his oracular admonitions. Junior, perhaps daunted by the tasks facing the General Education Board, expressed some doubts about the scale of the philanthropies that Gates envisioned. Where, he wondered, could "able and efficient" trustees be found to oversee such vast undertakings? But Gates swept aside this objection by asserting that the very size of these "endowment funds" would attract "the attention and the intelligence of the world, and the administration of each would command the highest expert talent."

As for the specific social needs that these philanthropies might address, Gates had a few suggestions: "the promotion of a system of higher education in the United States," "the promotion of medical research throughout the world," "the promotion of the fine arts and refinement of taste in the United States," "the promotion of scientific agriculture," "the promotion of Christian ethics and Christian civilization throughout the world," and "the promotion of intelligent citizenship and civic virtue in the United States."

This admittedly "cursory, tentative and illustrative list" offered a characteristic mix of missionary presumption (clearly, Gates already had some ideas about what constituted civic virtue) and obeisance before the new gods of science and research.

Apparently, John D., Sr., was quite comfortable with this mix. His purely personal philanthropies were largely religious in nature and weighted heavily toward missionary work, which was customarily divided between "home" and "foreign" fields. His interest in "improving" the South, for example, had predated the General Education Board by many years; his wife's father had been active in the Underground Railroad, which helped runaway slaves escape to Canada before the Civil War, and he was an early supporter of a Baptist seminary for Negro girls in Atlanta that evolved, with Rockefeller help, into Spelman College.

At the same time, his experience in the oil business had taught him to appreciate the practical benefits of scientific research. If anything, Rockefeller's belief in "research" was more broadly based than Gates's somewhat mystical faith in "scientific medicine." Rockefeller seems to have viewed scientific research as an extension of his own penchant for detailed fact-finding. In a speech he made in 1909 (which he included in his *Random Reminiscences of Men and Events*), he remarked, "My interest in the University of Chicago has been enhanced by the fact that while it has comprehensively considered the other features of a collegiate career, it has given so much attention to Research. . . ."

His very predilection for "Research," however, led him to move cautiously in the establishment of the great corporate philanthropies that Gates proposed. In such matters it was not his style to impose his preferences on others. (As we have already noted, he was a confirmed believer in homeopathy at the time he agreed to finance the Rockefeller Institute for Medical Research.) But one thing is clear: whether he admired bigness for its own sake (as his detractors claimed) or whether he saw centralized control as a means of reducing waste and inefficiency (as he himself insisted), the architect of the Standard Oil Trust did not need Frederick Gates to convince him of the advantages of thinking big. By the end of 1906 he was ready to take Gates's argument one step further by turning over "considerable sums of money" to one large trust, which would be devoted to "philanthropy, education, science and religion."

His son still had misgivings about philanthropic gigantism. After consulting with Gates and a family attorney named Starr J. Murphy, Junior came up with a counterproposal: three trusts, with interlocking boards of trustees, each trust to be authorized to dispense money in a particular field of philanthropy, with John D., Sr., retaining a "veto power" over the decisions of the boards during his lifetime. In a letter to his father dated December 31, 1906, Junior listed the specific functions of the three trusts "in the order of their importance as we see it." The first would specialize in promoting "Christian civilization" abroad; the second would assume the same task at home; and the third would be primarily a holding company for funds earmarked for eventual use by the University of Chicago, the GEB, and the Rockefeller Institute.

What happened between the writing of this letter in 1906 and the decision, in the spring of 1909, to seek a broad-based charter for a multimillion dollar foundation can be briefly summarized as a victory for bigness—and flexibility. The elder Rockefeller's clear preference was for a single, large philanthropic trust, but all efforts to spell out in advance the responsibilities of this trust were abandoned in favor of a charter containing some fancy phrases that left its future course of action entirely open: "to promote the well-being and to advance the civilization of the peoples of the United States and its territories and possessions and of foreign lands in the acquisition and dissemination of knowledge, in the prevention and relief of suffering, and in the promotion of any and all of the elements of human progress." (In this proposed charter, even Junior's qualifying "Christian" has been stripped from "civilization.")

If the Rockefellers and their advisers thought they could disarm suspicion by creating a philanthropic trust with no legal constraints on its ability to act for the public good, they were wrong. Although the General Education Board had been granted a federal charter in 1903, that was before scandalous revelations about Standard Oil executives and the government's successful prosecution of its antitrust suit had further tarnished the Rockefeller name. A bill to charter the Rockefeller Foundation was held up in Congress for three years. After numerous revisions and much overheated rhetoric (in which Rockefeller was accused of trying to give away his cake and have it too), it became clear that Congress was in no mood to place its stamp of approval on any kind of Rockefeller philanthropy.* Early in 1913 the more compliant New York State legislature passed an act of incorporation, which gave the Rockefellers exactly what they wanted: carte blanche to practice philanthropy on the grandest of scales without prior restrictions.

The stated purpose of the new Foundation was simply to "promote the well-being of mankind throughout the world." In other words, all details of the program—including the balance between the disseminating of knowledge missionary-style and the discovering of new knowledge through research—were to be worked out later.

The first Board of Trustees to tackle this problem consisted of nine men. Besides the two Rockefellers, there were Gates, Starr Murphy, Simon Flexner of the Rockefeller Institute, Wickliffe Rose, Charles O. Heydt (a longtime Rockefeller aide), Harry Pratt Judson, president of the University of Chicago, and Jerome D. Greene, who had been general manager of the Rockefeller Institute for two years (and secretary to the Harvard Corporation before that), At the first meeting, John D. Rockefeller, Jr., was elected president of the Board, and

*In an effort to convince Congress that he had no ulterior motives, Rockefeller at one point proposed that the trustees of the Foundation be selected subject to approval by a panel consisting of the president of the United States, the chief justice of the U.S. Supreme Court, the Speaker of the House of Representatives, and the presidents of Yale, Harvard, Columbia, Johns Hopkins, and the University of Chicago.

Jerome Greene, the secretary, was made responsible for day-to-day operations.

By the end of 1914 John D., Sr., had endowed the Foundation with securities worth more than $100 million. But although he remained a trustee for ten years, Senior never attended a meeting, in keeping with a policy he had articulated several years before: "I have not had the hardihood even to suggest how people, so much more experienced and wise in those things than I, should work out the details even of those plans with which I have had the honor to be associated."

This is not to say, of course, that the founder took no interest in the new venture that bore his name. The Board of the Foundation was made up entirely of men of similar tastes and values to his own (many were on his payroll in one way or another); and although the charter did not give Rockefeller the "veto power" that his son had originally proposed, it was highly unlikely that such a group of men would take any action that violated the deepest beliefs of the founder.

Beyond that, John, D., Sr., had reserved to himself the right to spend up to $2 million of Foundation income each year on "such specific objects within the corporate purposes of the Foundation as [he might] from time to time direct." In practice, this special category of gifts (which came to be known as Founder's Designations) represented Rockefeller's personal charities transferred to the Foundation's books. In the beginning such gifts made up a substantial proportion of the monies dispensed by the Foundation.

In 1915, for example, the RF spent a total of $3,862,444.40. Of this sum, $66,542.48 went to administration; $441,301.23 went to Rose's International Health Commission; $582,339.59 went to war relief; and $1,314,561.11 went to organizations designated by the founder. The largest of these Founder's Designations was a $500,000 gift for new construction at the Rockefeller Institute. The smallest was a $50 donation to the "Working Women's Protective Union." The list also included $4,050 to Charles B. Davenport's Eugenics Record Office, in Cold Spring Harbor; this gift was specially earmarked to pay the salaries of six field workers who were studying the family backgrounds of inmates in state institutions ("the feeble minded, the insane and others"), with an eye toward persuading the state to "take more vigorous steps to diminish the number of them who are reproduced thru bad heredity."

But more typically, the Founder's Designations supported missionary work carried out under religious auspices. In 1915 Rockefeller gave $250,000 to the American Baptist Foreign Mission Society, $50,000 to the Foreign Mission Board of the Southern Baptist Convention, $55,000 to the International Committee of the YMCA, and so on.

While all these gifts could be seen as promoting "the well-being of mankind throughout the world," it soon became apparent that the kind of philanthropy to which Mr. Rockefeller was personally drawn made an awkward fit with the larger purposes of the Foundation. Although the trustees moved slowly to develop a program of their own (and especially slowly after the beginning of the First World War), a memo presented to the board by Greene on October 2, 1913, was quite explicit in its recommendations about what the Rockefeller Foundation should *not* do.

The list of no-nos included aid to individuals or purely local enterprises; the permanent support of any institution, no matter how worthy, that the Foundation did not itself control; gifts "in perpetuity," especially those with narrow restrictions placed on their use; gifts to community enterprises that were not matched in some way by donations from other sources; and support for projects "of an immediately remedial or alleviatory nature, such as asylums for the orphan, blind or cripples." (War relief, of course, was a rule-confirming exception.)

*Jerome D. Greene*

More positively, Greene, whose mother and father were both missionaries, suggested that the Foundation concentrate on seeking out solutions "which go to the root of individual or social ill-being and misery."

Many of the Founder's Designations seemed to violate the spirit, if not the letter, of this memo. On July 19, 1917, John D. Rockefeller, Sr., publicly waived all future rights to allocate Foundation income to projects of his own choosing. The reason he gave was the "increasing demand" being made on the Foundation as a result of the war. In 1917 the Board voted nearly $6 million for war relief, which amounted to more than half the total disbursements for the year. To meet all their obligations, the trustees dipped into principal for $5 million and accepted another $5.5 million as a special gift from Senior.

But ending the category of Founder's Designations served another purpose as well. One benefit of corporate philanthropy (which the Rockefellers certainly appreciated) was the disassociation of the donor from the social consequences of his benefactions. As long as a significant portion of RF income was under Rockefeller's personal control, confusion between his private charities and the Foundation's more ambitious experiments in "scientific giving" was inevitable.

When the Founder's Designations ended, in 1917, so did the Foundation's direct involvement in strictly religious undertakings. Responsibility for supporting such activities was transferred to the business office of the Rockefeller family, at 26 Broadway. But this did not end the RF's involvement in missionary enterprises of a nonreligious nature. The trustees continued to weigh the advantages (and drawbacks) of two very different kinds of programs: those that supported the search for something new and those that supported the dissemination of already-established ideas and techniques to people who had not yet been exposed to them.

As we have seen, modern science had an important role to play in both kinds of programs. The campaign against hookworm in the American South marshaled weapons that had been developed by scientific researchers, but the campaign itself was conceived by Rockefeller, Rose, and their colleagues as a missionary endeavor. What had appealed so strongly to John D., Sr., about the original Sanitary Commission was precisely that "everything" was already known about hookworm. The eradicating of hookworm (like the improving of education in the South) was a problem of getting proven goods to the unenlightened. This took imagination, as Rockfeller had reason to know. Before Standard Oil representatives could sell kerosene in China's huge market, they had had to give away millions of cheap kerosene lamps to the Chinese masses. But the imagination called for was of a different order from that which characterized the research scientist.

Nevertheless, it was Rose, pursuing his missionary campaign against disease on a global scale, who came to the conclusion that the Foundation had better make the search for new knowledge its primary concern. What led him to this conclusion was a simple discovery: the knowledge that he needed to bring his campaign to a successful conclusion did not yet exist.

# 10

# The Greening of American Science

Nineteen thirteen, the year the Rockefeller Foundation was finally chartered by the New York State legislature, was a critical year for the factions contending for financial control of the scientific community in the United States and abroad.* In England, Parliament created the Medical Research Committee to look into the question of how government funds should be disbursed to qualified researchers for maximum impact. This body, later known as the Medical Research Council, would play a significant role, together with the Rockefeller circuit riders, in the development of penicillin as a life-saving antibiotic in the Second World War (see chapters 31–41).

On October 24, 1913, Edward Pickering, who two years earlier had dunned Andrew Carnegie without success for a $100,000 donation, asked the new Rockefeller Foundation to set aside funds for the purpose of aiding "men of genius" in the sciences. Secretary Jerome Greene's reply was not encouraging; he told Pickering that public health and social welfare were the trustees' primary concerns.

But this rejection was only the beginning of a long struggle for access to the Foundation's purse. In November, George Hale, Pickering's archrival in astronomical entrepreneurship, decided to raise his sights

*In Berlin the Kaiser-Wilhelm-Gesellschaft, a society for the promotion of research that had been established with money from private industry in 1911, was in the process of building its first research institutes independent of the German universities.

to include the entire spectrum of modern science. Like Pickering's, Hale's goal was to create large endowment funds whose income could be used to finance scientific research on a permanent basis. But his vision extended beyond the universities, museums, and learned societies of the academic establishment. His father, a manufacturer of elevators, had built a solar observatory near Chicago for Hale's personal use after his graduation from MIT in 1890; six years later the elder Hale acquired a 60-inch telescopic mirror that was eventually installed (under Hale's direction and with funds supplied by the Carnegie Institution) in the new Mount Wilson Observatory in 1908.

Not surprisingly, Hale was eager to involve industry in the financing of science. (One historian has called him "the J. P. Morgan of the scientific community.") And while he wanted to avoid government control, he was savvy enough to see that he needed a permanent fund-raising "instrument" that would inspire confidence in corporate donors while ensuring him access to levers of power within the government. His chosen instrument was the 130-member National Academy of Sciences.

Hale's plan was to turn to the moribund NAS into an elite body—not unlike Britain's Royal Society—that would stand above the narrow interests of universities and specialized institutes. This revitalized NAS would not only direct research funds to worthy recipients but also have its own building with its own research laboratories and a lecture hall for the education of the public—not unlike London's Royal Institution, where Michael Faraday had found a home in the early nineteenth century. Hale even played with the notion of stationing NAS representatives as "scientific attachés" in London and Paris.

But Pickering was not about to surrender the field without a fight. In December 1913 he countered with a plan to set up a national "clearing house" of science within the American Association for the Advancement of Science. Known as the Committee of One Hundred on Scientific Research, this group would be far more broadly based than Hale's elitist NAS, welcoming medical men, engineers, and even nonscientists to its deliberations. (Between 1900 and 1914 the membership of the AAAS itself jumped from just under two thousand to over eight thousand.)

Speaking for the new committee, Pickering wrote again to Jerome Greene in 1915 and once more in 1916, asking for an appropriation of up to $50,000, to be parceled out "in small grants, not as a rule exceeding $500," to investigators chosen by Pickering and his colleagues. In a covering letter Pickering noted that during his forty years as director of the Harvard Observatory, he had mastered "the application of business methods to scientific research on a large scale," while overseeing expenditures of nearly $2 million.

Greene's reaction, as expressed in memos to the trustees and letters to his colleagues, reveals the confusion that the early leaders of the Foundation felt when the question of supporting science came up. The secretary was known to his colleagues as a man with a "crisp, clear, incisive mind." Tall, dignified, a bit aloof at times, he could argue a case with "a high degree of persuasive charm" when he so chose. Yet on the issue of science patronage he waffled.

Greene recognized that "the promotion of research through subsidy has not been very satisfactorily worked out in this country"; both the Carnegie Institution and the Rockefeller Institute had abandoned the support of outside investigators in favor of in-house research, which was much easier to monitor. He was especially concerned that the United States was not doing enough to foster scientific development in light of the wartime devastation of the European scientific community.

One study had shown that American universities were spending $100 million a year on teaching and virtually nothing for "research in pure science." No wonder the prewar scientific output of the United States had compared so "unfavorably" with that of far smaller and less wealthy European nations—the rule-proving exception being astronomy, which had benefited from the extraordinary fund-raising efforts of Pickering and Hale.

In addition, Greene was convinced that scientific research in general was a sound investment that normally paid "very great" social dividends. But when he thought about trying to persuade the trustees of the Rockefeller Foundation to undertake a major commitment to science patronage, he got cold feet.

Preoccupied as they were with emergency war relief and with missionary-style programs in public health, the trustees were not disposed to look beyond the well-known fact that "a considerable proportion" of investigations in basic science produced no tangible results, at least in the short run. Greene knew this from personal experience; at one of the very first Board meetings he had suggested that the foundation create "a sort of University of Human Need," which would devise philanthropic programs based on "scientific research" into such problems as alcoholism, venereal disease, and mental illness. But the trustees were not interested.

The frustrated Greene confided to Simon Flexner at the Rockefeller Institute, "I cannot help thinking that a benevolent plutocrat who didn't have to bother with precedent and policies, as you and I have to do . . . could do a lot of good for science by arbitrarily selecting now and then a man or a group of men to be aided by the subsidy method." Even if as many as three-quarters of the subsidized investigations proved "futile," the quarter that bore fruit would undoubtedly "justify the whole expenditure."

This is not what Greene told the trustees on May 24, 1916, when he passed along Pickering's latest proposal with an explanatory memo of his own. After noting that "the great argument for aiding research is that knowledge breeds knowledge," he concluded, "It would be premature to present at this time a definite proposition for the aiding of research by the Rockefeller Foundation"—especially since the specific request from the AAAS involved "some serious difficulties of administration." Instead, Greene recommended bouncing the whole question of the "needs of scientific research" back to a special committee that would produce a report by the beginning of 1917. The trustees, with an almost audible sigh of relief, agreed; a few months later Greene left the Foundation to join an investment firm, and no committee was ever formed.

As it turned out, the manner and timing of the Foundation's entry into the complex field of science patronage was determined by events beyond the bounds of the RF boardroom.

The major event was the war in Europe, into which the United States was being inexorably drawn. In April 1916 a delegation from the National Academy of Sciences led by Hale met with President Woodrow Wilson to offer the services of America's scientific community in the defense of democracy. The president accepted the offer, and two months later the National Research Council (NRC) was officially established as the operating arm of the Academy, charged with coordinating research in time of war. Although the Committee of One Hundred was consulted, Hale had in effect stolen a march on Pickering; the NRC, not the AAAS, had become America's scientific clearinghouse. Hale was not modest about his achievement: "I really believe this is the greatest chance we ever had to advance research in America."

The NRC (which relied on a mix of public and private funding)

*George E. Vincent*

consisted of delegates from universities, industry, government agencies, and professional societies. But Hale, who pulled the strings, chose as executive director his own candidate, the University of Chicago physicist Robert A. Millikan, whose measurement of the charge of the electron was to win him a Nobel Prize in 1923. Scientists recruited through the NRC made major contributions to the Allied victory, especially in the development of sophisticated submarine and artillery detection devices. Even while this work was going on, Hale kept an eye on the future. If the NRC was to continue in its role as national coordinator of scientific research after the war, it would have to secure some kind of permanent funding. To this end, Hale set about forging an alliance with the Rockefeller Foundation.

In January 1917 John D. Rockefeller, Jr., relinquished his title as president of the Foundation and assumed the newly created post of chairman of the Board, which seemed more in keeping with his chosen role as grand strategist of the family philanthropies. His successor as president was George E. Vincent, the fifty-three-year-old head of the University of Minnesota. With Junior's change of title and Jerome Greene's resignation, Vincent in effect became executive officer of the RF, responsible for carrying out policies adopted by the Board. But Vincent, like Greene before him, also wanted a hand in setting Foundation policy.

An early proponent of social psychology, Vincent had been named to a series of increasingly important administrative posts at the University of Chicago by President William Harper, before Harper's own administrative failings led to his ouster by the Rockefellers in 1904. That same year Vincent was made full professor. Three years later he was appointed dean of the arts, literature, and science faculties; his success in boosting the academic prestige of the young university led to his call to the University of Minnesota in 1911. The job he did there (and as a member of the General Education Board, to which he was appointed in 1914) convinced John D., Jr., and his inner circle of advisers that Vincent was the right man for the RF presidency at this critical juncture.

# 11

# Battle Scars

Vincent was one of the most popular speakers on the circuit of the Chautauqua Institution, the adult-education center that had been founded by his father. It is no coincidence that the Rockefeller Foundation's public image in 1917 was in need of a little polishing. The Board's generous contributions to war relief were making a favorable impression. Nevertheless, suspicions persisted that this most ambitious of Rockefeller philanthropies was an attempt by John D., Sr., to hide his control of the American economy behind the façade of charity. The new president had the rhetorical skills to explain to skeptical audiences the Foundation's philanthropic mission; he also transformed the dry-as-a-ledger *Annual Report* into a readable document by adding the sprightly *President's Review.*

Yet, not long after his arrival in New York, Vincent became embroiled in a fierce internal struggle for control of the Foundation's destiny. His antagonists were the powerful directors of the Foundation's operating units, who resented the new president's efforts to check their free-spending, free-wheeling administrative style.

At this time the Foundation consisted of the Board of Trustees, the Central Office of Administration, and two operating units: Rose's International Health Board and the China Medical Board. The latter, created for the purpose of "developing a system of modern medicine in China," was unabashedly missionary in its outlook and methods. The notion of building a medical school in China had come from

Gates, after an earlier proposal to establish a major university there had run into opposition from entrenched church schools and Chinese officials. After much handwringing and several fact-finding expeditions to the Orient, the Foundation erected a medical school in Beijing, at a cost of more than $8 million.*

Construction of a vast campus, comprising fifty-nine buildings on a twenty-five acre site, was not completed until 1921. But by the time Vincent took office, in January 1917, both the China Medical Board and the International Health Board were functioning as virtually autonomous agencies, answerable not to the president of the Foundation but only to the Board of Trustees. Of course, as long as John D., Jr., had been president, his opinions, however diffidently offered, carried great weight. But this authority was not necessarily transferable to his successor. In fact, Junior had originally conceived of the Foundation as a "great holding company" that would fund, and possibly spin off, subsidiary boards, but would not itself become involved in operational details. His reasoning was that no layman, however accomplished in his own specialty, could possibly oversee day-to-day operations in a number of different technical fields.

As it happened, it was not lack of technical training but ignorance of the trustees' ambivalent attitude toward science (a subject Jerome Greene could have informed him about) that most seriously hampered Vincent's ambitions during his years as president of the Rockefeller Foundation.

In December 1917 Vincent presented the trustees with a "Forecast of Policy" that looked forward to at least one major postwar project, the creation of an institute of physics and chemistry. This independent facility would conduct "pure research" into the nature of matter and energy, much as the Rockefeller Institute for Medical Research tackled basic problems in the biological sciences. Indeed, the suggestion

*In 1928 the China Medical Board was severed from the Foundation and set up as an independent corporate entity with an endowment of $12 million. In 1947, when the Communists seized control of Beijing, all formal ties between the medical college and the China Medical Board were dissolved. But this drastic step may only have accelerated a process of Sinification that was already well advanced. Although the faculty and administration were overwhelmingly European in the early years, most responsible posts had passed into the hands of ethnic Chinese by the 1940s—evidence that the gospel of modern medicine, as preached by Frederick Gates, had indeed taken hold in pre-Communist China.

had originally come from Simon Flexner of the Rockefeller Institute, who felt that fundamental discoveries in physics and chemistry would benefit his own researchers. For the more practical-minded trustees, Vincent pointed out that the new institute would be of "enormous value" to the United States in the stiff industrial competition that was bound to follow the war.

However, opposition to the proposal immediately developed on two fronts. Although George Hale himself was strongly in favor of the new institute, Robert Millikan could not see how a suitable staff could be recruited without stripping the physical science departments of American universities of all their best researchers. A sample of eminent scientists polled by Millikan and Vincent divided on the issue; while no one wanted to remove himself from consideration in case the proposal went ahead, many expressed fear that a centralized institute would come under government influence—which all agreed would be a bad thing.

As Simon Flexner was well aware, similar arguments had been advanced when the Rockefeller Institute for Medical Research was proposed, and they had proved groundless. It is not difficult to imagine the trustees of the Rockefeller Foundation forging ahead with the new institute if they had been eager to create a counterpart to the internationally renowned Institute for Medical Research. But a great deal had happened since 1901 (and especially since 1913) to change the thinking of the men into whose hands John D. Rockefeller, Sr., had consigned the bulk of his fortune.

During its first two years of operation, the RF Board had seriously considered the establishment of a whole network of research institutes and had actually set up machinery to look into such controversial topics as "mental hygiene" and "industrial relations." But before any of these plans could mature, the fledgling Foundation suffered an almost mortal blow to its confidence. The problem began in 1914 when the RF announced that it was funding an "Investigation of Industrial Relations" under the direction of a former Canadian minister of labor, Mackenzie King.

To say that the timing of this announcement was spectacularly bad would be an understatement. Only a year earlier a bloody confrontation between state militia and striking workers at a Colorado mining camp owned by the Rockefeller family had left dozens dead and wounded, including women and children. To restore order after this so-called Ludlow massacre, President Woodrow Wilson was forced to call in federal troops. The largest stockholder in the mining company was John D. Rockefeller, Jr., whose Foundation now proposed to look into the causes of "industrial unrest and maladjustment . . . in a purely objective way and with scientific accuracy."

Public outrage at what appeared to be a plutocratic shellgame—using an ill-gotten fortune to whitewash capitalistic oppression—found an outlet in a congressionally chartered commission of inquiry headed by U.S. Senator Frank P. Walsh. Under relentless questioning by Walsh, Junior disclaimed responsibility for the day-to-day conduct of mining-company officials during the Ludlow strike. His attempt to portray himself as a disinterested philanthropist who just happened to have some mining stocks in his personal portfolio was greeted with scorn and derision. John Lawson, a leader of the United Mine Workers, spoke for more than his own membership when he declared, "It is not their money that these lords of commercialized virtue are spending, but the withheld wages of the American working class."

Nevertheless, following the advice of King and the publicist Ivy Lee, Junior managed to turn this humiliating experience into a personal triumph. He appeared at a series of open meetings in the mining

*John D. Rockefeller, Jr., 1915*

camps, where he sold the miners on King's compromise plan: the workers agreed to give up representation by the militant United Mine Workers and to join a newly organized company union, with which management promised to negotiate fairly.

Meanwhile, the RF trustees pondered the consequences for the Foundation of their poorly timed foray into industrial relations. One lesson seemed clear. In the future—the near future, at least—any actions that tended to blur the distinction in the public mind between the Rockefeller Foundation and the Rockefeller family were to be avoided. Among other things, this meant steering clear of direct involvement with economic issues. A proposed "Institute for Economic Research" was the first casualty of this change in attitude.

But an even more important caveat had been impressed on the policymakers of the Foundation. This lesson was summed up in a history of the RF written by Raymond B. Fosdick after he stepped down from a twelve-year term as president in 1948. According to Fosdick, who began representing Rockefeller interests as a young attorney in 1910, the RF trustees were so spooked by the industrial-relations fiasco that they decided to alter their basic strategy for spending the Founder's money. "Except for a narrow range of noncontroversial subjects, notably public health, medicine, and agriculture," Fosdick wrote, ". . . the Foundation [would] become primarily not an operating agency but a fund-dispensing agency." Only in this way could both giver and recipient be freed from any "suspicion of ulterior interest."

In essence, the trustees were giving up the authority to monitor the day-to-day performance of their beneficiaries; as compensation, the Foundation would be insulated from direct responsibility for the use of its funds. As we will see, this insulation did not always work as intended, and future boards did not always adhere to a strict hands-off policy. But in 1919 the trustees made it clear that they had no enthusiasm for a highly visible institute of physics and chemistry, especially in light of objections from a significant number of scientists and university spokesmen.

A long round of negotiations between Hale, Millikan, and Vincent brought about a complete shift in emphasis. Early in 1919 Vincent conveyed to the trustees a series of proposals from the NRC, all aimed at strengthening the American scientific community while enhancing the role in that community of the National Research Council. The Foundation would build and equip new laboratories at five major

American universities (to be selected by the NRC) and provide money to release from teaching duties a small group of faculty members (whom the NRC would certify as accomplished researchers). In addition, the Foundation would fund fifty fellowships a year to enable promising young scientists (nominated by the NRC) to study at the five selected schools.

Having just been freed from the drain of war relief (a total of $22.5 million between 1914 and 1918), the RF trustees were ready to take on new challenges in the postwar world. Yet they were wary of getting too deeply involved in academic politics. The zeal with which departmental chairmen and tenured professors divided up knowledge into so many disciplines and then defended their own intellectual turf struck some members of the RF Board as counterproductive.

By contrast, Hale's National Research Council projected a solid image of patrician professionalism. Not only did the NRC impress Vincent and his colleagues as a tightly run operation; it offered the Foundation the same kind of buffer from controversies over the selection of grant recipients that the Royal Society provided to the British Parliament.

Following another round of negotiations between Vincent and the NRC, a compromise emerged that proved acceptable to everyone. Direct aid to universities was out. So was the research institute that Simon Flexner had argued for and that Vincent had seen as an opportunity to bolster the authority of his office vis-à-vis the International Health Board and the China Medical Board. On April 9, 1919, the Rockefeller Foundation announced a grant of $500,00 to the National Research Council to administer a five-year program of postdoctoral fellowships in the natural sciences. Recipients could go anywhere in the world to advance their studies. In the first year of the program, thirteen fellowships were awarded, six in physics and seven in chemistry.

The funding of fellowships on a global scale was not absolutely new. The precedent had been set at the beginning of the century by the Rhodes Scholarships, as provided for in the will of Sir Cecil Rhodes. The first Rhodes scholars, chosen from all parts of the English-speaking world, arrived at Oxford University in 1902. That same year the General Education Board announced grants (of $300 each) to schoolteachers in the South who wanted to further their professional training. Among the first actions taken by the trustees of the Rockefeller

Foundation in November 1914 were awards to four Chinese for the study of medicine. And in 1917 Wickliffe Rose's International Health Board began giving fellowships to Latin Americans for advanced training in medicine and public health at American or European universities.

But until 1919 all Rockefeller-funded fellowships had been designed as adjuncts to other programs that could be justified on more traditional philanthropic grounds. The Chinese fellowships anticipated the trustees' decision to build a first-rate medical college in Beijing. The International Health Board fellowships were part of an effort to build up a cadre of medical investigators who could carry on, in their home countries, the battle against disease that Rockefeller men and money were waging worldwide.

The grant to the NRC was a departure in that the Foundation, for the first time, publicly embraced the support of scientific research by independent investigators. In fact, the program was structured to emphasize the professional, rather than the academic, status of the candidates. Although university scientists dominated the selection committees, they sat as representatives of the scientific community and not of their own schools. And while the fellows were expected to return to their home institutions, they were to be selected for their potential contributions to science in general rather than to medical research or to any other discipline with immediate practical applications.

There was no intent to nudge young researchers along any particular path or toward any particular goal. The unspoken assumption behind the RF-NRC fellowship program was that, simply by pursuing scientific truth, scientists would automatically promote the well-being of mankind, as if some "invisible hand" were yoking progress in the sciences to social progress.

# 12

# Triumph—and Disappointment

The word "philanthropy" derives from the Greek for "love of man." In our time, of course, "philanthropy" is reserved exclusively for a love that is expressed with money. When Frederick Gates first understood that John D. Rockefeller had to become a wholesale philanthropist, he realized that this could be best accomplished through what he called "scientific giving"—by which he meant a rationally organized system of getting goods and services to large numbers of needy people. Since their needs exceeded even Mr. Rockefeller's resources, it was necessary to concentrate the philanthropists's buying power where it would do the most good. Without apparent hesitation Gates chose to invest in knowledge.

The principal purpose of the Sanitary Commission was to distribute knowledge that already existed to those who needed it. The principal purpose of the General Education Board was to give people access (through better schools) to already existing knowledge. The principal purpose of the Rockefeller Institute was to wrest from nature new knowledge. As long as this new knowledge was assumed to be closely connected with the healing arts, there could be no quibbles about its social value. Even a small chance of curing yellow fever or cancer was clearly worth a considerable investment. But once the philanthropoids who administered the Rockefeller benefactions accepted the notion that knowledge of *any* kind was worth pursuing because it *might,* at some future time, advance the well-being of mankind, knotty issues of selection and accountability arose.

These were issues that Jerome Greene had worried over in his memos to the Board of Trustees, and that John D. Rockefeller, Jr., had addressed in his letters to his father before the Foundation was chartered. How could laymen support the search for knowledge in a rational way when their comprehension of the search was necessarily limited to superficial summaries and paraphrases? Even without professional credentials a patron of the arts might come to be accepted as a connoisseur, even an authority, in certain genres and periods. But how could a science patron who was unfamiliar with advanced mathematical notation or the intricacies of quantum mechanics ever hope to make an intelligent allocation of limited resources between competing grant proposals in physics or chemistry?

The NRC fellowship program offered one way out of this dilemma: find surrogates with impeccable credentials in the scientific community and pay *them* to make the actual decisions. This prudent if somewhat bloodless approach could easily have become the model for the Foundation's subsequent ventures into science patronage. But all counsels of caution, however sensible, were about to be swept away by the indefatigable Wickliffe Rose, who was already formulating plans for science patronage on a much grander scale.

Although Rose was a professor of philosophy, his interest in science owed little if anything to philosophical considerations. Rather, it would seem that Rose was led to science patronage by his involvement in the "missionary" side of the Rockefeller philanthropies.

When he launched the RF's International Health Commission, in 1913, he began with the same strategy that he had used earlier in the Peabody Fund and the Sanitary Commission and that Gates, Buttrick, and others had pioneered on the General Education Board.

In the best Rockefeller tradition, he made no move to dispense funds, goods, or knowledge to the less fortunate until he and his staff had conducted extensive fact-finding expeditions. Rose personally toured England, the British West Indies, and the Near and Far East (Egypt, Ceylon, Malaya, the Philippines). Then, in cooperation with local governments (a point he insisted on), he organized antihookworm campaigns in more than fifty countries where the hookworm larvae thrived because winter temperatures remained above fifty degrees Fahrenheit.

In many places these campaigns produced results as gratifying as those obtained in the American South. But new problems emerged as well. As early as 1914 it became clear to Rose that the expanded work in public health was being held back by a lack of trained workers. The skills required to carry out large-scale programs of preventive medicine—dealing with populations numbering in the millions—had little in common with the skills that an ordinary physician needed to treat patients on a one-to-one basis.

With Rose as the driving force, the Rockefeller Foundation built and endowed the School of Hygiene and Public Health at Johns Hopkins University. This institution (which opened in 1918) offered courses in such metamedical disciplines as sanitary engineering, epidemiology, administration, biostatistics, and bacteriology. The concept won quick acceptance; the Johns Hopkins school became known as the "West Point of public health." A similar RF gift went to Harvard in 1921; and during succeeding decades the Foundation spent more than $25 million to create schools and institutes of public health around the world, from Zagreb to Calcutta, from Copenhagen to São Paulo.

But the creating of institutions did not guarantee a supply of properly qualified students or a flow of trained graduates to countries where they were most needed. Accordingly, Rose and his associates developed a program of international fellowships. The first award was made on May 22, 1917, to Dr. Carlos P. Chagas, a Brazilian pathologist who studied in the United States, France, and Germany and went on to a distinguished career in his home country. Among other things, he identified the parasite responsible for a debilitating tropical disease (spread by the bite of a beetle) now know as Chagas's disease. By 1923 the number of fellows appointed annually had risen to over a hundred; and a survey conducted in 1927 showed that the original purpose of the program—to build up an international cadre of public-health workers—had been achieved: nine out of ten returning fellows made their careers in public health.

There remained other problems, for which Rose had no immediate answers. The mission of his International Health Commission (which became the International Health Board in 1916) was to conquer disease by the systematic application of knowledge that *already* existed. Research per se remained the province of the Rockefeller Institute. At first Rose saw no reason to question this division of labor. But at almost every turn he and coworkers came up against gaps in existing knowledge that hampered their efforts. Fact-finding expeditions were no help if the "facts" required were nowhere available. Slowly, Rose came to realize that to fight ignorance and disease on a worldwide scale, which was his ambition, he would have to prime the pump—by supporting researchers whose job was to wrest new knowledge from nature.

Not even the hookworm problem was as thoroughly understood as Charles Wardell Stiles had supposed. Eradicating the disease in any one country provided no more than a temporary victory, since immigrants from a still-infested area could quickly reinfect a "cleared" zone. Furthermore, the conditions under which the larvae thrived (or perished) were found to vary from locale to locale, and Rose's agents had to adapt their campaigns accordingly. They also had to learn to distinguish between a comparatively harmless hookworm infection (in which the parasites and the host's antibodies achieved a stable equilibrium, with little long-term effect on the victim's health) and the ravages of full-blown hookworm disease. The need for new information became even clearer when Rose and his associates extended their efforts to other diseases, like yellow fever and malaria.

On his first swing through the Orient, in 1914, Rose had found public-health officials, especially in British-controlled ports like Singapore and Hong Kong, preparing for a new epidemic of yellow fever. They feared that the imminent opening of the Panama Canal would bring their harbors into direct contact with shipping from the Gulf of Mexico, where the disease was a major health hazard. This was somewhat ironic since the building of the Canal itself had been made possible by the work of U.S. Army Major Walter Reed, who had discovered that yellow fever was transmitted by mosquito bite, and Colonel W. C. Gorgas, who had virtually eliminated the disease from Panama by vigorous antimosquito measures.

When Rose returned from the Far East, he went to Washington to ask Colonel Gorgas what could be done to block the spread of yellow fever. At their first meeting Gorgas told him just what he wanted to hear: the facts on yellow fever were in. Not only had the carrier of the disease been identified, but its habits had been well established. There were only a few "endemic centers" from which periodic epidemics erupted. Wipe out those "seedbeds." Gorgas said, and yellow fever could be "eradicated from the face of the earth within a reasonable time and at a reasonable cost." The Foundation hired Gorgas to do just that—and within a few years the incidence of yellow fever in South and Central America had indeed been greatly reduced.

But once again eradication proved a more elusive goal. Other "carriers" and other "seedbeds" were discovered. After many false starts and many personal tragedies—at least thirteen RF staff members contracted yellow fever during the course of their investigations, and six died—the New York laboratory of the Rockefeller Foundation developed the first effective vaccine against the disease in 1937. During the Second World War alone, more than thirty million people received this protection. Today a combination of vaccination and mosquito control has almost realized Gorgas's original dream of eradication. But for Rose and other pioneers in public health back in the 1920s the moral of the story was otherwise: it was never safe to assume that you knew all you needed to know about a given disease.

Similar learning experiences awaited RF workers who set out to conquer malaria, typhus, influenza, tuberculosis, and other age-old scourges of mankind. In most cases it was not enough to apply existing knowledge; a disease could not be controlled, much less eradicated, without fundamental advances in bacteriology, parasitology, epidemiology, and so on.

This recognition kindled Rose's growing interest in basic scientific research. Sometime around 1920 he began to jot down his thoughts about the pivotal role of science in modern life. Originally intended for his eyes alone, these notes ring with the optimism of the man and his time. The War to End All Wars was over. Now the War to End All Ignorance could begin. And the supreme weapon in that war was Science.

"All important fields of activity," Rose noted, "from the breeding of bees to the administration of an empire, call for an understanding of the spirit and technique of modern science." Nations unwilling to learn this truth were bound to be "dominated" by those that did: "Promotion of the development of science in a country is germinal; it affects the entire system of education and carries with it the remaking of a civilization.

Characteristically, Rose moved quickly from the general to the specific:

> Begin with physics, chemistry and biology. Locate the inspiring productive men in each of these fields, ascertain of each of these whether he would be willing to train students from other countries; if so, ascertain how many he could take at

> one time; provide the equipment needed, if any, for operation on scale desired.

Rose's goal was nothing less than an "international migration of select students" who would be not only trained but also "inspired" abroad, before returning to a life of "service" in their own countries.

To illustrate what he had in mind, Rose referred not to the fellowship program of the International Health Board but to an earlier episode in American intellectual history:

> Half a century ago America was doing but little for the development of science. Earnest young Americans studied with leading scientists abroad, particularly in Germany; returned to plant the seed in American soil. The seed germinated; the plant took root; science is now thoroughly transplanted in America and is propagating itself both at home and abroad. What has taken place in America may be repeated in other countries, and by systematic fostering the development may be hastened.

# 13

# Trickle Down

If one had to sum up in a single word the key to Rose's success as a philanthropoid, that word would be "confidence." From the early days of the Sanitary Commission, he had enjoyed the complete confidence of the Rockefellers, especially John D., Jr. To give Rose the scope his new vision of science required, Junior asked him to assume the presidency of the General Education Board in 1922. When it became clear that the charter of the GEB could not be amended to include work in other countries, Junior created the International Education Board to Rose's specifications.

The IEB began in January 1923 with no endowment fund as such. Junior simply pledged to give $200,000 per year, for ten years, to projects that Rose and his fellow directors selected. In a letter to Rose elaborating on this commitment, Rockefeller wrote,

> It is my thought that this new organization should feel its way along very modestly and carefully. . . . I am sure you know me well enough to feel that you would not want me to contribute to any enterprises or projects along this line which you could not convince me were important, timely and much needed.

Whatever Junior may have thought, Rose was not a man to attack a problem "modestly." Ten months later Rose sent Rockefeller a rough estimate of expenditures for the IEB's first fiscal year. The total was $348,000. Junior's response was measured. After requesting "a little more specific information" on the listed items (which ranged from support of eminent scientists like Niels Bohr to fellowship grants for young researchers), Junior deposited another $150,000 in the IEB account. Together with his earlier pledge of $200,000, this just covered the first year's budget.

After a similar experience the following year, Junior decided that the arrangement was unnecessarily cumbersome; in 1925 he handed over securities worth $20 million to the Board of Trustees, which was authorized to use both income and principal as it saw fit. At the same time, he declined to serve as a trustee, explaining, "It is a technical field, and I have no qualifications as an adviser."

Rose himself had no such qualms. The notes he had jotted down about science and education became the blueprint for the International Education Board. Actually, there was more to Rose's vision than the "promotion of science." He also hoped to export some of the "missionary" programs that the General Education Board had pioneered in the South, including demonstration farms and boys' and girls' clubs. But the main focus of his stewardship of the IEB was science—a focus that became clear during the first month of his first European fact-finding tour when, as he put it, "work started with a bang."

During December 1923 he traveled to London, Oxford, Cambridge, Manchester, Rothamstead, Dublin, Edinburgh, and St. Andrews to see the leading scientists of Great Britain in their laboratories. At each stop he laid out what he always referred to as his "scheme" for supporting science through grants to major institutions and university departments and through an international fellowship program.

On a typical day in London, he spoke to the chemist Frederick G. Donnan of University College; to the Dutch physicist Paul Ehrenfest of the University of Leiden, who happened to be passing through; to the physiologist A. V. Hill, who had just won a Nobel Prize for his work on muscle metabolism; and to Grafton Elliot Smith, a physical anthropologist whose new institute had been built with Rockefeller Foundation funds as part of a large grant to London's University College.

He found them to a man enthusiastic: "I have never been afield with a scheme that commanded such instant, intelligent and keenly interested response." J. J. Thomson, the physicist who won a Nobel Prize in 1906 for his discovery of the electron, assured Rose, "This is another example of the fact that Mr. Rockefeller has solved what has hitherto seemed an insoluble problem: namely, how to spend large sums of money wisely."

In his campaign against disease, Rose had circled the globe bringing hope and money to hard-pressed people. He was accustomed to bring greeted by supplicants with outstretched hands. But this was different. After a month of talking to Britain's leading scientists, Rose wrote Buttrick, "Visiting with these men, seeing their laboratories, and getting an insight into the things they are working at is a liberal education. It is the most interesting, the most stimulating experience I have ever had."

After long days of travel, interviews, and "working" dinners, Rose found time to continue his education in scientific matters, reading introductory texts on such subjects as anthropology and physics. Of one such book, *The Structure of the Atom,* by Professor Edward Andrade, he commented, "There are mathematical formulas, to be sure, that are beyond the layman in these matters; it remains, nevertheless, that one finds it tremendously fascinating."

Rose the missionary was undergoing a change of vocation. Instead of bringing enlightenment to those in darkness, he was begging permission to aid in the search for new knowledge—and he was enjoying every minute of it. Not that he entirely doffed his missionary hat. In less developed countries he could still sound like a dispenser of proven goods to the less fortunate. In Belgrade, for example, he described the Serbians as "backward in science and administration. Keen for improvement." From Czechoslovakia he wrote, "You know I had in mind rather important plans for this country. So far as I can see we shall realize all I had in mind and more.

But most important to Rose was a sense of a new beginning. As he put it, "All doors are open." This was due partly to meticulous advance planning, partly to the potent Rockefeller name, and partly to the presence of International Health Board fellows who had returned from their RF-sponsored training abroad and who often held positions of influence in their homelands. "If one wanted an example of the value of fellowships properly administered it is to be seen in Czechoslovakia and Poland," Rose wrote on March 9. "To see them in action is an inspiration. One of these men is now head of the health service in Poland and has plans for a staff of trained young men."

Rose could not wait to get his "scheme" off the ground. While the power to inaugurate programs actually lay with the trustees back home, Rose knew that they would approve virtually any actions he took. So he began to grant fellowships and commit Rockefeller funds to larger projects "on the run," as it were. In Zurich, which had not suffered physical damage during the war, he found a lively center of scientific activity, full of young men doing research "with a zest one finds infectious." Evidently, Rose himself caught the bug: "I was glad to recommend three travelling fellowships for research workers here."

A steady stream of nominations and applications poured into the New York office. On February 29, 1924, a telegram arrived from Rose, datelined Prague: "Approve Gottlieb fellowship proposal. Have Board act." The trustees had no objections; they had just set aside $6,000 for fellowships and had authorized Rose "to make definite commitments for fellowships under this appropriation subject to final confirmation by the Board." The trouble was, no one in New York had ever heard of the "Gottlieb fellowship proposal" or, for that matter, of Gottlieb. Further correspondence cleared up the confusion, and Dr. Hans Billroth Gottlieb, a chemist from Munich, went on to three years of RF-sponsored study at the University of Chicago.

Rose, who repeatedly stressed the "experimental" nature of the IEB's work, did not hesitate to make up his own rules as he raced from one country to the next. In February 1924 he spent six days in Rome, where he met a group of "outstanding mathematicians." One of these, Professor Vito Volterra, offered to arrange a little get-together in his apartment with "a few of the most interesting scientists in Rome." At this meeting Volterra mentioned "a young mathematician, Mr. Fermi," who was out of the country on a one-year government grant but who Volterra felt could benefit from further study abroad. This suggestion was immediately seconded "by several others present," and Volterra agreed to submit Fermi's name to Rose's board. On the basis of this informal procedure, Enrico Fermi was awarded an IEB fellowship, which he spent at the University of Leiden, working with Ehrenfest.

All told, Rose's trip generated some sixty nominations for fellowships. He also negotiated a number of financial arrangements with senior scientists. In Copenhagen, for instance, he met with the Nobel Prize winners Niels Bohr and August Krogh, both of whom were planning to expand their laboratory facilities. On the basis of proposals received in New York, the IEB trustees had already agreed to help finance this expansion. With larger quarters the two men would be in a position to accept many more students from abroad, including those on IEB fellowships. Rose promised Bohr $40,000 for his Institute of

Theoretical Physics and agreed to cooperate with the Rockefeller Foundation's Division of Medical Education in funding Krogh's Institute of Physiology. It was, Rose felt, "the three best days of this long journey."

While Rose continued to talk about the IEB's mission of bringing science to backward countries, the grants to Bohr and Krogh showed where his heart lay. In his own words, the important thing was to "make the peaks higher." Support the best—and the benefits would trickle down to everyone else. This strategy had the advantage of requiring little or no scientific expertise on the part of the grant giver. Indeed, it would have been an impertinence for a layman like Rose—no matter how many introductory texts he studied—to discuss the details of specific research projects with a Bohr or a Krogh. What such men needed was freedom from financial worries and adequate space to work in, both of which the IEB could provide.

In all, during its five-year existence, the International Education Board appropriated just over $16 million in support of the natural sciences around the world. Of the total, some $1.5 million went for fellowships and traveling professorships. All but a small fraction of the rest went for construction, equipment, endowment, and maintenance of laboratories and institutes—that is, for physical plant rather than for research projects per se.

Not surprisingly, most of this money was disbursed in the form of large grants. Of the institutions that got any aid, more than half received in excess of $100,000. The three largest grants went to the University of Cambridge, Harvard University, and the California Institute of Technology. Cambridge got $1 million to build a new library and another $2 million to build and equip laboratories for its departments of agricultural science, botany, physiology, and zoology. Harvard received $2 million to consolidate its widely dispersed biological laboratories in one building—a move that led to the creation of a unified Department of Biology with an entirely revamped curriculum.

The most expensive project undertaken with IEB funds was the construction of the 200-inch reflecting telescope at Mount Palomar. The project, initiated by George Ellery Hale and supervised by the California Institute of Technology, began in 1928 with a $6 million pledge from the trustees of the International Education Board. When the IEB was terminated the following year, responsibility for the commitment was assumed jointly by the General Education Board and by the RF's newly established program in the natural sciences. Mount

Palomar was Hale's last major project; he died in 1938, ten years before the work was completed. By the time the big telescope (named in Hale's honor) began operating on a regular schedule, in 1949, the three Rockefeller boards had contributed a total of $6,550,000.

Rose also served as president of the General Education Board from 1923 to 1928; during these years the activities of the GEB were closely coordinated with those of the IEB under the banner. "Make the peaks higher." In private, Rose assured Fosdick, "Emulation is the greatest force in the world. The high standards of a strong institution will spread throughout a nation and will even cross oceans."

Before Rose's tenure the General Education Board had contributed some $60 million to the general endowment funds of nearly three hundred American colleges and universities. Rose ended this program and concentrated the GEB's resources on building up strong science departments in the few selected institutions. Between 1923 and 1929 nearly $12 million went to bolster research and teaching in the natural sciences at such institutions as Harvard, Princeton, the University of Chicago, Cornell, and Stanford.

Conspicuously absent from Rose's grandiose vision was any formal

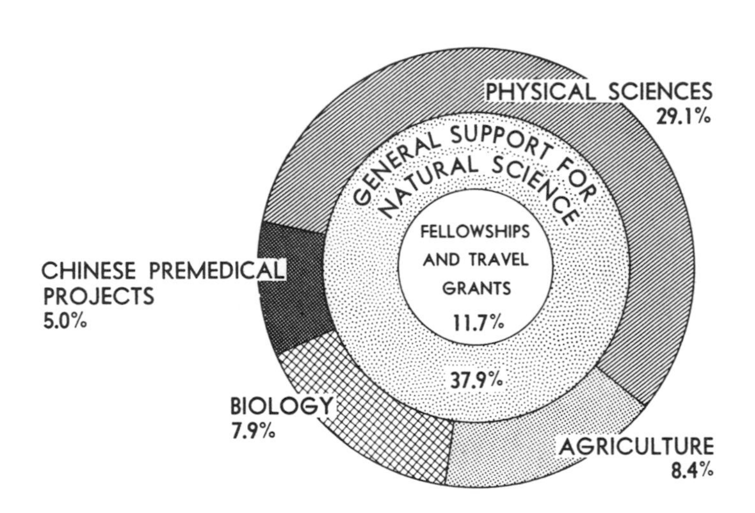


*How the Rockefeller Foundation, the General Education Board, and the International Education Board divided their support for the natural sciences in the pre–Warren Weaver years (prior to 1932). Total appropriations for the natural sciences: $43,815,359.*

mechanism for investing small amounts of money in specific research projects by younger investigators. Not until 1929, when the Rockefeller Foundation instituted its own grants-in-aid, was such a mechanism recognized as a useful adjunct to larger philanthropic ventures. And not until the 1930s, when a modest RF grant to a pathologist named Howard Florey paved the way for the development of antibiotics (see chapter 37), did the world take notice of how much could be accomplished in science patronage with very small sums. But even Rose had reason to know that the results of a grant were not necessarily commensurate with its size.

The smallest of all IEB grants was also one of the few targeted to a specific research project. Early in 1924, while Rose was still traveling in Europe, Jacques Loeb of the Rockefeller Institute informed the trustees of the IEB that Albert Einstein needed the help of a mathematician in undertaking some new studies in quantum mechanics. Although Einstein had indicated that he could hire a suitably trained man for $30 a month ($360 a year) the trustees felt that $500 a year would be more appropriate. (This was later increased to $700.)

Rose cabled his approval; and the first check—$250 for six months—was mailed to Einstein on February 2. When Rose arrived in Berlin in early March, he called at the Einstein home on the Haberlandstrasse and found the physicist apologetic and embarrassed over the fact that he had mislaid the check. Apparently, Einstein had placed it in a book, and now he could not remember which one. Or possibly the housekeeper had done something with it. But he told Rose, "I am sure it is somewhere in the house."

A search of his library proved fruitless. Rose not only assured him that a duplicate check could be issued; he asked if there was anything else the IEB could do for him. Recovering his composure, Einstein replied that the rest of his needs were being met. A duplicate check reached Berlin on March 28. Einstein cashed it promptly, as he did all succeeding checks until the end of 1925, when he completed the project and no longer needed the services of the mathematician. The total IEB outlay was $1,400, not counting postage and the bank's stop-payment fee. The lost check was never found.

In Rose's IEB, flexibility was always more important than precedent or protocol. When necessary, Rose would authorize a grant that anticipated the more narrowly focused programs, built around specific research projects, that his successors in the Rockefeller philanthropies initiated. One of the IEB's smaller "equipment" grants—for $50,000—

went to the Swedish chemist Theodor Svedberg in 1928. Two years earlier Svedberg had won a Nobel Prize for work on the chemistry of colloids. To aid in his investigations, he had invented the high-speed centrifuge, a device that separated different materials in a mixture according to their densities. Now he wanted to improve it. But although the Swedish government had built him a new $300,000 lab, he insisted that he needed more money.

The IEB came to the rescue. In time the "ultracentrifuge" became a standard tool in all chemical and biological laboratories; indeed, it is one of the technical advances that helped move biology from a basically qualitative to a largely quantitative science. Other scientists, in Sweden and the United States, played important roles in perfecting the ultracentrifuge; they, too, were supported by Rockefeller money, channeled through both the Rockefeller Institute and the Rockefeller Foundation's Division of Natural Sciences.

Moving from success to success (and explaining away his rare misjudgments with disarming candor), Wickliffe Rose had a virtually blank check from John D. Rockefeller, Jr., and the IEB and GEB trustees until 1929, when, in a kind of palace coup, his dazzling one-man show was brought to a halt and his administrative flamboyance replaced by the more analytic, bureaucratic style of Raymond B. Fosdick.

# 14

## A Scientific Education

Of course, Rose did not actually run his boards single-handed. Shortly after returning from his first European trip for the IEB, he noted, "It is extremely important that we get representatives in the field to guarantee the quality of the work we are undertaking; we are making some progress in this direction." In building his administrative machinery, Rose naturally built on his experience with the Sanitary Commission and the International Health Board. But those were both operating agencies. Although cooperation with local authorities was the rule, the brunt of the work was actually carried out by people who were recruited, trained (often on fellowships), and paid as Rockefeller employees. The IEB was a different institutional animal. Its field representatives would only assist the recipients of grants and fellowships; they would perform no research themselves. They were to be observers, listeners, enablers—the first scientific circuit riders, after Rose himself.

On June 13, 1924, Rose asked Albert R. Mann, dean of the College of Agriculture at Cornell University, to serve as "Director for Europe of the Board's work in agriculture." In November of the same year, Augustus Trowbridge, professor of physics at Princeton, was named "Director for Europe" in the field of science. The fifty-five-year-old Trowbridge, who had distinguished himself in the wartime effort to develop artillery detection devices for the American Expeditionary Forces, had the scientific stature to meet with Europe's leading researchers on a more or less equal basis.

These early circuit riders kept on the move. In 1925 the IEB established an office in Paris next to that of Selskar (Mike) Gunn, who represented the International Health Board in Europe. In the two years that Mann was on the job (he returned to Cornell in 1926 and was replaced by Claude B. Hutchison, a professor of plant breeding at the University of California), he visited every country in Europe except Russia, Albania, Lithuania, and Portugal. Trowbridge, picking up where Rose had left off, canvassed scientific leaders in France and other countries, explaining the IEB fellowship program. Unlike the International Health Board, which was looking for trainable public-health workers for specific projects in specific countries, the IEB wanted nothing but the best. Only exceptional research workers who had already earned doctoral degrees need apply. In addition, every candidate had to be personally interviewed by an officer of the board.

Some scientists found these criteria too strict. Madame Curie, for instance, complained that the requirement for a personal interview discriminated against candidates from her native Poland and other European countries that were visited only infrequently by IEB representatives. But Trowbridge, backed by Rose, held out for what he called a "hard-boiled" policy. Except in the rarest of circumstances, the IEB would not relax its high standards.

This in no way implied an arm's length attitude toward the scientific community of which the IEB was rapidly becoming a part. Quite the contrary. For example, the French had a strong reputation in mathematics, but their universities were considered somewhat parochial in outlook. To stimulate cross-pollination among the mathematicians of Europe, the IEB spread the word that it was looking for young mathematicians who were willing to go abroad for a year or two. This discreet "advertisement" came to the attention of a twenty-five-year-old native of Poland, Szolem Mandelbrojt, who was studying at the Collège de France.

Mandelbrojt's professor asked him whether he would accept a fellowship if offered one. Attracted more by the chance to travel than by the urge to study at a particular institution outside France, he said yes. As Mandelbrojt recalls, his interview with Trowbridge at the IEB office was "rather comical":

> He asked me where I wanted to go, and how much money I wanted. There were no fixed stipends then. I was very modest at the time, not very rich, and I asked him for forty dollars a

> month in living expenses. I thought it was a lot. At first I wanted to go to Sweden, but my professor persuaded me to go to Rome to work with Volterra. I don't regret going to Rome at all. When I arrived, I spoke a very poor Italian. Volterra asked me to replace him on some lectures he was scheduled to give on the theory of functionals; and I had to improve my Italian very quick. Volterra was like a father to me. I'd go to his house almost every day, eat with his sons. The whole experience was very important to me. It was good to be in contact with many different kinds of mathematicians; and it was good to be a young man and free to come and go, paid to do just what I wished to do.

The size of the stipend, however, was a problem. Mandelbrojt found that $40 a month was not enough to live on, despite the hospitality of Volterra. So he returned to Paris, and in the company of a Hungarian fellow named Francis Krbek, who was having similar problems, went to the IEB office again. Somewhat to his surprise, his monthly stipend was raised to $60 a month, and then to $90: and when he got married, in 1926, it was increased again, to $162 a month.

Later, when Mandelbrojt became a professor at the Collège de France, his opinions on young fellowship candidates were often solicited by Wilbur E. Tisdale, a physicist who came to Paris in 1926 to take charge of the IEB fellowship program, and who continued to perform the same function, in much the same fashion, for the Foundation after it absorbed the IEB's science interests in 1929. During these years each circuit rider began keeping a daily "log" or "diary," filled with useful information, subjective impressions, speculations, and occasional gossip. Copies of these journals were forwarded to the New York office, where they came to form an invaluable record of the "state of the sciences" on the European continent during this period.

Between 1924 and 1929 the International Education Board awarded a total of 509 fellowships in the natural sciences, the majority going to students of physics, biology, mathematics, and chemistry, in that order. More than a few recipients went on to high achievement. In addition to the young Enrico Fermi and Werner Heisenberg, there was the American wunderkind J. Robert Oppenheimer, who arrived in Europe in 1928, at the age of twenty-four, to study the new quantum mechanics. Oppenheimer, never one to extend praise lightly, later had this to say about his European mentors: "Some of the excitement and wonder of the discoverer was in their teaching."

Once Wickliffe Rose became a convert to science, his own confidence in the scientific community never flagged. The strength that he drew from this faith allowed him to dominate the Rockefeller philanthropies in the 1920s in a way that no one did before or has since. Raymond Fosdick (who served on the IEB board from its inception to its termination) wrote, "[Rose] was a formidable man in debate and few of us trustees ever cared or dared to oppose him. Occasionally, when one or two of us were not quite convinced and ventured a contrary argument, we were overwhelmed with a logical exposition that was so complete and convincing that there was nothing to be said in reply."

Especially memorable was the meeting at which Rose presented his proposal for building up the mathematical faculties at the University of Göttingen and the University of Paris. The board was being asked to appropriate more than $500,000. To justify this expense, Rose lectured the trustees

> with the aid of elaborate charts and diagrams, not on mathematics at Göttingen or Paris alone, but on mathematics in every leading institution around the world. He was reporting on where man had arrived in his mathematical thinking, and where the opportunities for progress seemed brightest. It was characteristic of the immense pains and thoroughgoing analysis with which he scanned every recommendation he brought before his trustees. Göttingen and Paris were preferred in his judgment because of all the places in the world at that time they represented the peaks in mathematical science.

It should come as no surprise that Rose, a man who inspired confidence in others, had an unshakable confidence in himself. He was averse to personal publicity, regularly rejected honors bestowed on him by grateful governments and academic institutions, and discouraged any mention of his work in the press. At the same time he suffered from no false modesty about his abilities.

He pursued his self-education program in the sciences with such assiduity that within a few years the scientists he consulted seemed to accept him as one of their own. (Of course, considering the funding source he represented, it was clearly in their best interests to do so.) His progress in the comprehension of physics, for example, can be gauged by two items taken from his official correspondence. On July 8, 1924, he wrote to thank Simon Flexner of the Rockefeller Institute for sending him a book entitled *The New Physics:* "It is a remarkable summary of recent achievement in this field. The chapter on relativity takes me quite beyond my depth. Having got thus far, however, I must go further."

A year and a half later, Rose read in *Scientific American* about the "hard" radiation from outer space being studied by Robert A. Millikan, the Nobel Prize–winning physicist who was then at the California Institute of Technology. Millikan had coined the phrase "cosmic rays" for this phenomenon, but the rays were also known, for a time, as "Millikan rays." On January 13, 1926, Rose sat down to share with Millikan an idea he had had. "A layman has no business troubling a busy scientist about scientific matters," he began. But then he went on,

> It is probably very foolish, but as I have no scientific reputation at stake, you will overlook it. The point is this: As I understand it, the cause of radioactivity in such substances as radium, thorium, etc. is not known. I also understand that these Millikan rays are much harder even than the gamma rays from radioactive substances. I am wondering if the radiation from radioactive substances may not be brought about by the action upon these substances of the Millikan rays.

Rose even suggested an experiment that might test his hypothesis: place some radium in a chamber "behind more than six feet of lead," and wait to see whether the radioactive emissions cease. If you find the suggestion "too foolish," Rose told Millikan, "throw it into a waste basket and don't trouble yourself to reply."

But Millikan did reply, a week later:

> Dear Mr. Rose, I was very glad to see from your letter of January 13th how much a real scientist you are. The same suggestion had come to me from the best of scientists, both here and abroad. I have pretty good evidence that radioactive

> substances screened entirely from the penetrating rays continue their activity unabated. The experiment has not yet been made wholly quantitative, but I hope to make it so in the near future. I am really sorry that it looks now as though we could not explain radioactivity in this way, but at any rate I will tell you very definitely about it in the near future. Sincerely yours, R. A. Millikan.

Rose's theory was incorrect. As it turned out, Millikan's own theory about "his" rays—that they were very energetic photons—was also mistaken. But what is significant about this interchange is that it occurred at all. Rose was a layman who dared to take as his philanthropic "specialty" science as a whole. William Whewell might have approved, but it is hard to imagine anyone today—even a respected scientist in some other discipline—sticking out his neck to pass along to a Nobel laureate like Millikan some half-baked theory about Millikan's own field. It is equally hard to imagine a nonmathematician daring to summarize for an audience of laymen the state of modern mathematics, concept by concept, institution by institution.

In truth, Rose's approach to science patronage—work with the best in every field, from physics and mathematics to chemistry and biology—was an anachronism even in the 1920s. The IEB succeeded only because Rose had such a remarkable intellect and because there were so few sources of support for research scientists that Rose could hardly fail to do good things if he gave money to a needy few at the top of their profession—the Bohrs, the Hales, the Millikans, the Einsteins.

But this era was quickly coming to a close. Increasingly, governments and quasi-governmental agencies like the Kaiser-Wilhelm-Gesellschaft in Germany were taking on the care and feeding of world-famous scientists. There were only so many institutes to build and libraries to endow. And there was only so much glory to be gained from strengthening the strongest. Another, more selective approach to science patronage was in the wind. The bywords of this approach were "selectivity" and "concentration." Pick an area that is ripe for development but lacking in vital resources—money, equipment, trained investigators—and supply what is lacking. Then stand back and see what develops.

Selective science patronage is bound to be riskier than a strategy of "making the peaks higher." Like investment in penny stocks, support of individual researchers in specific "underdeveloped" sciences can return enormous dividends, in terms of social benefits and prestige. It can also produce catastrophic losses, for both society and the investor. This was the kind of gamble toward which the Rockefeller Foundation was now moving.

# 15

# A Cautionary Tale

The Rockefeller philanthropies were not the only players in the postwar game of science patronage. In 1925 the newly created John Simon Guggenheim Foundation began giving fellowships in the arts and sciences. Among the first candidates to get a Guggenheim was the twenty-four-year-old Linus C. Pauling, who had just received a doctorate from the California Institute of Technology; in his application he said he wanted to help create a "mathematical chemistry" that could eventually elucidate the molecular nature of chemical bonding. During the foundation's first quarter century, more than a third of all Guggenheim fellowships went to scientific researchers like Pauling.

One rationale for these fellowships was that they gave scholars who were still undecided about a scientific career a taste of real research before they became bogged down in routine teaching assignments. Enrollment in American universities was booming; between 1900 and 1930 the proportion of the college-age population attending institutions of higher learning rose from 4 percent to nearly 13 percent, and the number of graduate students rose from 6,000 to 47,000. One mathematician told Rose in 1923 that most of the scientific investigators in America were "so burdened with teaching responsibility that but little time [was] left for research."

In daring to select fellowship recipients on its own, instead of delegating this authority to a well-credentialed buffer like the National

Research Council, the Guggenheim Foundation had taken a leaf from Rose's book. Any mistakes made would be the responsibility of the foundation. The man who directed the Guggenheim Foundation in its early years was Henry Allen Moe, an attorney with a scholarly interest in Elizabethan literature. He created several tiers of unpaid juries and advisers to evaluate candidates, and he usually took their advice. Despite the foundation's worldwide reputation for excellence, Moe worried about the big ones that got away—or almost got away, like the one-armed researcher in immunology whose application was borderline but who went on to win a Nobel Prize. Looking back, Moe remarked, "We hooked that one and he did us proud, but I get nervous when I think how close we came to losing him. We make mistakes."

There is no doubt that the post–World War I "boom" in fellowships did much to change the status of scientific research in the academic community. One historian of science has gone so far as to call the shift toward supporting individual scientific researchers directly, rather than indirectly through the institutions where they worked, "one of the most significant events in the history of learning in the United States."

Nevertheless, evaluating the success of a specific fellowship program is more difficult than it might appear. On the one hand, any fellowship program that cannot boast a few big names on its roster—the Paulings, Fermis, Heisenbergs, and Oppenheimers who go on to greatness—would surely be deemed a failure; and fellowship officers rightly take pride in helping young geniuses early in their careers. On the other hand, despite Moe's fretting, the mistakes of a fellowship officer tend not to be very visible. Thwarted geniuses who fail to make great discoveries because they are rejected by this or that foundation do not, of course, show up on any tally sheet. As for the odd rejectee who goes on to win a Nobel Prize, few people will remember or care which foundation failed to come through with a stipend years earlier.

Indeed, the very best fellowship candidates, the future Nobel laureates, usually stand out so far above the crowd that they have little trouble finding support; and it can be argued that little credit accrues to the particular foundation that happens to land them. According to this argument, the real test of a fellowship officer's acumen is not how many geniuses he helps but how well he chooses among the lesser candidates.

The very difficulty of evaluation makes a fellowship program a peculiarly safe philanthropic investment. By contrast, a foundation that directly subsidizes specific research projects leaves itself open to all kinds of second-guessing. If a subsidized discipline fails to make progress while other fields that were passed over for grants flourish, the error in judgment cannot be hidden. Similarly, a foundation's reputation is bound to suffer if it continues to back a well-known researcher long after his or her laboratory has ceased to be productive. Even worse, grantees can behave in ways that damage society and reflect badly on the granting agency; such misbehavior can range from misuse of funds (a risk in any kind of philanthropy) to fraudulent research that leads to harmful initiatives in public policy or undesirable shifts in public opinion.

In Wickliffe Rose's vision of science patronage, support of large institutions came before support of specific research projects; and the IEB was careful to work only with the strongest, most reputable institutions ("Make the peaks higher"). When Rose did give money to scientists for specific projects, he took some obvious steps to reduce the risks: he dealt only with internationally renowned investigators, like Albert Einstein and Theodor Svedberg, and he hired capable scientists like Augustus Trowbridge as circuit-riding assistants, to serve as his eyes and ears in the scientific community.

But with his boundless self-confidence, Rose disdained another obvious strategy for reducing risk. Instead of concentrating on one or two disciplines that he and his small staff could hope to gain some intimate knowledge of, he dared to support science right across the board ("Begin with physics, chemistry and biology . . ."). If he ever worried about his ability, or the ability of his advisers and colleagues, to recognize excellence in all the diverse fields that came under the scrutiny of the IEB, no evidence of such second thoughts can be found in the archives. Rose got away with it. But, understandably, few cared to follow in his footsteps.

Once science patronage entered the high-risk era of project-by-project subsidies, most philanthropoids tried to contain the risk by adopting, in one form or another, a policy of concentration. The Carnegie Institution of Washington understood the advantages of such a policy from the beginning. Over the years the Carnegie has added and dropped departments as its officers and trustees sought to identify areas of science where its resources would have the maximum impact. Yet so seductive is the appeal of scientific "objectivity" that science patrons constantly run the risk of empowering "experts" whose hidden agendas are at odds with the stated goals of both science and philanthropy. As an illustration of the rewards, and the very real dangers, that science patronage entails, we can take a brief look at a field of concentration that for many years was almost synonymous with the Carnegie name: genetics.

The Carnegie Institution's support of genetics began in 1904 with the establishment of the "Station for Experimental Evolution" at Cold Spring Harbor, on Long Island. One of its first staff members was George Harrison Shull, who developed hybrid corn at the research station in 1908. The facility was home for many years to Barbara McClintock, winner of a Nobel Prize in 1983 for her work in plant genetics. The last director was Alfred Hershey, who won a Nobel Prize in 1969 for his contributions to molecular biology. Two years later the Carnegie's Department of Genetics turned its buildings and equipment over to its longtime neighbor the Cold Spring Harbor Biological Laboratory and ceased operations, in fulfillment of the Carnegie Institution's self-proclaimed role as an "explorer who helps to open a new land. Once the area has been surveyed, the explorer moves on and leaves its development to others."

One fact that goes unmentioned in the Carnegie's current promotional literature is that the longtime director of its research efforts in genetics was Charles B. Davenport, whose doctrine of eugenics was used to justify native American racism in the first half of the twentieth century.

A graduate of Harvard College who never let anyone forget that he could trace his own ancestry as an "American" back to the seventeenth century, Davenport was teaching zoology at the University of Chicago when he persuaded the Carnegie Institution to create the Station for Experimental Evolution. Davenport's scientific interests extended from the genetics of canaries and chickens to the hereditary causes of antisocial behavior in man. In 1910 he founded the Eugenics Record Office (ERO), also in Cold Spring Harbor, to look into such matters as the family histories of the "feeble-minded, the insane and others and to lead the state to take more vigorous steps to diminish the number of them who are reproduced thru bad heredity."

Initial funding for the ERO came from Mrs. E. H. Harriman, with smaller contributions from John D. Rockefeller and other wealthy donors. In 1918 the ERO became part of the Carnegie Institution; two years later, the ERO and the Station for Experimental Evolution were combined into a single Department of Genetics.

That same year (1920) the U.S. Congress was considering a revision of the immigration laws. A longtime associate of Davenport's, Harry

H. Laughlin, appeared before a House committee as an expert witness. In presenting his credentials, Laughlin stressed his position as a "member of the Carnegie Institution of Washington." The purpose of his testimony (and of a detailed report he later submitted) was to provide a scientific rationale for legislation designed to "protect the blood of the American people from contamination and degeneracy." The Immigration Restriction Act of 1924, passed by an overwhelming majority, incorporated the eugenicists' explicit bias in favor of "Nordic" types. After 1924, immigrants would be admitted to this country on the basis of national quotas that frankly discriminated against southern Europeans, Orientals, and blacks.

Many prominent geneticists—including some on the staff of the Carnegie's Department of Genetics—were uncomfortable with Davenport's insistence that bad genes were responsible for a broad spectrum of bad behavior, ranging from "feeble-mindedness" to criminality. But there is no question that the support of the Carnegie Institution throughout the 1920s gave the Davenport approach to social problems a respectability it would not otherwise have had. The influence of this doctrine of genetic determinism was not limited to debates over immigration. By the early 1930s thirty states had passed compulsory sterilization laws that gave panels of "experts" the power to sterilize individuals who fell into such undesirable social categories as "sexual perverts," "drug fiends," "drunkards," and "epileptics." Through 1941 a total of 36,000 sterilizations were authorized, approximately half of them in California, where zeal for "cutting off defective germ-plasm in the American population" (Laughlin's phrase) ran especially high.

Davenport directed the Carnegie Institution's genetics programs until his retirement, in 1934. A blue-ribbon committee of appraisal appointed the following year found the eugenics work scientifically unsatisfactory. Henry Laughlin was eased into retirement in 1939, whereupon the Institution terminated all support for eugenics.

The Carnegie's belated change of heart about Davenport and eugenics can be traced in its official publications. The 1934 *Year Book*, which appeared just after Davenport's retirement, described his three decades of research into "experimental evolution, genetics and eugenics" in these terms:

> The natural close relation among these activities was maintained under the leadership of Dr. Davenport. It was thus

> possible to bring to bear upon the problem of human development and betterment a wealth of data arising out of studies on heredity derived from a wide range of original sources. The contribution made by these interlocking researches marks one of the most important advances in biological research by the Institution.

The same *Year Book* praised the "important contribution . . . made by Dr. Laughlin in application of the principles of heredity to the problem of immigration control as a practical question facing the United States and other countries."

By the 1970s philanthropic history had been rewritten. In its seventy-fifth-anniversary report, issued in December 1977, the Carnegie Institution pointed with pride to its long list of accomplishments as a patron of science. Under the heading "Genetics" more than twenty researchers were cited by name, beginning with George Harrison Shull, of hybrid-corn fame. Understandably, the names of Henry Laughlin and Charles Davenport were nowhere to be found. Nor was so much as a passing reference to eugenics.

Today there is universal agreement that Davenport's eugenics was bad science, that is, work that cannot be replicated, because it proceeded from incorrect premises by way of basic procedural flaws. In retrospect it seems clear that the Carnegie Institution should have withdrawn its support of Davenport earlier than it did. The question then becomes, how could such an institution, with access to the best scientific minds of the day, have had so much trouble distinguishing between "good" and "bad" science being conducted in its own laboratories by its own salaried employees?

Davenport had his public critics almost from the beginning. As early as 1913, English geneticists were objecting to the ease with which he leaped from meager data to conclusions that happened to match his preconceptions, while ignoring competing hypotheses. One particularly egregious example: a 1919 Carnegie Institution publication, intended for use by the U.S. Navy, asserted that "thalassophilia," or love of the sea, was almost certainly a genetic trait, inherited as a sex-linked recessive factor. In other words, what drove men, rather than women, to run away to sea was something in their genes.

But American geneticists who found fault with eugenics were slow in expressing their doubts in public. Only one scientist (the zoologist

Herbert S. Jennings of Johns Hopkins University) came forward to counter Laughlin's congressional testimony in the early twenties. Some of this reluctance can be attributed to the uneasiness that members of any professional community feel about airing their dirty linen. In science this attitude is reinforced by the assumption that nonreplicable work will eventually be exposed as worthless. Short of proven fraud, bad science is usually allowed to die a natural death as it becomes clear that other members of the profession give it no credence. But this winnowing process takes time, which (as in the case of the eugenics movement) may result in tragic social fallout.*

Caution would seem to be appropriate in the drawing of conclusions from the cautionary tale of eugenics, since this particular set of circumstances may be unique in the history of science. Ideology, which is supposed to have no place in the laboratory, clearly warped the judgment of many people in and out of the scientific community. Davenport never tried to hide his belief that eugenics was a necessary tool in the struggle against the barbarism that threatened the "American" values that he held dear. It is also apparent that the Carnegie Institution's long involvement with Davenport created a sense of mutual dependency that made objective evaluation increasingly difficult.

Nevertheless, if one lesson can be drawn from the rise and fall of the science of eugenics, it is this: any program of science patronage that dares to become identified with a specific research project or scientific discipline is an inherently risky business, with possibly serious consequences for both patron and society.

Warren Weaver, director of the Rockefeller Foundation's science patronage programs from 1932 to 1950, understood the risks involved. He liked to tell a story about a prominent businessman on the Board of Trustees who, mistaking science for certainty, thought that science patronage ought to be simple. Weaver had just finished explaining to the trustees his reasons for recommending a seven-year grant to an outstanding chemist with a superlative record of past accomplishments. The businessman shook his head and said that he still did not understand "*just what* Professor X will do " with the Foundation's money.

*In the 1930s the English psychologist Cyril Burt "supported" his own views on racial and class distinctions with phony identical-twins studies. He was not exposed as a fraud until the 1970s—after an entire generation of British schoolchildren had been "tracked" through a rigidly hierarchical educational system whose design had been heavily influenced by Burt's views.

Another trustee—Dr. Herbert Gasser, a Nobel laureate who was at that time head of the Rockefeller Institute for Medical Research—came to Weaver's aid with this comment: "If Professor X knew just exactly what he was going to do, he wouldn't need to do it."

To minimize the risks inherent in project-by-project science patronage, Weaver adopted a two-fold strategy. (1) He chose a narrow field of concentration in which he had at least some hope of becoming knowledgeable; and (2) he borrowed from Wickliffe Rose the notion of a corps of circuit riders who would keep in touch with that portion of the scientific community whose cooperation was essential to the success of his program. To warn him of the dangers that lay in store if he made a mistake, he had before him the examples of the Carnegie Institution's painful association with Davenport's eugenics, and of Edwin Embree's abortive attempt to lead the Rockefeller Foundation into large-scale science patronage in the 1920s.

Embree, who had immediately recognized the need for concentration, flirted with eugenics before backing off just in time. What finally brought him and his program down, in a series of contretemps that compounded comedy with tragedy, was his failure to understand the importance of the circuit-riding tradition.

# 16

# Choosing Up Sides

Edwin Embree was his own worst enemy and, with benefit of hindsight, his severest critic. Years after he had left the Rockefeller Foundation and gone on to a notable career as director of the Rosenwald Fund, he had this to say about his performance as a patron of science:

> If I had that job today I think I could build up a program that would be not only brilliant but would command support even from the conservative. But for three years I sweat blood on the job in New York. On the whole I think the results were those that would be expected from a very immature person. The things I got done were not bad, but I did not get the philosophy of the thing for years and years and was therefore unable to formulate a really statesmanlike program.

The Rosenwald Fund was created in 1917 by Julius Rosenwald, a longtime RF trustee, to aid Negro schools and farmers in the South. Embree's achievements in this "missionary" field, where men of goodwill could agree on the definition of the problem and even on the broad outlines of the solution, were substantial and warmly applauded. Yet between 1923 and 1927, when he headed the RF's Division of Studies, his efforts to find a field of concentration worthy of philanthropic attention failed to impress the Board of Trustees; and even

*Edwin R. Embree*

when he persuaded the Board to back one of his initatives, he failed to create the machinery necessary to monitor the results. This was an invitation to disaster. In fact, Embree's brief tenure at the Foundation can be taken as a case study in how *not* to conduct the business of science patronage.

To begin with, Embree's first loyalty was not to science but to his mentor the RF's President, George E. Vincent, who was still trying to enhance the power of the central administration at the expense of the International Health Board and the China Medical Board. It was Vincent's contention that the RF was an organizational nightmare, a loose confederation of semi-autonomous divisions lacking central direction and tight budgetary control. The fault, he argued, lay in the concept of the Foundation as a philanthropic "holding company" that delegated policy-making responsibility to its operating units.

Vincent's argument was seconded by Raymond Fosdick, who had joined the RF Board of Trustees as the "family's representative" in February 1921. Disturbed by what he heard, Junior acted with the deliberate speed for which he was famous. On October 18, 1923, he invited a number of trustees and family advisers, including Vincent,

Rose, Gates, and Fosdick, to meet him at lunch at the Whitehall Club in Manhattan to discuss "several questions of policy which [had] arisen in connection with various foundations established by [his] father—such questions as interlocking directorates, a common treasury department, etc."

This luncheon was the beginning of a complex round of discussions and negotiations that led to the complete restructuring of the Rockefeller philanthropies in 1928. Although Fosdick was the principal player, Vincent had a hand in the eventual restructuring. But he also had a more pressing goal—to carve out for himself an area of undisputed authority within the Foundation.

The powerful directors of the operating boards were constantly jockeying for a bigger share of the Foundation's income. The budget of the International Health Board grew from just over $800,000 in 1915 to more than $3 million in 1924. Annual outlays by the China Medical Board jumped from $157,623 to over $1 million during the same period. To keep pace with these expenditures, John D., Sr., donated securities worth $50,438,768.50 to the endowment, bringing his total gifts to the Foundation to over $180 million by 1919.

But in that same year a third operating unit of the Foundation was created, the Division of Medical Education. Its mandate was to improve the teaching of medicine around the world, as the General Education Board was empowered to do in the United States. The director of the new division was Dr. Richard M. Pearce, a professor of pathology at the University of Pennsylvania, who had been recruited by Rose three years earlier to look into the problems of medical education on a global scale. Pearce was soon running his division in an independent fashion, very much in the manner of Rose. Like his mentor, he was a strong-minded administrator who quickly won the confidence of the trustees. Within a few years, he had put into place a multimillion-dollar building program to create a center of scientific medicine at the University of London.

It was not Vincent's style to mount a direct challenge to the entrenched satraps of the operating boards. He was by nature a "moderator" who sought consensus on major policy decisions before taking action. Some found his administrative style infuriatingly slow. (Even Embree later characterized his presidency as a time of "shilly-shally indecision.") But when he saw an opening, he moved quickly to create a new center of power within the Foundation, headed by the man he had brought to New York with him, Edwin Embree.

Embree was yet another recruit to philanthropy from the ranks of academic administrators, but he was not an educator in the sense that Rose and Vincent were. A Nebraskan by birth, he was the youngest of seven children of a railroad telegrapher. Descended on both sides of his family from southern abolitionists, he spent much of his childhood in a region of Wyoming where his family's "nearest neighbors" were Blackfoot Indians. He graduated from Yale in 1906, with a B.A. in philosophy, and went to work as a reporter on the *New York Sun.* A year later he returned to Yale as assistant editor of the *Alumni Weekly,* a post that led to other assignments in the alumni office. In the course of this work, he met Vincent, the up-and-coming college president from the Midwest. When Vincent joined the Foundation, in 1917, he brought Embree with him as secretary.

In his first *President's Review,* Vincent sounded a theme that he would keep returning to during his twelve years in office: private foundations must do only what they do best and avoid fruitless competition with large government agencies. Given its limited resources, the Foundation's best chance of making a significant contribution was "by concentrating its funds upon a few convincing demonstrations and statesmanlike programs." At the same time, since private philanthropies had more freedom to act than government agencies, it was imperative that the RF remain lean and flexible and ready to seize any opportunities that came its way.

In truth, the Foundation that Embree joined in 1917 was long on concentration but short on flexibility. Except for the NRC fellowships, most of its work was in public health and medical education. A few grants outside those fields had been voted now and again, but it was never easy for the trustees, with the limited time available at Board meetings, to choose among a myriad of proposals concerning areas in which they had no competence.

To make it easier for the Foundation to explore new fields—and to give the president and his protégé a power base of their own—Vincent asked the Board of Trustees to create a permanent "Division of Studies." The new division was officially added to the Foundation's books on January 1, 1924. It assumed responsibility for all Foundation activities unconnected with public health and medical education; these ranged from the National Research Council fellowships in physics, chemistry, and mathematics to a $500,000 contribution to the building and endowment fund of the Marine Biological Laboratory in Woods Hole, Massachusetts.

From the beginning Embree's philosophy at the Division of Studies was opportunistic. Only when the trustees ignored his impassioned plea for a major RF effort in the humanities and the fine arts did he and Vincent turn to the sciences. After some thought they concluded that their prospects were brightest in the general area of "human evolution." Early in 1925 Embree came to the trustees with a proposal for "cooperative aid to studies under university auspices in the sciences of Human Biology"—a discipline that he defined to include "those subjects commonly designated under (a) anthropology, (b) physiology, and (c) physiology and psychiatry, and their application in mental hygiene."

This proposal was certainly not lacking in ambition. Embree, with Vincent's blessing, was staking a claim to an important area of science patronage that remained outside the jurisdiction of the other Rockefeller boards. Unfortunately, what had emerged from Embree's yearlong investigation and deliberation sounded like a parody of the incisive analyses and persuasive calls to arms that the trustees were accustomed to hearing from Wickliffe Rose.

Embree tried to offer something for everyone. For the more adventurous-minded trustees, he pointed out that governments had already begun to grapple with "practical problems of human biology," not excluding such eugenical matters as "group or racial hygiene." Although he took no stand one way or the other on compulsory sterilization of the feebleminded or restrictive immigration laws, Embree argued that the Foundation's natural role was to help develop "an adequate basis of well documented fact upon which . . . new procedures may rest." This could be done by the tried-and-true method of strengthening "permanent centers" of teaching and research and by awarding fellowships.

For those trustees who might have doubts about eugenics, Embree described his new program as no more than a logical extension of the Foundation's traditional concern for "individual health." There was no need for direct involvement in "controversial problems," such as population control or the physical basis of racial differences, since "it need not be the function of the Foundation to select specific researchers." Nevertheless, he concluded, "The ultimate, potential possibilities of satisfactory answers to such questions stimulate the imagination."

RF trustees like Simon Flexner and Rose himself, who prized precision in the use of language, could only have winced at such locutions as "potential possibilities"—and wondered if Embree's stylistic confusion indicated a deeper confusion about subject mater.

In cobbling together his something-for-everyone package, Embree miscalculated. The trustees, only too aware of their inadequate credentials as laymen sitting in judgment on "complicated technical proposals," responded best to the aggressive, even arrogant confidence of a Wickliffe Rose. Tentativeness, no matter how honest, put them off. After hearing Embree's presentation, they decided against committing the Foundation "to a comprehensive program in the field of Human Biology."

Instead, the Division of Studies was authorized to continue its survey of new fields and to bring to the Board's attention "from time to time specific projects to be considered on their merits." Since this was essentially the mandate that the division had been operating under for more than a year, the trustees' action was hardly a vote of confidence in Embree. A temporary decline in Foundation income (due to a drop in the yields of certain securities in the RF portfolio) undoubtedly put the Board in a cautious mood. But the message was clear enough; despite the support of President Vincent, Embree's Division of Studies would remain on a short leash.

# 17

# The Vanishing Eugenicists

The social policies so avidly promoted by American eugenicists like Charles Davenport had no particular appeal for Edwin Embree. What drew Embree to eugenics was its relevance to the mission of the Rockefeller Foundation: "To promote the well-being of mankind throughout the world." Eugenics offered itself as a science of human well-being. This was an important consideration to a Foundation officer trying to persuade his Board of Trustees to support a program of science patronage.

While Jerome Greene had believed that the Foundation was empowered by its charter to support scientific research—since the pursuit of truth necessarily led to human betterment—the Board itself had never embraced this view. The National Research Council fellowships, the grant to the laboratory at Woods Hole, and even the far-reaching science patronage of Wickliffe Rose could all be justified (at least in name) as aid to "education." Not until the reorganization of 1928 would the trustees formally equate "the advancement of knowledge" with "the well-being of mankind."

Out of the shambles of his "Human Biology" proposal, Embree salvaged a project whose social utility was indisputable. This was a study of the "vanishing" aboriginal cultures of Australia. As Embree explained it to the trustees, not only were the cultures in question dying out—so the study had to be done quickly if it were going to be done at all—

but the information to be gained had a significance to our society that went beyond a merely scholarly commitment to "knowledge for its own sake." This significance was affirmed by no less eminent authorities on the social uses of science than the leaders of the American eugenics movement.

The first impetus for a study of Australian aborigines had come from a New York attorney named Madison Grant. In 1916 Grant had published a book called *The Passing of the Great Race,* which unabashedly espoused the genetic superiority of the "white race" over all other subdivisions of humanity. Two years later he helped organize the Galton Society for the Study of the Origin and Evolution of Man. This was one of many impressively titled organizations spawned by the American eugenics movement; like the others, its principal purpose seemed to be the promotion of solidarity in the movement by marshaling, on one letterhead, the names of scientists and nonscientists who shared a concern about the effect of "bad" genes on the prevailing political, economic, and moral order.

Prominent on the letterhead of the Galton Society was the address of the American Museum of Natural History, in New York. Among the charter fellows listed, along with Grant, were Edward L. Thorndike, professor of psychology at Columbia University's Teachers College, William K. Gregory, curator of comparative anatomy at the American Museum, and Charles B. Davenport of the Carnegie Institution.

It was on such a letterhead that Raymond Fosdick received, in late December of 1923, a document written by Dr. Gregory explaining the importance of a study of "savage races . . . from the viewpoint of modern biological anthropology."

Gregory argued that the prolongation of life by modern medicine was not an unmixed blessing, since without the checks of bacterial and viral diseases, "hereditary diseases and mental and moral inferiority are increasing with shocking rapidity." He went on, "Unless the danger is recognized soon . . . the ruthless processes of Natural Selection must inevitably reassert themselves and a civilization based on physical and moral inferiority may fall into ruin from within." Gregory offered a way out. Civilization *could* be saved "by substituting intelligent guidance for the ruthless, wasteful selective methods of nature." But first it was necessary "to find out how Natural Selection operates within some of the more primitive races of mankind."

Just in case Fosdick missed the connection between the rapid deterioration of the human gene pool and the work of the Rockefeller Institute, the International Health Board, Johns Hopkins University, and other Rockefeller-supported bastions of modern medicine, Madison Grant added a note: "I trust this plan will appeal to Mr. Rockefeller because the study of primitive man is the best approach to a proper understanding of the artificial conditions of selection now operating in civilized countries."

Addressing this appeal to Fosdick was a shrewd move for a number of reasons. Between 1911 and 1915 the Rockefeller family had contributed nearly $25,000 to Davenport's Eugenics Record Office in Cold Spring Harbor. While the Rockefellers' contributions never came close to matching the generous support available from other private donors like the Harriman family (which gave Davenport $500,000 between 1910 and 1918), there was no reason to believe that Senior or Junior had rejected the underlying philosophy of the eugenics movement—or that their interest might not be rekindled by the proper approach at the proper time.

As for Fosdick himself, he was known as a Wilsonian liberal, deeply involved in many social issues of the day. In 1921 he had accepted membership on the advisory council of the newly founded Eugenics Committee of the United States of America. Among the council's one hundred members were many luminaries associated with the Rockefeller philanthropies, including Dr. William H. Welch of Johns Hopkins University; Frank R. Lillie, a zoologist at the University of Chicago; and President Charles W. Eliot of Harvard University. Fosdick's brother, the well-known preacher Harry Emerson Fosdick, was a member, and both Charles Davenport and Madison Grant sat on the board.

The chairman of the Eugenics Committee was Irving Fisher, a Yale professor of political economy and a leading public-health advocate, who spelled out the group's philosophy in this way:

> The time is ripe for a strong public movement to stem the tide of threatened racial degeneracy following in the wake of the War. America in particular needs to protect herself against indiscriminate immigration, criminal degenerates, and the race suicide deplored by President Roosevelt. Our work will consist of eugenics research and education, and its gradual appli-

cation, through legislation and otherwise, to economic and social betterment.

Despite Fosdick's personal involvement and the family's previous generosity, the Rockefeller philanthropic boards had so far proved unresponsive to the importuning of the eugenicists. The trustees showed no inclination to take up a burden that Senior had relinquished with his last Founder's Designation in 1915.

In 1920 and again in 1921, Harry Laughlin tried to get a Rockefeller subsidy for publication of his book *Eugenical Sterilization in the United States*. Each time he was turned down. The Rockefeller advisers, including Fosdick, found nothing wrong with the bulk of the manuscript, which they described as "material relative to the history of sterilization, legislation in various states, court decisions, etc." There is some indication that Laughlin might have got his subsidy if he had agreed to delete, as requested, several chapters containing "direct propaganda favoring sterilization." But Laughlin refused, and the Rockefeller advisers backed away.

It was not so much the eugenicists' ideas that cost them Rockefeller backing as their tendency to confound research with advocacy. Nothing was more highly valued in the Rockefeller camp than fact-finding pursued without conscious bias. Only after all the facts were known could the proper action be taken. The Rockefeller model for bias-free social research was Abraham Flexner's 1910 report on American medical schools, which had been funded by the Carnegie Foundation for the Advancement of Learning. In 1913, after completing a similar report on European medical schools, Flexner was asked to join the General Education Board, to organize a campaign to bring about the reforms that his reports had so eloquently called for.

There was irony in the fact that Madison Grant and Charles Davenport has chosen to name one of their research-and-advocacy groups after Francis Galton, a British geneticist (1822–1911) who coined the word "eugenics" but specifically warned against mixing research and advocacy. Eugenics in both England and America offered two faces to the public: as an objective science dedicated to finding out the facts about heredity, and as a social movement dedicated to improving the human race through influence on its breeding practices. Although Galton himself believed in "improving the breed," he had a healthy respect for the complexity of this task, declaring research to be the immediate need, with social application as a distant goal. As a prophet of the dangers inherent in eugenics, he found more honor in his own country than in the United States, where his name was appropriated while his admonitions were ignored.

Since American eugenicists like Laughlin continually let their policy recommendations get ahead of their research, their chances for Rockefeller backing were dim. But a purely objective study of Australian aborigines was something else again. While Fosdick did not feel qualified to evaluate the scientific validity of the proposal, he could make sure that it reached the right desk within the bureaucratic maze of the Rockefeller philanthrophies. The first desk that he chose belonged to Beardsley Ruml, the recently appointed director of the Laura Spelman Rockefeller Memorial.

This was yet another Rockefeller philanthropy, named in honor of Senior's wife. Established in 1918 (five years after Mrs. Rockefeller's death), the Memorial had a capital fund of $74 million but no particular direction until Ruml persuaded the trustees in the fall of 1922 to launch an ambitious program of support for the social sciences, through the creation of university-based research centers. Studying Australian aborigines in their native habitat was hardly what Ruml had in mind, but he agreed to join forces with Edwin Embree to look into the matter.

The original budget submitted by the Galton Society called for $50,000 for a three-year study. Dr. Gregory thought this a fair estimate, considering the high cost of white labor ("one cannot hire natives as in Asia and Africa") and the aggressive behavior of certain groups of aborigines. Clark Wissler, curator of anthropology at the American Museum, had recently received the following news from a correspondent in the Crocodile Islands, off Australia's Northern Territory:

> The natives that inhabit the country between Darwin and Arnheim Bay are friendly but some of them still practice promiscuous cannibalism. (Within the last two months two bodies had been eaten by natives who are at present about my camp.) Several white men had been killed in these districts but it has usually been due to their own indiscretion. So far as we know the natives have never killed a white man for food.

Far from being put off by the rigors of the proposed investigation, Embree was enthusiastic. This was just the sort of opening the Division of Studies had been created to exploit. At a meeting with Wissler, Gregory, and Davenport, he talked about creating a "Rockefeller Commission on the Study of Primitive People" that would spend at least five years gathering data on everything from the head sizes and basal metabolic rates of native Australians to their birth and death ceremonies, as well as something that Davenport described as their "gregarious instincts."

Before submitting a formal proposal to the RF trustees, however, Embree consulted with a number of scientists and scholars associated with the Foundation, among them Simon Flexner of the Rockefeller Institute (a longtime RF trustee), President James R. Angell of Yale University, and the Princeton biologist E. G. Conklin. Although all concurred that the work itself was important, there were reservations about the Galton Society.

The two unwritten rules of the Rockefeller philanthropies were (1) get the facts first and (2) work only with the best. Despite eugenics's high standing with the American public and with Congress, the scientific reputation of Davenport and his colleagues had been under attack by the English school of geneticists for more than a decade. In America scientists like Herbert Jennings and Raymond Pearl (a colleague of Jennings's at Johns Hopkins) had begun to question the methods of the Davenport school. What Embree heard was enough to cool his enthusiasm for his would-be collaborators.

Without informing the eugenicists, Embree put together a proposal for a study of Australian aborigines to be carried out under *Australian* auspices—possibly through a group called the Australian National Research Council, which Beardsley Ruml had told him about. When the RF trustees expressed interest, Embree got them to subsidize a fact-finding trip to Australia by Dr. Grafton Elliot Smith, who had been recommended by Raymond Pearl. Elliot Smith (as he was known) was an Australian-born scholar whose Institute of Anatomy had been a major beneficiary of a $2 million RF grant to the University of London a few years earlier. Coincidentally (but conveniently), he was also a corresponding member of the Galton Society of New York.

Meanwhile, Davenport, who was unaware how far Embree had strayed from the Galton Society's original proposal, kept trying to influence the composition of the "Rockefeller Commission." What could be more important, he wrote Embree, than a scientific comparison of the intellectual development of "the lowest human race" with that of "the average European"? And he had just the right man to oversee this work: an Australian-born psychologist named Stanley D. Porteus, who was teaching at the University of Hawaii after spending several years as director of research at the Vineland (New Jersey) Institute for Feeble-Minded Girls and Boys. During the 1914–18 war Porteus had administered nonverbal intelligence tests to Australian aboriginal children; and he had found that hardly any of these "human anachronisms" (Davenport's phrase) scored above a twelve-year-old level. Porteus himself had seen this as evidence of widespread "mental retardation" among the aboriginal population—a conclusion that obviously appealed to Davenport, although he conceded that "investigation of the mentality of the Australians is practically an untouched field."*

On May 7, 1924, the Executive Committee of the Foundation authorized Elliot Smith to travel to Australia, at the Foundation's expense, to consult with scientists there about field studies of aboriginal cultures.

When Davenport learned that the Foundation was proceeding without further input from the eugenicists, he was furious. He demanded that some representative of the Galton Society be included in any preliminary talks with the Australians. In the ensuing exchange of letters between the Foundation and the eugenicists, Embree's murky style—which often hobbled him in his relations with the Board of Trustees—turned out to be an advantage. The leaders of the Galton Society withdrew their objections to Elliot Smith's trip on the assumption that the society would later be invited to cooperate in forming an "Australian-American Commission for the Study of Primitive Peoples." What Embree had actually promised was this: "If a commission or an officer of the Foundation later visits Australia in order to mature definite proposals, at that time, I should imagine rather than at present the question of representation by the Galton Society would arise."

Despite the high seriousness of all the participants and the importance—even urgency—of the work proposed, the genesis of the Aus-

*Porteus's predecessor at Vineland was H. H. Goddard, the man who coined the term "moron." It was Goddard's family histories of the "feeble-minded" that served as the principal documentation for Davenport's assertion, in 1910, that feeblemindedness was an inherited trait, carried in the germ-plasm as a Mendelian recessive, and responsible—through the mating of unwitting carriers—for much of the pauperism and degeneracy found among the lower social orders.

tralian project had a comic-opera air. Embree's motives for getting into science patronage were primarily administrative; the time had come for the Foundation to blaze new trails, and Vincent had given him the role of advance scout. Having made the decision to focus on science (after the trustees turned a cold shoulder to the arts), he had to find some science to support, and the suggestion from the Galton Society arrived at just the right time, with the imprimatur of one of the most influential RF trustees, Raymond Fosdick.

While providing on-the-job training for Embree, the Australian project also held an important lesson for Embree's successors. Even within the scientific community, public reputations can be misleading; behind the most impressive letterhead may lurk the danger of embarrassment—or worse—for the unwary science patron. It was a lesson that had to be relearned from time to time throughout the Foundation's fourteen-year association with the Australian project.

# 18

# Anthropology in the Antipodes

The power of the Rockefeller name—and Rockefeller money—was evident from the moment Elliot Smith stepped off the boat in Sydney in August 1924. Just the preceding year a conference of scientists from countries interested in the Pacific region had met in Sydney and Melbourne. In an effort to stimulate research while avoiding costly duplication of effort, this Pan-Pacific Conference had voted to give Australian reseachers sole responsibility for the study of aboriginal peoples both in Australia proper and in the "Mandated Territories" (including Papua and New Guinea) that Australia governed under a mandate from the League of Nations.

But there was a slight catch. Not only did Australia have no researchers trained in anthropological fieldwork, but no Australian university was equipped to supervise such work. So the conferees asked the Australian government to establish a department of anthropology at the University of Sydney, with an annual subsidy of £2,000. For this sum the legislators would get more than the satisfaction of supporting homegrown science; the professional anthropologists could also help prepare colonial administrators for their work in the Mandated Territories. Edwin Embree had been apprised of this development; in fact, it was the Australian government's favorable response to the initiative of the Pan-Pacific Conference that had given him an excuse for proceeding without the Galton Society.

But on his arrival in Sydney, Elliot Smith found that the political climate had changed. Before committing itself to an annual subsidy for anthropology, the government had asked the advice of a British colonial official who had been sent out from London to prepare a report on the administration of Papua. This official, a man of strong opinions, told the Australians that it was ridiculous to spend good money giving future colonial administrators a university education: what counted most in the "bush" was character, and this could be guaranteed by careful selection of graduates from good "Public Schools."

Impressed by this frank declaration of class bias, the Commonwealth government reneged on its promised subsidy, an action that the secretary of the Australian National Research Council called a "stab in the back." But the arrival of Elliot Smith, bearing letters of introduction from the Rockefeller Foundation, turned the situation around. After meeting with the prime minister, the press, and academic leaders from the major universities, he reported to Embree that the government had changed its mind again; the subsidy was once more on. Although Elliot Smith had been careful not to give the impression that Rockefeller backing was a sure thing, the mere hint of dollars from abroad had rekindled the enthusiasm of local politicians for anthropological research.

On November 7 the Board of Trustees of the Rockefeller Foundation pledged $100,000 toward the support of Anthropological studies in Australia over a five-year period.

In a letter of congratulations to Elliot Smith, Embree could hardly contain his enthusiasm. He looked for "far-reaching results" from the enterprise that Smith's "sympathetic and tactful assistance" had initiated. Then he added, in an instructive parallel to Wickliffe Rose's scientific ambitions,

> I hope that our common interest in anthropological work may lead our paths to cross many times. I have thought a little occasionally of the possibility of obtaining a little more intelligent knowledge of these problems by some regular work for a few months in such a laboratory as, for instance, that of yours in London. I imagine, however, it would be difficult for a layman to get anything from brief periods of study, even under the most stimulating environments. At any rate, I shall hope that I may have many opportunities of considering with

you personally problems that may come to our attention in the field of human biology.

When Elliot Smith responded with an open invitation to his lab at the Institute of Anatomy, Embree quickly backed off. He explained that he had been expressing a "hope" rather than a definite request and that other commitments made it impossible for him to acquire firsthand experience of the scientific life at this time. Among his other commitments (although he did not mention this to Smith) was the preparation of a proposal for a wide-ranging "Human Biology" program, which the trustees politely but firmly shelved at their meeting in February 1925.

Disappointed as Embree was by this action, he treated it as no more than a temporary setback. The trustees were still on record as approving "the principle of keeping Foundation policy flexible" (that is, not concentrating all efforts on public health and medical education), and the Division of Studies still had a mandate to explore new fields and to bring "mature" proposals to the Board's attention. And the Australian project was still on track, although events on that distant island continent were moving somewhat slower than expected.

One of the key people at the University of Sydney (the professor of anatomy) had died; and until his successor was named, the appointment of a professor on anthropology could not proceed. Back in New York, Ruml recorded his impression that the trustees had gotten "cold feet on the question of anthropological research." But Embree still had visions of involving the Foundation in the study of "vanishing" primitive cultures throughout the Pacific. In the spring of 1925 he made plans to visit the area with Dr. Clark Wissler of the American Museum of Natural History.

Wissler was a fellow of the Galton Society, but this connection was only incidental to Embree's decision to travel with him. A former student of Frank Boaz's at Columbia, Wissler (1879–1947) was curator of anthropology at the American Museum, former chairman of the Division of Anthropology of the National Research council, and a consultant to the Bernice Bishop Museum, in Honolulu—as solid a figure in the anthropological establishment as one could ask for. He was an authority on the American Indian, having done extensive fieldwork among the Sioux and Blackfoot tribes, but his interest extended to other primitive peoples who were coming into contact with the white man. Although Wissler had no doubts about the ultimate result of such contacts (in one book he referred to Nordic Americans as carriers of "the lamp of civilization"), he was eager to record in a scientific way the beliefs and customs of the less fortunate races. He had been recommended to Embree as the "best man" in the Galton Society, by Dr. Vernon Kellogg, the Stanford biologist who was elected to the RF Board of Trustees in February 1922.

All told, Embree and Wissler were away from New York for six months. Their most important stops, from the viewpoint of Foundation business, were New Zealand, Australia, and Hawaii. The academics they met in the Antipodes—which Embree noted was "an ungodly distance from all other inhabited spots—were hardly the first-rank scientists Rose had hobnobbed with on his European tour. But in his reports back to his "dear chief" (President Vincent) in New York, Embree radiated enthusiasm. Traveling with Wissler, he wrote, was "a fine education." The situation in Australia was "encouraging" for anthropology, with both federal and state governments contributing to the budget of the newly established department in Sydney. Everyone seemed pleased with the appointment of the new professor, A. R. Radcliffe-Brown, a Cambridge-trained man who had in 1910–13 conducted a brilliant investigation of the kinship and marriage systems of the Australian aborigines. After the war Radcliffe-Brown had established the School of African Studies at the University of Cape Town, and it was from Cape Town that he had been called to Sydney, principally on the recommendation of Elliot Smith.

In a letter that he dispatched to Vincent while en route to Hawaii from Australia, Embree allowed himself a small pat on the back: "If I do say it myself, outside one or two of those early South American surveys of Pearce's, I doubt if any country has been more thoroughly canvassed than Australia and New Zealand have by the two of us on this trip." He assured Vincent that the "recommendations for work in the Pacific" that he would submit on his return to New York were "likely to be of far-reaching importance," although, he hastened to add, they would "involve very small sums as compared with most of those of other Boards and Divisions." He tempered his enthusiasm with a single caveat: "One of the handicaps of Australia in all branches of work is its remoteness and the resulting difficulty men find in keeping in touch with work and workers in Europe and America." To overcome this handicap, he suggested a strong program of fellowships for "advanced training abroad of the most promising men."

What Embree did not realize, evidently, was that the difficulty of keeping in touch was a handicap in both directions. Between the end of 1925, when Embree sailed for Hawaii, and June 30, 1938, when the Foundation closed its books on anthropological research in the South Pacific, no representative of the Foundation set foot in Australia. If Embree had thought more deeply about the consequences of Australia's isolation from the mainstream of Western scientific life, he might have had second thoughts about committing the Foundation to a long-term project there.

On May 26, 1926, the trustees reaffirmed the five-year, $100,000 commitment to anthropological research in Australia that they had voted two years earlier, and named the Australian National Research Council as the sole channel through which these funds were to be distributed. In keeping with RF policy at the time, all monies released to the ANRC for fieldwork had to be matched, dollar for dollar, by funds raised locally. The trustees also appropriated an additional $10,000 a year to allow professors from abroad to teach and work in Australia; finally, they authorized a fellowship program in "Human biology," with candidates to be nominated by a special committee of the ANRC.*

In his own detailed report on the Antipodes trip, Clark Wissler was unstinting in his praise for the Australian National Research Council. Despite its name, the council had more in common with the U.S. National Academy of Sciences than with its war-born offspring, the National Research Council. The ranks of the ANRC were limited to one hundred members, "chosen according to merit" and "high scientific attainment." Although hampered by lack of funds since its founding, in 1919, the council was, in Wissler's view, "the only body able to handle scientific responsibilities on a national level."

As if to justify this assessment, the ANRC responded promptly to the Foundation's vote of confidence. A newly formed committee on anthropological research, headed by Professor Radcliffe-Brown, began organizing expeditions to several parts of the Australian continent and nearby territories; and the first request for RF funds arrived over

*At Embree's request, the Board also pledged $100,000 to the University of Hawaii for studies of the "biological, mental and social conditions of the people of Hawaii," and another $50,000 to a sister institution, the Bernice P. Bishop Museum, for research in "Polynesian anthropolgy." Like the Australian appropriations, these grants were contingent on locally raised matching funds.

the signature of the honorary treasurer of the Australian National Research Council, Henry G. Chapman, professor of physiology at the University of Sydney and director of cancer research there.

But the smooth start of the program was misleading, to say the least. Although Embree and the officers of the ANRC had exchanged numerous letters trying to define the terms of the grant, the potential for long-distance misunderstanding surfaced before the year was out. On December 22 the honorary secretary of the council, A. J. Gibson, wrote to Embree to ask whether the sum of £1,265, which had been advanced to two researchers at the University of Sydney by a local "corset manufacturer," could be counted as "matching funds" against the Foundation's $20,000 appropriation for 1926.

The two researchers, Professor Henry G. Chapman and S. A. Smith (a brother of Dr. Grafton Elliot Smith) were both members of the council's anthropology committee. The research sponsored by the corset manufacturer was a "national census of anthropometric measurements on Australian women." Up to September 1926 some eleven hundred women had been carefully measured to determine "the real shape of the body of a woman by the use of numerical data, in a form admitting of analysis of the measurements." In time, the researchers hoped to measure a total of twenty thousand women. Although the letter was not explicit, Chapman and Smith were apparently supervising the census but not doing the actual measurements themselves.

After consulting with his associates, Embree sent back a carefully worded letter, in which he regretted to inform the Australians that the corset manufacturer's census was not what the trustees had had in mind when they voted funds for "field work among primitive peoples" in the South Pacific. But the Australian National Research Council—or, to be more precise, Henry G. Chapman—did not accept this decision lightly. Six months later, the honorary treasurer of the ANRC was still trying to convince the Foundation that this work should be considered a serious anthropological investigation, whether or not the RF found the corset manufacturer's money acceptable as "matching funds."

By the time this point was resolved, Embree was no longer making decisions about the Australian program at the Foundation. On February 23, 1927, the Board of Trustees, acting on a recommendation from Fosdick, abolished the Division of Studies, transferred administration of its programs to Richard Pearce's Division of Medical Education, and kicked Embree upstairs to a post as vice-president. These changes did not take effect until April 1, and the Australians, whether by design or oversight, were not immediately informed. Embree, answering a letter from Radcliffe-Brown on March 30, rather casually remarked that "the Rockefeller Foundation had somewhat reorganized the administration of its work and that "official correspondence . . . should hereafter be carried on with Dr. Pearce." In case further proof was needed of the difficulty of communicating halfway round the world via steamship mail, letters from Radcliffe-Brown and other members of the ANRC continued to arrive in New York addressed to Embree as late as the middle of June.

When the Australians finally got the word, the news emboldened Professor Chapman to seek a new—and possibly more favorable—ruling on the corset issue from Pearce. On August 1, Pearce responded, reaffirming Embree's decision that the corset manufacturer's census did not come within the terms of the RF grant and indicating that as far as the Foundation was concerned the issue was closed.

# 19

# Double Downfall

The reorganization that cut the ground from under Embree was primarily the work of three men. John D. Rockefeller, Jr., had never been comfortable with the structure of the Foundation. It was neither the somewhat aloof "holding company" that he had once argued for nor a tightly organized operating unit with a clearly expressed mandate, like the Rockefeller Institute, the original Sanitary Commission, and so many other projects that bore the Rockefeller name. It was, in a word, too messy for Junior's tastes.

For George Vincent the problem had to do with authority. With the various operating boards pulling in different directions, there was a costly and confusing duplication of effort, an inability to respond to new challenges—and no clear line of command that would allow the president to impress his own mark on the institution.

The man who headed the reorganization effort—the choice of both Junior and the trustees—was Raymond Fosdick. He was a social reformer who felt most comfortable working inside society's power centers, a man who believed that nothing could stand in the way of reason in the service of conscience.

As we noted earlier, Junior had begun inviting members of the various boards to discuss their mutual problems in an informal setting as early as 1923. These meetings evolved into a regular "Monday luncheon" group attended by representatives of all the Rockefeller philanthropies: the Foundation, the Rockefeller Institute, the General

*Raymond B. Fosdick*

Education Board, the International Education Board, the Laura Spelman Rockefeller Memorial, and Mr. Rockefeller's personal advisory committee (which he sometimes referred to as his "committee on benevolence"). In addition, the trustees created an interboard budget committee in February 1925.

When it became apparent early in 1926 that the discussions were not leading to the reforms that Vincent felt to be essential, he persuaded the trustees to take an even more comprehensive look at the "organization, policies and programs of the Foundation"—with special emphasis on the powers of the president. "Until the precise relations of the boards to the Rockefeller Foundation and the responsibilities and duties of the Central Administration—i.e., the President and Secretary, have been clearely defined," declared Vincent, "the conditions for effective administration will be lacking."

A three-man "committee on reorganization" (Fosdick, Vincent, and Rose, with Fosdick doing most of the work) solicited the views of all the trustees and officers by letter and, when possible, in personal interviews conducted in New York and Paris. Junior had already written to Fosdick about his fears that the operating boards were too independent. Success within the Foundation, he said, tended to be measured "by the volume of business done by a Board or by one department of a Board" rather than by the RF as a whole. Junior's description of what might happen to a Foundation that failed to pull together could not have been more succinct: "Unless we keep ourselves clear-eyed and fresh and keep the machinery elastic, we run the risk of dry rot."

A few months later John D., Sr., gave his blessing to his son's efforts to rationalize the administration of the Rockefeller philanthropies. "If the whole thing were to be done today," wrote Senior to Junior, "you have rightly understood me as feeling that it should be done and doubtless could be done through a single organization."

With the founder, the chairman of the board, and the president in favor of change, a reorganization was inevitable. But what shape would the new Foundation take? The first hint came in a conversation between Fosdick and Rose toward the end of 1926. Rose suggested that the "science work" of the IEB might become the germ of a new division of the Rockefeller Foundation—"a division of fundamental science." The moment Junior heard about the suggestion, he wrote to tell Rose how appealing he found it. Not only was Rose's idea "in line with the principle of consolidation recently enunciated" in correspondence

*John D. Rockefeller, Sr., and John D. Rockefeller, Jr., 1921*

between the two Rockefellers, but it would also "round out the Foundation along what seems to be a logical line of development"—by placing all the scientific work of the various boards under one roof.

As for the International Education Board, Junior tactfully suggested that once its science programs were removed, the IEB might concentrate its efforts in the "general field of education throughout the world," doing the same sorts of things that the General Education Board was doing in the United States. In other words, the IEB could return to the work for which it was ostensibly created, before Rose became caught up in the excitement of supporting scientific research on a grand scale.

The termination of the RF's Division of Studies in early 1927 was only one step in the general overhaul of the Rockefeller philanthropies that Fosdick's committee set in motion. The China Medical Board was dissolved as an operating unit of the Foundation, and steps were taken to reconstitute it as an entirely separate entity, with its own $10 million endowment. In addition, the International Health Board lost its privileged status as a virtually autonomous operating unit and was restructured as a "Division" of the Foundation, under the close supervision of the RF Board of Trustees.

But these were mostly negative steps, designed to shake off the dead hand of past decisions and ingrained habits. After many more months of discussions and consultations. Fosdick presented the trustees with a comprehensive plan for reorganizing programs and policies, which was formally approved on January 3, 1929.

According to this plan the remaining assets of the Laura Spelman Rockefeller Memorial ($63 million) and the International Education Board ($16 million) were turned over to the Foundation, and their programs in support of the social sciences and the natural sciences were transferred to RF divisions that the trustees had created for just this purpose several months earlier. The Foundation also absorbed the natural sciences and humanities programs of the General Education Board. At the same time, the mandate of the RF's Division of Medical Education was changed; instead of providing general support for medical teaching institutions, it would henceforth concentrate on funding research in the medical sciences. Indeed, this was the principal rationale offered by Fosdick for the whole reorganization. From now on the Foundation would seek "to promote the well-being of mankind" by contributing to "the advancement of knowl-

*John D. Rockefeller, Sr., 1924*

edge." This meant, in Fosdick's own words and emphasis, that "*research* became the chief tool by which our newly acquired program was promoted."

In essence, the reorganization of 1928 represented the final triumph of the science patron over the missionary in the battle for control of the Rockefeller philanthropies. It is certainly ironic, then, that the two Rockefeller philanthropoids who had been most closely associated with science patronage should have terminated their Rockefeller connections at the very moment when the concept they had championed was being adopted as the basis for all future RF programs. But despite the coincidence in time, the circumstances of their going were quite different; again, it is the contrasts between Rose and Embree that are most instructive in understanding the era that came next.

On November 19, 1927, Rose turned sixty-five. He was due to retire as president of the GEB and IEB the following June. Although no one in the Rockefeller philanthropies had enjoyed such success, and no one (except possibly Fosdick) had such easy access to the Rockefellers, he had been around long enough for his autocratic habits to rub more than a few people the wrong way. As we have seen, Junior had personally thanked him for the suggestion that the science programs of the IEB might eventually be incorporated into the Foundation; and in the reorganization of 1928 that is exactly what Fosdick proposed to do. Yet Rose was not at all pleased. He had not intended that the transfer would take place so quickly, nor had he expected that the IEB would disappear entirely just because its science programs had found a new home.

When he realized that Fosdick had taken his suggestion and fashioned it into a devise for undermining both of his education boards, Rose fought back. On February 10 and again on March 1, 1928, he wrote to Fosdick expressing dismay over the proposal that the General Education Board "exhaust its capital within a period of five years" and that the International Education Board "close its work within a very short time by contributing $10,000,000 toward the Peking Union Medical College."

In the first place, Rose said, "The General Board has something like sixty millions to spend. I haven't the remotest idea that the administration would be able to spend that sum soundly in five years." As for the IEB, it was plainly "illogical" to ask the board to contribute to a medical college, since its programs had never so much as touched on medical matters.

Rather than rush the education boards out of existence, Rose suggested a gradual transition period of "three to four or more years" to allow the new officers in the Foundation to learn the ropes. When Fosdick showed no signs of budging, Rose took his case to John D. Rockefeller, Jr., who unreservedly backed Fosdick. Once Junior's position was clear, Rose simply withdrew. On May 22, 1928, the day on which Fosdick submitted his recommendations for a completely reorganized Foundation to the trustees, Rose submitted his letter of resignation as a trustee of the International Education Board, to take effect not later than June 30, 1928, the day on which he was due to retire as president of both the IEB and the GEB and as a trustee of the foundation.

Rose's formal connection with the Rockefeller philanthropies, which he had served with such distinction for three decades, was at an end. The institutions he had created or shaped to his purposes—the International Health Board, the International Education Board, the General Education Board—were being altered to suit the purposes of other men and other times. But as far as the natural sciences were concerned, the "new" Rockefeller Foundation would begin where Rose had left off, taking over his emphasis on Europe and, with it, the administrative machinery he had developed, including the central office in Paris and its dedicated corps of circuit riders.

The legacy that Edwin Embree left the newly reorganized Foundation was of a very different order. As early as October 1926 Fosdick had personally evaluated Embree's science program and found it wanting. He wrote to Simon Flexner,

> I have come to the conclusion that the only thing to do with the biological aspects of Embree's department will be to transfer them temporarily to Pearce's division, with the understanding that they are to be terminated as soon as possible. I find that they do not amount to anything anyway.

Leaving aside for a moment the propriety of this evaluation by a layman with a long-standing fondness for the eugenics movement, there is no question that the Foundation was wise to reject the administrative precedents that Embree had established. In the absence of anything like Rose's circuit riders, the Division of Studies had involved RF officers in the funding of complex scientific projects whose progress they could not possibly monitor.

Richard Pearce of the Division of Medical Education grasped this point immediately. His remedy was drastic. Since the project could not be properly administered from New York or canceled without hardship to the grantees, Pearce transferred all decision-making responsibility to the Australian National Research Council. Every year for the duration of the grant (through 1931), the Foundation deposited lump sums as promised in the ANRC account. In effect, Pearce was telling the Australians to take the money and spend it however they wished. The rationale he gave was unexceptionable: "So great a time is required for correspondence between New York and Sydney that it has seemed wise for us to take this action without consulting specifically in advance."

Pearce's letters to the ANRC reveal his determination to make the best of what he obviously considered to be a bad thing. Although the tone is correct and formal, it is hard not to read a sigh of relief into sentences like these, addressed to Professor Radcliffe-Brown on October 21, 1927: "If we can assist by counsel, or in other ways, please feel free to call upon us. The responsibility, however, for selection, for invitation and for arranging salaries and expenses [of fellow and visiting professors] should rest entirely with you and the Council."

This enforced reliance on a distant scientific body whose leaders Pearce had never met was entirely out of keeping with the precedents that Rose had established in Europe. It is true that the IEB transferred funds (to be used for fellowships and research grants) to the British Medical Research Council in London and the Notgemeinschaft der Deutschen Wissenschaft in Berlin. But unlike the ANRC, these were internationally recognized bodies, which drew on the talents of world-class scientists in their respective countries. Furthermore, the IEB circuit riders remained in close contact not only with the executive leadership of these bodies but also with the scientific communities they represented, so that any signs of political or personal interference with what were supposed to be purely scientific matters quickly came to the attention of Rose and his lieutenants. Embree's failure to provide similar safeguards would have unfortunate consequences in the years ahead, despite Pearce's attempt to insulate the Foundation from just such difficulties.

# 20

# Scandals

Administrative problems notwithstanding, the scientific work carried out under the auspices of Embree's human-biology program turned out to be far more significant than Fosdick's brisk appraisal suggested.

The Foundation was responsible for the creation of a new quarterly journal, *Oceania,* which became one of the most respected publications in its field. And the enormous mass of information gathered by RF-funded workers among the native populations of Australia and the Pacific islands had a lasting impact on the way these "primitive" peoples were viewed in the West. To cite just one example, Stanley D. Porteus, a principal beneficiary of the RF grant to the University of Hawaii, traveled to central and northwest Australia in 1929, at the invitation of the ANRC, to "determine the mental status of the aborigines."

This was the very study that Charles Davenport had proposed five years earlier. But what Porteus learned in the field must have made unpleasant reading for Davenport and his fellow eugenicists. On his 1929 expedition, which lasted seven months, Porteus tested over three hundred subjects, more than half of them children. Although he used a variety of measurement tools, he relied mainly on the nonverbal Maze test, which he had devised. His most significant finding was that the "mentality" of aboriginal children showed "very little difference from white children's norms."

He also saw right away that "one cannot apply tests to a group of primitive people, such as these aborigines, without finding out a great deal about the tests as well as the subjects." In fact, the more he tested native populations in Australia, Africa, and Hawaii, the less comfortable he became with the adjective "primitive," and the more emphasis he put on the "courage, hardihood, mechanical ingenuity, acuity of vision [and] social foresight" of these people, who could not possibly have survived in their inhospitable environments if they were (as portrayed by earlier observers, including himself) "inherently dull or stupid."

Porteus's change in attitude toward the Australian aborigines—from condescension to appreciation and even admiration—cannot be ascribed solely to his RF-sponsored fieldwork in Australia in 1929. He had been moving toward a more generous interpretation of their mental abilities even before this trip. Nevertheless, the publications that came out of his 1929 Australian expedition demonstrate vividly the impact of on-the-scene fact-finding on an open-minded investigator. And the writings of Porteus and other RF-funded investigators (often published in the RF-funded *Oceania*) contributed to a growing awareness among anthropologists that aboriginal cultures are most fruitfully viewed not as anachronisms illustrating "early" steps in the development of Western civilization but as alternative solutions to the general problem of human survival.

It is not clear whether Fosdick or anyone else at the Foundation recognized this shift of attitude as an important consequence of the Australian project. An evaluation of the Foundation's relationship with the ANRC, prepared for the RF trustees in July 1938, makes no mention of it. The evaluator simply cites the "scientific" importance of the information gathered, points to the establishment of *Oceania,* and then goes on to emphasize the "practical" value that such information might have for "colonial administrations faced with such problems as the breakdown of tribal authority, the conflict between native and European law, the educability of natives, etc."

For Fosdick, whose principal concern between 1925 and 1935 was to bring administrative order to the Rockefeller philanthropies, none of these benefits could have justified the serious risks to the Foundation's public image that Embree had incurred with his haphazard methods of operations. The problem is best summarized by the first sentence of the 1938 appraisal: "No [RF] representative . . . has visited Australia during the eleven year period when support of research in anthropology was provided under Foundation grants."

Despite Pearce's decision to shift administrative responsibility from the Foundation to the Australians themselves, RF officers could not wash their hands of events in Australia as long as Rockefeller money was being spent there. Before the RF's original five-year commitment could expire, in December 1930, the deepening world economic crisis had placed the entire project in jeopardy. The Australian government threatened to cut off support for the Department of Anthropology at Sydney; and there was no hope of making up the difference locally. According to a letter that Radcliffe-Brown sent to the Foundation in June 1930, the Australian public had shown little interest in anthropological studies, and even if private donors did come forward, the Australian National Research Council, unlike its American namesake, had no machinery for accepting funds from such sources.

While pleading for an immediate extension of RF support, Radcliffe-Brown complained about the way the grant had been administered:

> You will remember that when I was in New York, I discussed very fully with Dr. Embree the plans for our work and the policy that was to be adopted. Since Dr. Embree has left the Foundation, there has been no one there who is fully acquainted with our plans and our work and I feel that this is somewhat of a disadvantage.

This letter was addressed to "Dr. Vincent" as president of the Foundation. But Vincent had already retired (as of December 1929), and Max Mason, the former president of the University of Chicago, had succeeded him. Under the new administration, oversight of the Australian project had been shifted to the director for the social sciences, Edmund E. Day.

So uncertain were conditions in Australia that Radcliffe-Brown could not even forward a formal proposal until the spring of 1931. Since RF aid was contingent on locally raised matching funds and since the government's subsidy could not be assured beyond the end of 1931, the ANRC limited its request to $20,000 for a single year—"to tide anthropological research over this difficult period without any break in the continuity of the work."

Although the Foundation's new administrative team was just then in the process of shaping an entirely new program for the next decade, the Executive Committee immediately granted the ANRC request. Furthermore, the retirement for matching funds was dropped, in recognition of the near-bankruptcy of the Australian Commonwealth; between 1928 and 1930 roughly half of the RF allocation of $60,000 had gone unclaimed because the Australians could not come up with the funds to match it.

The Foundation's gesture of support once again persuaded the Australian government to renew the subsidy to Sydney's Department of Anthropology, although on a much reduced scale. In response the Foundation pledged another $60,000 over the next three years.

By the time all the arrangements for this grant were completed, there had been yet another change in personnel—this time on the Australian side. Radcliffe-Brown had accepted a call to the University of Chicago, where, in the words of the historian Adam Kuper, he spent the next five years re-creating American social anthropology "in his own image." In 1932 he became a professor at Oxford University, where he proceeded to do much the same thing.

There can be no doubt that, in terms of academic achievement, Radcliffe-Brown had been an outstanding choice as Australia's first professor of anthropology. Along with Bronislaw Malinowski of the London School of Economics, Radcliffe-Brown was a dominant force in academic anthropology for three decades (1925–55), and his students, often supported by Rockefeller-funded fellowships of one kind or another, carried his influence into the next academic generation as well. But because his personality was so out of joint with the place and the time, his four years at Sydney were hardly an unqualified success.

Radcliffe-Brown was a self-made man in every sense, with all the strengths and weaknesses that go with such an accomplishment. Born Alfred Reginald Brown in 1881 in Birmingham, England, he lifted himself out of his family's not-so-genteel poverty by distinguishing himself as a scholar, first at Birmingham University, then at Trinity College, Cambridge. At Cambridge his personal eccentricities were legendary. Politically, he was a declared anarchist, but socially he adopted "the manner of an English aristocrat, rather romantically conceived." He dressed with great elegance, even when circumstances might have dictated otherwise. For example, in the resolutely "egalitarian milieu of Sydney in the twenties," he affected a cloak and opera hat.

He believed that by careful planning a man could turn every moment of every day to his best advantage. He refused to talk to anyone who did not interest him. People he admired or needed, he cultivated with a single-minded ruthlessness that even his friends noted. His name itself was a carefully calculated gesture; he added the hyphen when he had his mother's name legally incorporated into his own in 1926.

It is not surprising that such a man should have had trouble wheeling and dealing with local politicians—academic and otherwise—during the financial crisis of 1929–31. According to Kuper, Radcliffe-Brown managed to antagonize many of the legislators whose support was essential to the department's annual subsidy. In fact, considering "his overbearing ways and political maladroitness," it is possible that his departure from Sydney in 1931 was at least partly responsible for the government's about-face in renewing its subsidy.

Even after the crisis in Sydney was resolved, the political and personal tensions that had precipitated it remained a mystery to Foundation officers thousands of miles away. When Frederick Wood Jones (professor of anatomy at the University of Adelaide) visited New York in the fall of 1932, he told Alan Gregg, RF director for the medical sciences, "[The Foundation] made a considerable blunder in dealing with the so-called Australian National Research Council, since state rivalries are acute in Australia and [the] Sydney faction is in control of RF grants for fellowships." Despite Wood Jones's obvious personal stake in the matter, Gregg noted in his diary, "There is something in this probably and we should be on our guard against any abuses or continuation of any arrangement without review of the situation."

A few days later Radcliffe-Brown, now ensconced in Chicago, sent the Foundation his own assessment of the situation he had left behind:

> Only a very small minority of persons in Australia recognize the value of scientific research except when it has immediate utilitarian results. . . . The Australian National Research Council itself hardly lives up to its name. Apart from its funds for anthropology derived from the Foundation it has only the scantiest possible funds for any kind of research and in the last six years it has made no effort to increase them.

In 1935 the Foundation agreed to provide one last grant, of $30,000, to help the Department of Anthropology at Sydney complete the work that had been initiated under RF auspices. Since the trustees had already decided against including anthropology in the RF's new natural-sciences program, a terminating grant of this size seems more than generous. In light of the bizarre events that had occurred in Sydney the preceding year, the Foundation's action might be considered positively heroic.

On the morning of May 25, 1934, Henry G. Chapman, professor of physiology at the University of Sydney and honorary treasurer of the Australian National Research Council, was found unconscious in his rooms at the university. He died without regaining consciousness, early the next morning. The cause of death was first thought to be a stroke, but a postmortem examination revealed that Chapman had committed suicide by taking poison. A noted toxicologist, he had often been heard to boast that his knowledge of deadly compounds gave him the power to take life without any possibility of discovery. When police detectives searched his rooms at the university, they found a cache of poisons and equipment for their administration. The police also uncovered evidence that Chapman had been spending considerable sums of money on his "private affairs, mostly associated with women." In addition, hundreds of letters addressed to Chapman in his capacity as officer of numerous Australian scientific societies (including the ANRC and the Royal Society of New South Wales) were strewn around the rooms unopened.

Chapman's family and acquaintances (he had few friends) had always thought of him as well-to-do. He had maintained at least three residences and several cars, and he had spoken of a large pastoral property that he owned somewhere in Australia. In a will prepared less than a month before he died, he named his wife as sole beneficiary of an estate estimated at approximately £80,000. The police found no such assets; if these had ever existed, Chapman had gone through them before his death.

In the course of their investigation, the police discovered that £17,900 in government bonds—which had been entrusted to Chapman for safekeeping and which represented the entire net worth of the Australian National Research Council and a substantial portion of the assets of the Royal Society of New South Wales—had disappeared.

Apparently, Chapman had cashed the bonds surreptitiously and spent the funds on his own "affairs."

These revelations did not come as a complete surprise to Chapman's colleagues in the Australian scientific community. A year earlier, in the spring of 1933, apparent irregularities in the ANRC books had prompted a call from other officers for an outside audit. Chapman at first resisted, accusing the other officers of personal animosity. When the executive committee of the ANRC formally requested an audit, Chapman agreed to present the books for examination on May 25, 1934—the day on which he was found unconscious in his rooms. The missing bonds and unopened letters told part of the story. But it was months before all the details could be worked out, since Chapman had destroyed all the ledger books of the Australian National Research Council.

Right after Chapman's death, A. J. Gibson, the secretary of the ANRC, cabled Edmund Day in New York, to inform him of the distressing news. The key sentence in the cable read, "SERIOUS DEFALCATIONS NOW APPARENT INVOLVING ALMOST TOTAL IMPAIRING OUR ASSETS." Without records there was no telling how much of the RF grant was missing, but for the moment the ANRC would be hard-pressed to find any money to support its "FELLOWS IN DISTANT FIELDS"—of whom there were at least six.

Day cabled back to inform the ANRC that he was forwarding the dollar equivalent of £700 immediately, to be charged against the unspent portion of the current appropriation.

As auditors unraveled the confusing state of affairs, the Australians were able to report some good news to Day. RF funds on deposit in local banks had somehow survived Chapman's "defalcations," although a £5,000 gift received several years before from the Carnegie Corporation was among the missing. Along with profuse apologies and expressions of gratitude, Sir George A. Julius, the president of the ANRC, expressed hope that the anthropological research "made possible in the first case chiefly through [the Foundation's] generous contributions" could continue in the future, "though perhaps on a more modest scale than in the past."

Day himself was sympathetic to the Australians' plight, and to anthropology in general. But after June 30, 1938, when the last installment of the Foundation's terminating grant ran out, it would be up to others, in Australia and elsewhere, to support research into the "vanishing" cultures of the South Pacific.

The entire history of the Australian project may be seen as a cautionary tale for science patrons—and for philanthropists in general. Of course, the Chapman scandal was a special case; even his closest collegues had no idea of how deep a hole he had dug for himself or how far he had gone in betraying the trust he had earned during a career of more than three decades in the highest echelons of Australian science. But there were other serious problems with the Australian project that even the most desultory monitoring system might have been expected to register: the precarious political support for the government subsidy (which Radcliffe-Brown was precisely the wrong man to oversee); the jealousies that divided the "Sydney faction" from scientists at other Australian universities; the somewhat shabby reality behind the impressive facade of the Australian National Research Council. Perhaps the most important piece of information that a science patron should have weighed before getting involved in the Australian project was the transparent "thinness" of the Australian scientific community when compared to its American and west European counterparts.

The problem was not merely a scarcity of qualified anthropologists. This was, after all, a relatively new discipline. According to Kuper, "as late as 1939 there were only about twenty professional social anthropologists, in the modern sense, in the British Commonwealth." And since one of Embree's goals was to build up anthropology through fellowships and seed money, it would have been inconsistent to expect to find a cadre of trained researchers on hand.

The real "scandal" of the Australian project (seen purely as an exercise in science patronage) was not Chapman's later defalcations but the failure of Embree and Wissler, when they visited Australia in 1925, to sound a warning over the fact that one man, a not particularly distinguished professor of physiology at the University of Sydney, should hold at the same time the following posts: honorary treasurer of the Royal Society of New South Wales, honorary general treasurer of the Australian Chemical Institute, honorary treasurer of the management committee of Sydney's Science House, and honorary treasurer of the Australian National Research Council.

It is hard to imagine the men who took control of the Rockefeller Foundation after the reorganization of 1928 mistaking the empty shell of a "National Research Council" for the real thing. The entire reorganization was predicated on the assumption that the best way to promote "the well-being of mankind" was to support basic scientific research. Fosdick's plan did not spell out how this was to be done. The ultimate responsibility, of course, lay with the Board of Trustees. But the more deeply the Foundation became involved in science, the less secure the trustees felt about their ability to guide and evaluate specific projects.

The solution was obvious: if scientific research was to be the focus of future programs, the officers would have to scientifically literate—not just self-educated polymaths like Wickliffe Rose or eager dilettantes like Edwin Embree but people trained in the methods and outlook of the modern research worker. In short, professional scientists.

In the spring of 1928 the Foundation named as its first director for natural sciences Max Mason, a former professor of physics who was then president of the University of Chicago. The new era had begun.

# 21

## But Is It Science?

Max Mason was the kind of person to whom the words "brilliant" and "genius" cling—early in youth as a promise and later in life as a kind of reproach. Born in Madison, Wisconsin, in 1877, he was a brilliant undergraduate at the University of Wisconsin, where he was noted for his talents in sports, chess, bridge, music, languages, and billiards as well as in science. He was a brilliant graduate student at the University of Göttingen, where he took a Ph.D. magna cum laude in mathematics in 1903 (and supposedly astonished his professor by solving a difficult thesis problem in just ten days). He was a brilliant teacher of mathematical physics at Wisconsin, where, together with his protégé Warren Weaver, he wrote a textbook on electromagnetic theory.

But his promise as an original thinker in math and physics was never fulfilled. He could never accept the new quantum theory. He had what Weaver later called "a phobia with respect to writing." He was better at generating ideas than at following them up. Typical of his intellectual style was a feat of creative engineering during the first World War; working with a team of scientists under the auspices of the National Research Council, Mason conceived of a major improvement on the best submarine-detection apparatus then available, built a crude prototype, and demonstrated its superiority, all within a week. Brought to combat readiness at the Navy's New London Experimen-

*Max Mason*

tal Station, Mason's device proved to be by far the best submarine detector in the Allied arsenal.

Perhaps because they sensed the gap between his potential and his achievements, other people kept trying to harness Mason's unused intellectual capacity in an administrative role. In 1925 he was named president of the University of Chicago; three years later he was lured to the Rockefeller Foundation to head the new Division of Natural Sciences, with the understanding that he would take over as president when George Vincent retired the following year.

On balance, Mason was no more successful as an administrator than he had been as a theorist and researcher. He did not suffer trustees or fellow officers gladly. His refusal to defend his proposals with the detailed and documented arguments that had become a hallmark of the deliberative process within the Rockefeller philanthropies soon alienated Chairman of the Board John D. Rockefeller, Jr. He gave only sporadic guidance to subordinates; Weaver recalled that Mason's idea of leadership was to "more or less let people do what they pleased."

His closest friends, Weaver among them, were all too aware of the personal problems that may have prevented Mason from living up to

*John D. Rockefeller, Jr., 1933*

his potential. His health was never the best; he was tormented by ulcers while still a young man. In addition, his private life was tumultuous, even tragic. His first wife gradually lost her mind, despite his Herculean efforts to find a cure for her. When she died, suddenly and unexpectedly, during his tenure at Chicago, he remarried quickly, but this relationship was stormy from the start, and ended, after the Rockefeller years, in a stormy divorce. (Mason was to become a widower two more times.)

Despite his sympathy for his mentor, Weaver could be quite candid about Mason's failings, which happened to mirror, as in a distorting glass, the personal strengths on which Weaver's own successful career as a philanthropoid was based. Mason "couldn't discipline himself intellectually or physically or morally or any other way," said Weaver, adding,

> He was so clever, so brilliant, so able—and so lazy. He simply could not devote himself, in any sustained way, to any task. . . . If he had [had] the capacity for sustained effort, I think he would have been one of the greatest scientists, or one of the greatest somethings, that the world has ever known. And

> as a result of the fact that he could not follow through, I have very reluctantly to write off his life as a comparative failure. In terms of what he was, as compared to what he might have been, I think the ratio is about as adversely low as it is for anybody I have ever known.

Nevertheless, the appointment of Mason began to pay dividends for the Rockefeller Foundation almost from the moment of his arrival in New York, in late 1928. The administrative restructuring presided over by Fosdick was virtually complete, but it would take several more years of hard deliberation to translate into effective programs the new policy that equated "the welfare of mankind" with "the advancement of knowledge." Not surprisingly, Mason first made his presence felt through decisions *not* to support certain kinds of science—in particular, eugenics.

Unable to pry any money out of the Foundation itself, American eugenicists had found a more receptive ear at John D. Rockefeller, Jr.,'s personal advisory committee (which Junior usually referred to as his "committee on benevolence"). Starting in 1925 this committee had contributed $5,000 of Junior's personal funds each year to the American Eugenics Society. Reviewing this commitment in 1927, Thomas B. Appleget, the secretary of Junior's advisory committee, wrote that he was "favorably impressed with the program and attitude of the Society." The advisory committee duly voted to renew Junior's $5,000-a-year pledge for two more years.

During these years Raymond Fosdick held membership on both Junior's advisory committee and the advisory council of the American Eugenics Society, while serving as the "family's" representative on the boards of all the major Rockefeller philanthropies. If the Rockefellers were willing to be publicly associated with the eugenics movement during this period, it goes without saying that Fosdick knew of—and approved— the connection.

When the matter of support for the American Eugenics Society came up again for review (in February 1929), the report of the secretary—still Appleget—sounded an entirely different note. Now Appleget was moved to write that the society, through its simplistic application of Mendelian principles to the complexities of human genetics, had "spread a gospel of predestination and despair, and neglected to consider the proper function of environmental influences." Tapering-off grants of $3,500 for 1929 and $2,000 for 1930 were recommended, with complete cessation of support thereafter. Although Appleget referred to recent criticisms of eugenics by respected biologists (such as "Dr. Jennings of Johns Hopkins"), the arrival of Max Mason at the Foundation undoubtedly had much to do with this turnabout by Junior's personal advisers.

Just a month earlier the advisory committee had looked with much favor on a $20,000 request from yet another eugenical organization. This was the Eugenics Research Association, which had been created by Frederick Osborn, a friend of the Rockefeller family. Like his famous uncle Henry Fairfield Osborn (who served as president of the American Museum of Natural History from 1908 to 1933), the forty-year-old Frederick Osborn was passionately interested in the study of human heredity. His motive in founding the Eugenics Research Association was to join forces with Charles Davenport in an ambitious five-year program to study such vital matters as "the rules which govern the inheritance of human traits," "race mixtures," and "differential fecundity." (The last was a reference to the often-noted—and invariably deplored—fact that Negroes, Jews, and other non-Nordic races tended to have more children than did Nordic types.)

Appleget's investigation of this proposal convinced him that it was "the best one which has come before us in the field of heredity" and that it promised to be "infinitely more scientific in its approach than the American Eugenics Society." The participation of Frederick Osborn was a definite plus—and not just because he was a successful Wall Street banker whom Junior knew and respected. As an amateur of science, Osborn was notably more open-minded in his approach to human heredity than his uncle, whose views on the innate superiority of the white race were virtually indistinguishable from those of Madison Grant. Another obvious plus, or so it seemed, was the proposed connection with the well-known eugenics research program of the Carnegie Institution.

To Appleget's surprise, Dr. Charles Merriam, director of the Carnegie Institution, was "very circumspect" in his appraisal of Davenport's work. In strict confidence he told Appleget that the Institution had been

> engaged for some time in an attempt to re-state the problem of our researches in relation to eugenics . . . to make sure that these investigations are on a basis of fundamental research

corresponding to that in the well established fields of scientific investigation.

From this tortuous circumlocution, it could be inferred that the Carnegie Institution was unhappy about its long-standing contractual relationship with Charles Davenport and his associates. Appleget may not have understood all the scientific issues, but he knew that if there was disarray at the very center of the eugenics establishment, outsiders had better tread with care. "The safest step," he concluded in his report to the advisory committee, "would be a declination." But in view of "Mr. Osborn's relationship to Mr. Rockefeller," Appleget was uneasy about a summary rejection.

The just-completed reorganization of the Rockefeller philanthropies provided Junior's advisory committee with a graceful way out. The main purpose of the reorganization was to concentrate in a single agency all the Rockefeller expertise in science. So the advisory committee referred Frederick Osborn's proposal to the Rockefeller Foundation, where it came to rest on the desk of the director for natural sciences, Max Mason—who promptly turned it down.

The next time the executive secretary of the American Eugenics Society, Leon Whitney, came to the Foundation's offices to talk about a grant, it was Mason whom he saw. The interview was brief and not particularly friendly. Mason raised the question of whether the society's "propaganda" was too often "in advance of scientific knowledge." Although Whitney denied the allegation, Mason told him that there was "almost no possibility of RF support for this type of work."

The severing of connections between the Rockefeller philanthropies and the eugenics movement was not yet complete. In the reorganization of 1928 the Foundation had assumed the assets and commitments of the Laura Spelman Rockefeller Memorial, which, under the leadership of Beardsley Ruml, spent nearly $50 million between 1922 and 1928 on research in the social sciences and related fields. Ruml, who was twenty-seven years old when he took over the Memorial, had a reputation for wide-open experimentation. In Edwin Embree's unflattering—and undoubtedly envious—assessment, "Ruml started his program and succeeded in giving away [millions of dollars] before he could be stopped."

One of the commitments that the RF inherited from Ruml was an $84,000 grant to the University of Vermont for a study of deteriorating rural communities in the state. This work had grown out of an earlier "Eugenics Survey," which examined the drain on state resources caused by some fifty "degenerate, deficient and delinquent families."

Among other objectives, the statewide study sought to determine what effect "the substitution of foreign race elements for the native stock" had had "upon the quality of life of the Vermont town and upon the ideals which have made the name of the Green Mountain state respected and loved throughout the nation." When the Vermont officials responsible for this study asked the Foundation for an additional $49,000 to complete the work, the request was denied.

With this action the last formal tie between the Rockefeller philanthropies, now dominated by Mason in scientific matters, and the eugenics movement had been broken—at a time when fears roused by the economic depression might have made the trustees especially receptive to the propaganda of the eugenicists. Indeed, the eugenicists kept trying. In February 1932, the same month in which Warren Weaver arrived in New York to join the RF staff, Harry Laughlin was in touch with yet another of Junior's private philanthropies, the Bureau of Social Hygiene.

The bureau had been created in 1911 to look into the causes of, and to seek a cure for, the epidemic of prostitution that reformers believed to be sweeping the Western world. Later, under the direction of Colonel Arthur Woods, a former New York police commissioner, it had conducted systematic surveys of police administration and drug addiction. When Laughlin asked for a $1,000 contribution to the "Third International Congress of Eugenics," scheduled for August at the American Museum of Natural History, the bureau declined, after first ascertaining that neither Max Mason nor Edmund Day, the RF's director of social sciences, regarded the eugenics movement as "scientific or having anything to offer at this time."

Mason's unequivocal rejection of eugenics in the early thirties can hardly be considered prescient. Enlightened scientific opinion had long since turned against the more extreme contentions and slipshod methodology of Davenport and his associates. But the eugenicists were still very much in the public eye, and for all the private reservations that Charles Merriam expressed to Appleget, the Carnegie Institution continued to lend its name and prestige to the movement until 1940, when the Department of Genetics at Cold Spring Harbor was entirely restructured.

In the absence of a firm anti-eugenics policy, there was always the chance that the blandishments of the eugenicists would find a sympathetic hearing somewhere in the Rockefeller philanthropies—with painful, not to say tragic, consequences. The great danger of eugenics was that it cloaked baleful doctrines in a guise of scientific respectability that few nonscientists could be expected to see through. For example, who would have sensed anything untoward in the presentation of Mrs. Cora Hodgson, an English lady who called at the Bureau of Social Hygiene on October 4, 1932, to discuss the work of the International Federation of Eugenic Organizations? This was yet another umbrella group, whose aim was to foster eugenics research on a worldwide basis.

Both the Carnegie Institution and the American Eugenics Society were associated with the federation, which was headed by Ernst Rudin, director of the Kaiser Wilhelm Institute for Psychiatry in Munich. In November, Professor Rudin himself wrote to the bureau, asking for the dollar equivalent of 40,000 reichsmarks to support an investigation of "the probable heredity prognosis of offspring in the most frequently occurring types of mental disease with hereditary taint"—including hysteria, epilepsy, and feeblemindedness. Once again, the bureau declined.

Two months later Hitler was named chancellor of Germany. With the Nazis in power, eugenicists like Rudin no longer needed to solicit funds from abroad. Working closely with Heinrich Himmler, German scientists formulated the Eugenical Sterilization Law, which went into effect on July 14, 1933. Special Heredity Health Courts, consisting of two medical doctors and a judge, were set up to decide which "defectives" in the German population would be sterilized. In the first year alone, 56,244 people were forced to undergo this operation; within three years, the number of sterilizations had reached 225,000.

The leaders of the American eugenics movement, seeing their ideas put into practice on such a large scale in so enlightened a country, were overjoyed. Leon Whitney assured American Jews that the Nazi programs in *Rassenhygiene* (race hygiene) were being conducted in an impeccably scientific manner. In 1936 Harry Laughlin accepted an honorary degree from the University of Heidelberg, where the Nazis had staged a virulently anti-Semitic "rededication" ceremony the preceding December.

Three years later *Rassenhygiene* in Germany was taken to its logical conclusion. On September 1, 1939, while Hitler's armies invaded Poland, two of his most trusted aides were given the power to implement a semisecret euthanasia (mercy-killing) campaign. The goal was to cull from the German nation the incurably insane and other "lives unworthy to be lived." During the next two years 50,000 such lives were "suppressed" (the official term) by lethal injections, and by the newly developed method of gassing in rooms disguised as shower baths. Then the program was suspended, because of a popular outcry led by prominent churchmen.

Apparently, even in wartime the German people were not ready to see their own ranks culled by government order, no matter how "scientific" the principles used to justify the operation. But the experience that had been gained in the euthanasia program was not wasted. It was simply transferred to a larger and more important project that Hitler set in motion early in 1941 and that went ahead without significant protests from the German people: the systematic destruction of European Jewry. Even while they were constructing the death camps in Poland, the Nazi veterans of the euthanasia program continued to refer to themselves by the title that had been adopted as an innocuous cover for their earlier operation: "The Charitable Foundation for Institutional Care."

American eugenicists were not so quick to comment publicly on the developments in Germany between 1939 and 1942. For the record, therefore, it is perhaps not unfair to quote from a letter that Charles Davenport wrote to Madison Grant in 1925:

> Our ancestors drove Baptists from Massachusetts into Rhode Island but we have no place to drive the Jews. Also they burned the witches but it seems to be against the mores to burn any considerable part of our population.

# 22

# The Search for Relevance

Changing the direction of a philanthropic foundation is not unlike changing the course of an ocean liner. The very nature of the enterprise rules out sudden turnabouts. Momentum must be reckoned with. Even when a philanthropist is not legally bound to support a particular project for more than a year at a time, there are moral obligations that cannot be ignored. A recipient whose otherwise worthwhile project is excluded from a foundation's newly redefined "area of concentration" needs time to seek funding elsewhere.

In the early 1930s the slow process of phasing out old RF programs and designing new ones to promote "the advancement of knowledge" was complicated by the worldwide economic upheaval that came to be known as the Great Depression. In 1929 the annual income of the Rockefeller Foundation hit an all-time high, representing a 6.59 percent yield on its investments; that year the officers and trustees had $14,139,743.40 to spend. By 1933 the yield on the Foundation's investments was down to 4.21 percent, and its income had fallen to $8,248,673.97; furthermore, the severe drop in market values made the trustees reluctant to sell securities to raise cash for current outlays, as they had done during the First World War.

Of course, compared with the rest of the scientific community (not to mention the world at large), the Foundation was not badly off. Congress slashed the budgets of all federal scientific agencies by an average of 12.5 percent from 1931 to 1932. State legislatures took similar aim at research funds allocated to public universities like those of California, Wisconsin, and Michigan; private universities either had to curtail major fund-raising drives (as Stanford and MIT did) or had to watch their once sizable endowments for research dwindle away in the stock-market retrenchment (as Caltech and Cornell did). Unable to raise money to equip its new physics building, Washington University in St. Louis converted the third floor of the building to a children's skating rink.

Employment opportunities for young (and even established) researchers kept dwindling as well. While Ph.D.'s poured out of America's institutions of learning at an ever-increasing rate, the Bureau of Standards in Washington and the major industrial research labs (like those financed by General Electric and AT&T) were laying off half their technical staffs. Under the circumstances the budget of the RF's Division of Natural Sciences—which averaged about $1.5 million a year during the thirties—"constituted a very substantial fraction of the free funds available for research in the whole United States." As Warren Weaver recalled years later, it was a time when "one could have a great intellectual leverage with a relatively small sum."

For this very reason the trustees scrutinized with great care the budgets of the new and still somewhat shaky Division of Natural Sciences. With the world economic system in disarray, with millions starving, with the specter of social revolution haunting the rich and powerful, it was inevitable that the causal connection between the "advancement of knowledge" and "the well-being of mankind," so glibly asserted by Fosdick in 1928, would come under skeptical review by the RF Board in the early thirties.

The battle cry of "relevance" was in the air. Although no one wanted to see the Foundation bankrolling soup kitchens or otherwise getting involved in relief work on a large scale as it had in World War I, the trustees were now looking for programs that would pay measurable dividends—by alleviating the distress of individuals and/or bolstering the social order—sooner rather than later.

The measures advocated by eugenicists of the Davenport school—restricting immigration, sterilizing the "unfit," promoting marriages between the genetically advantaged—were specifically designed to combat evils like immorality and crime, while shoring up those twin pillars of the social order: the community and the family. No wonder the eugenical approach to society's problems had appealed to liberal reformers like the young Raymond Fosdick.

The Rockefeller Foundation's rejection of eugenics in the early thirties was based on the perceived lack of scientific merit in the work of the eugenicists. By no means did Mason's actions imply a rejection of "relevance" as a criterion for judging whether a particular scientific project was worthy of support. In October 1930 Mason and the man he had chosen as the RF's new director for natural sciences, Herman Spoehr, shared with the trustees their still-tentative thoughts about a new and more "concentrated" program of science patronage. Not surprisingly, since Spoehr's own research had been in photosynthesis, his presentation emphasized the importance of learning how plants convert the energy of sunlight into fuel and food, and how our bodies make use of what the plants produce.

The discussion coincided with a chorus of stories in the popular press concerning a "science moratorium." Originally proposed in 1927 by an Anglican bishop (who later confessed that he had been kidding), the idea that science should "take a holiday" for a few years gained adherents as the Depression worsened. The new antagonists of science were a mixed bag. On the one hand, there were liberals and Marxists who denounced the entire enterprise of modern science as elitist and (given the social background of its patrons) inherently conservative. For this group, nothing less than a restructuring of the academic and financial underpinnings of scientific research would suffice to bring researchers back into the circle of "human values" from which they had strayed. On the conservative side, there were prominent churchmen who believed that science was simply moving too fast and that the traditional institutions of society needed a breather to adjust to the torrent of new information flowing from the laboratories.

In the same month in which Mason and Spoehr met with the RF trustees to discuss the future of the Foundation's science programs, G. K. Chesterton wittily summarized the conservative argument for a science moratorium in an article in the *New York Times Magazine:* "There is nothing wrong with electricity; nothing is wrong except that modern man is not a god who holds the thunderbolts but a savage who is struck by lightning."

The RF trustees did *not* buy this argument. But the minutes of their October 1930 meeting reveal a new awareness of the risks they were running in supporting scientific research—the risk of squandering limited resources on work of dubious quality, the risk that otherwise legitimate work might be turned to antisocial purposes, the risk of creating a privileged class of scholars whose pursuit of knowledge for its own sake would fail to satisfy society's need for useful information. Without turning their backs on science patronage, the trustees began retreating from the uncritical "scientism" of the Rose years. Only science that could be justified as socially relevant had a chance of being funded. Among the disciplines that failed to meet this test was anthropology of the type that Edwin Embree had supported back in the twenties.

In the reorganization of 1928 it had been agreed that anthropology should come under the aegis of the director of social sciences. Despite Fosdick's cavalier dismissal of Embree's legacy, the new director of social sciences, Edmund Day, at first made a bid to *double* RF expenditures in this area—on the grounds that cultural anthropology fit in perfectly with the trustees' new emphasis on "the problems of human behavior and personality."

To bolster this argument, Day circulated, in the summer of 1931, a memo that he had received from Professor Radcliffe-Brown, now at the University of Chicago. In addition to repeating his old admonitions about the urgency of studying rapidly disappearing primitive peoples, Radcliffe-Brown claimed that such work would lead, in time, to the discovery of "universal sociological laws having the same validity as the laws of natural science." In his own pitch for a Foundation program in anthropology, Day pointed out that the RF was already spending some $150,000 a year on scattered projects of a more or less anthropological nature; he proposed that these projects be refocused along the lines laid down by the trustees, and that the level of support be increased to $300,000 a year.

Day's enthusiasm was reflected in the Foundation's *Annual Report* for 1931, where cultural anthropology was explicitly recognized as a "special field of interest" within the social sciences. But strong pressures, both from inside and from outside the Foundation, were already working against this program.

Max Mason thought that Day's proposal might have some merit if the rationale of the work could be defined with more precision. Day's colleagues were not sure exactly what "cultural anthropology" was, and Day himself had trouble drawing the boundaries that separated it from such related disciplines as history, sociology, social anthropology, and ethnology. It was decided that no firm decision could be reached without a survey of the entire field of anthropology.

Early in 1932 Leonard Outhwaite—a former staff member of the

Laura Spelman Rockefeller Memorial who happened to be out of a job—accepted the assignment. Outhwaite spent most of nine months on the survey, interviewing two hundred informants at more than sixty institutions in the United States and abroad. He had every reason to expect that an ambitious commitment was in the offing and that there would be a position for him in the new program. In his final report, which he submitted in the fall of 1932, he urged the Board to think big: "It is recommended that the Foundation accept in principle a plan for developing the anthropological field intensively for a period of fifteen years."

But time had passed Outhwaite by. The trustees' vague formula of concern about the "problems of human behavior and personality" had been considerably sharpened by events in the summer and fall of 1932. Franklin Roosevelt had been elected president, largely on the basis of his promise to do something about the continued economic slide. The League of Nations' ineffectual response to the Japanese invasion of Manchuria had blasted hopes that the world organization might prove a bulwark of world order. The Nazis were poised to take over Germany. Both at home and abroad, political, economic, and even moral debate had been reduced to shrill confrontations between partisans of one ideology or another.

When the future of the American way of life itself seemed in danger, it was hard to get excited about the vanishing aborigines of the southwest Pacific. The whole question of the relevance of Foundation programs was due for another airing at a special meeting of the trustees in April 1933, but when Outhwaite met with Day in the middle of March, he could see the handwriting on the wall.

Day was preparing a statement for the trustees in which he proposed to concentrate the resources of the Division of Social Sciences on such undeniably relevant areas as "economic planning" and "social control." An ambitious program in cultural anthropology had no place in this new set of priorities. Having lost his patron, Outhwaite had lost the battle before it began. Over his bitter protests his recommendations were rejected by Mason and the other RF officers at a staff conference in March. While breadlines formed and the armies of unreason mobilized, the Foundation could spare none of its already reduced income for a long-term commitment to such a "soft" and somewhat rarefied academic pursuit as cultural anthropology.

It was against this background of ardent concern for socially relevant science that Warren Weaver (with the backing of Max Mason) persuaded the trustees to start up a "concentrated" program in fundamental biological research, stressing the use of tools (often borrowed from the physical sciences) that might someday permit measurements of the very stuff of life.

# 23

# Betting on Uncertainty

In the last decade of the nineteenth century, the town of Reedsburg, Wisconsin, had a population of "something under two thousand persons." The estimate comes from the pen of Warren Weaver, who was born in Reedsburg on July 17, 1894. One can be sure that if a more precise figure had been available, Weaver would have used it. He was, as he himself noted later in his life, a man who always took "rather too much delight in precision."

In describing his own childhood in Reedsburg, Weaver recalled that the word "science" was "exceedingly little used in that village at that time." Nor were the products of science much in evidence. At the time of his birth, no houses in town were wired for electricity. A few businesses had telephones (the first had been installed in 1887), but a general service for householders would not be available until he was four years old. Since there were no automobiles, it didn't seem to matter so much that the streets were unpaved.

In many ways Reedsburg was typical of the small towns of nineteenth-century America that were soon to become the focus of so much uncritical nostalgia. Life in these towns came to be seen as life before the Fall—before the countryside was transformed by machines and by habits of thought that originated with the scientific enterprise of the new century. The popular newspaper cartoonist Clare Briggs, a Reedsburg native, captured this small-town nostalgia in his widely syndicated cartoon series featuring tag lines like "The Old Swimming

*Warren Weaver*

Hole," "Ain't It a Grand and Glorious Feeling?" "When a Feller Needs a Friend," and so on.

Weaver's own memories of growing up in Reedsburg hardly squared with these images of antediluvian bliss. Like many of his neighbors, he came from German Lutheran stock—farmers, shopkeepers, churchgoers. He remembered being bored at the endless prayer meetings; he had few friends and few opportunities to make them, since he had no talent for "the active games of childhood." He remembered being sick a lot.

His father ran a drugstore where much of the business was conducted in German. In addition to such staples as patent medicines (which promised cures for everything from tuberculosis to the common cold), soaps, and ice cream sodas, Isaiah Weaver filled his store with toys during the Christmas season. To build up his stock, he made a special buying trip to Chicago each fall; on his return he always brought something special for Warren and his older brother. On one memorable occasion Warren's present was "a one-dollar Ajax electric motor and a dry cell to run it."

The shy, sickly, introspective little boy fell in love with the toy motor. "Within a week," he recalled, "I had built from spools and suchlike all the little rotating devices which the tiny torque of the motor could manage. Then I began to penetrate its mysteries." He took the motor apart, piece by piece, and reassembled it.

> And—miracle—the motor would still run! I decided then and there that I wanted to spend my life building motors. . . . Some adult, sensing my enthusiasm, remarked that I would probably grow up to be an engineer. I at once completely adopted that word and idea, and through my grade school and high school years there never was the slightest doubt what I wanted to do. I wanted to be an engineer.

To hone his skills, he built ever more complicated radio sets.

It wasn't until his sophomore year at the University of Wisconsin, where he was enrolled in a five-year program leading to a degree in civil engineering, that Weaver realized he had been the victim of a "semantic error." This realization was triggered by his encounter with "a really poetic branch of mathematics"—differential calculus—and a great teacher, Charles Sumner Slichter, whose zest for the "excitement and power and logical beauty" of mathematics was contagious.

Under Slichter's tutelage Weaver realized that he didn't want to be an engineer at all but a *scientist*—a word not yet in common usage in Reedsburg. If Weaver's interest in "pure" math was ignited by Slichter, his conversion to science was confirmed by contact with another charismatic teacher, Max Mason, the leading theoretician in the University of Wisconsin physics department. Although Mason was twenty-two years older than Weaver, teacher and student quickly became close friends. They collaborated on a number of scientific papers; and for a while Weaver was quite comfortable in his role as Mason's protégé—first on the Wisconsin faculty and then on the staff of the Rockefeller Foundation in the early 1930s.

Weaver would not have said it, of course, but Mason's most brilliant stroke as president of the Foundation was probably his recruiting of Weaver as the RF's director of natural sciences. Rarely have a man, a job, and the time been so perfect for each other. Weaver was in charge of the Foundation's scientific programs from 1932 until his retirement in 1959. During these years, his support of quantitatively precise methods of biological research influenced a whole generation of investigators and played an important role in the development of such fields as molecular biology, a term Weaver coined in 1938, and plant genetics, which gave birth to the so-called green revolution in food production in the early 1950s.

According to the historian Robert Kohler, Weaver was one of the first, and certainly one of the most successful, of a new breed of "managers of science" who have come to dominate the international scientific community in a world increasingly dominated by the methods and products of science.

Ironically, Weaver was not Mason's first choice for the job. He might never have joined the Foundation at all if his predecessor as director of natural sciences, the biologist Herman Spoehr, had not called it quits after only one, unhappy year in New York. According to Weaver, Spoehr (who had been a staff scientist at the Carnegie Institution of Washington) "was unhappy for a very honorable reason. He was so good a scientist that he never felt himself in possession of a sufficient body of accurate information to justify his making the decisions that he had to make. . . . And this just tormented him."

Weaver's remark is significant because the whole thrust of the new RF policy was to put scientists in charge of dispensing money to scientists. As Weaver saw it, only certain kinds of scientists were fit to be science patrons. Spoehr's problem was that he lacked what Weaver called "a good administrative conscience." More often than not, a philanthropoid is called upon to make a decision based on insufficient information: "Once he's *made* it, [the good philanthropoid] puts it behind him . . . just as definitely as a good surgeon has to put the situation behind him when he walks out of the operating room." When Spoehr realized he couldn't do this, he returned to research at the Carnegie Institution. And the way was open for Weaver.

Behind Weaver's success as a science patron lay an unusual cast of mind in which a delight in precision coexisted with a high tolerance for uncertainty. It may come as a surprise to those who think of science as a realm of dry-as-dust calculations—and of scientists as cold-blooded devotees of the sure thing—but the pursuit of scientific truth typically involves a great deal of financial and personal risk. Investigators may spend years following false leads or bumping up against technological or conceptual dead ends without any assurance of ultimate success.

Like anyone who gambles on his own future, the modern scientist cannot afford to play the game as an amateur. The amateur gambler takes risks without first figuring the odds. Typically, he will rely on an untested but plausible-sounding "system," an unverifiable tip from a self-styled authority, or even a sentimental hunch (a horse whose name reminds him of his mother). The compulsive amateur gambler may even feel better when he loses.

The professional gambler hates to lose. He gets a kick out of taking risks; if he didn't, he would take up a safer occupation, like selling shoes or working for the Internal Revenue Service. But he is not moved by sentiment. Before placing a bet—whether at the track, in the financial arena, or in his personal life—he will gather as many facts as he can and then analyze them with a high degree of objectivity before acting. He knows he cannot eliminate all risk (and he would probably become bored if he could), but he will do anything within reason to increase his chances of winning. This desire to take risks while at the same time trying to keep those risks to a minimum may entail a logical contradiction; but in such matters there is no arguing with success. For both the scientist and the professional gambler, a basic prerequisite for success is a thorough grasp of the mathematics of probability—in other words, figuring the odds.

Not at all by coincidence, Warren Weaver numbered among his many interests the theory of probability and the statistical techniques that have been devised for dealing with situations in which, as he once wrote, "the individual event is . . . shrouded in mystery." He was referring to modern physics, modern genetics, and modern information theory, all of which are based on the notion that one can aspire to "dependable knowledge" only of the *frequencies* at which certain events occur. But what he had to say applies with equal, if not greater, force to science patronage.

Weaver was not only a scientist and a manager of science but also a lively popularizer of science (whose work in this area was rewarded by a UNESCO prize in 1965). In 1963 he published a little book called *Lady Luck: The Theory of Probability.* In introducing this "rich and practical and lovely" subject to a general audience, he pointed out that the theory of probability in Western culture began with attempts to figure the odds governing some popular card games in seventeenth-century France. As Weaver put it, "In the highly fashionable but slightly disreputable atmosphere of the gambling rooms, Lady Luck was born." Not until the twentieth century did it become clear that statistical reasoning was an indispensable method for "thinking about uncertainty"—a method without which we would be hard-pressed to think about reality at all.

Weaver was fond of noting that he was born "at almost precisely the right time." Attracted to science as a child without ever having heard the word, he helped create the Age of Science in which we all live. But precisely because he was born in another era, he maintained a clearer view of the limitations of science than did many of his intellectual descendants. Apropos of his lifelong habit of attending church, he once said, "It is rather surprisingly the case that the only time man is ever really sure is not when he is dealing with science but when he is dealing with matters of faith."

The program in science patronage that he presided over at the Rockefeller Foundation—which was known at various times by various titles, including "vital processes," "quantitative biology," "experimental biology," and "molecular biology"—was a gamble from the beginning. That was its appeal to Weaver. If he had wanted a sure thing, he could have stayed in Madison, Wisconsin, where, as chairman of the mathematics department, he taught a subject that he loved to students he respected in a university town that had become his home.

# 24

# Taking the Plunge

If there is a better life than that of an enthusiastic teacher in a good school, I don't know what it is!" Weaver wrote that sentence in his autobiography, and considering that he was a man devoted to precision in language, the uncharacteristic exclamation point must be taken at full value.

Weaver loved the challenge of interacting with "excited (and exciting)" young minds. He loved the stimulation of good talk with friends on the faculty of the University of Wisconsin, "men and women of wide experience, varied interests, great competence." He looked forward to the long summer vacations as a time of spiritual renewal, and he looked forward to "the first smokey tang of autumn," which brought "an almost compulsive yearning to get back on the campus" and plunge once again into his delicious round of duties.

He and his wife (the former Mary Hemenway) loved Madison itself. The town was small enough that the university "set the tone of the whole community," yet large enough to attract and hold eminent scholars, researchers, creative people of all sorts. The Weavers went to Chicago for an occasional shopping trip at Marshall Field's or for an academic convention. But they couldn't imagine living there, or in any big city. They had never been to New York, or to anywhere else on the eastern seaboard, other than the Washington area, where, in the last year of World War I, the twenty-two-year-old Warren had worked with a small group of scientists who were trying to devise new flight instruments for the nascent Army Air Corps.

So in the fall of 1931, when Max Mason phoned Weaver and asked him to come to New York to talk about a permanent staff position at the Rockefeller Foundation, Weaver had good reasons to say no. He was comfortable financially and socially; he was expecting a significant promotion from the University of Wisconsin; he and his wife had just settled into a new house, which they had designed and built; their young son was happy in a good school; and Mrs. Weaver was pregnant again. Looking back years later, Weaver summed up his mood at the time in a phrase: "Why should we leave Madison for anywhere?"

Nevertheless, he took a train to New York to talk to Max. He could hardly ignore a call from the man who had helped steer him toward the comfortable circumstances in which he now took so much pleasure. The train trip itself and the brief stay in New York were as discomforting as he had imagined they would be; and the conversations with his old mentor and with the other officers of the Foundation were, if anything, even more upsetting.

Asked if he had any thoughts about what the Foundation ought to be doing to aid science, he said that it might be wise to look more closely into the biological sciences. Many problems in biology were just being opened up to quantitative investigation, a development that could be expected to produce all sorts of important discoveries in the coming years. As one example, he mentioned the electron microscope. Although no working models had yet been built, the wave aspects of quantum theory indicated that such an instrument could reveal details of tissue structure at least a thousand times finer than those resolvable with an ordinary light microscope.

Weaver felt free to argue the case for the biological sciences with all the intellectual passion he could muster, since it was apparent that no one could possibly find his enthusiasm self-serving. Having no background in biology, he was clearly disqualified from heading such a program. On this point, however, he was mistaken. The officers and trustees, strongly influenced by Mason, not only bought Weaver's argument but also insisted that Weaver was the man to direct the realignment.

He returned to Madison to discuss the unexpected offer with his wife. The proposed salary was substantially larger than his current income, but the Foundation could not promise him the security he enjoyed at the university, and he regarded with a dismay bordering on panic the prospect of living in a big city—especially an eastern city. The opportunity to travel was a mixed blessing, since it would mean long absences from his family. Having toted up all the reasons against taking the job, Weaver uncharacteristically delayed a final decision. When he remarked to his wife that they really had to make up their minds one way or the other, Mrs. Weaver replied, "Of course, we *have* made up our minds." And (as Weaver tells the story) he immediately realized that she was right: "We could not go on living with ourselves unless we met the challenge."

As much as he loved the life of a university professor, it was clear to him that he had gone as far as he could go in his chosen career. He was a good teacher who could convey even the most difficult notions quickly and effectively to his students. But to his sorrow he had found that he lacked "that strange and wonderful creative spark that makes a good researcher." Alone or in collaboration with Max Mason, he had written some thirty mathematical papers. By his own estimation they were competent and workmanlike, but uninspired: "I never seemed to get a first class original idea. . . . To take a body of knowledge and advance it . . . off into some new imaginative direction—that I've never been able to do."

As for his interest in physics, there too Weaver found himself in a cul-de-sac. The 1920s had been one of the truly creative periods in the history of physical science. Building on earlier work by Max Planck, Albert Einstein, and Niels Bohr, a small band of daring theoreticians had virtually remade physics from top to bottom, replacing the certainties of electromagnetic field theory with the statistical formulations of quantum mechanics and, in the process, revolutionizing man's understanding of his place in the universe and of the limits of knowledge itself.

The principal actors in this intellectual drama (many of whom had received support from the International Education Board) were Europeans: younger men like Werner Heisenberg and Erwin Schrödinger and Wolfgang Pauli, slightly older figures like Bohr and Max Born. Later the torch passed to a brilliant group of slightly younger men, mostly Europeans but including a handful of Americans; many of these (like Enrico Fermi and Robert Oppenheimer) had pursued their studies of the new physics with the help of IEB or NRC fellowships.

Of course, there was still important physics being done outside of quantum mechanics. It was quite possible in the 1920s to be a physicist at a major American university—certainly a physicist who specialized in experimentation or teaching—and completely ignore what the quantum theorists were doing in Europe. Indeed, some American physicists of the older generation actively resisted the new ideas, either because they did not understand them or because they feared that their own expertise was about to become obsolete or because (like Einstein himself) they were repelled by the philosophical contradictions, ad hoc assumptions, and general messiness that marked the early versions of the quantum theory.

The dilemma facing this last group was eloquently expressed by John Zeleny, the fifty-seven-year-old chairman of Yale's physics department. Zeleny could not gainsay the success of the new theory in accounting for a number of experimental observations (such as those concerning the interaction between X rays and matter) that had posed insuperable problems for the old physics. But for Zeleny more was at stake than a close fit between theory and observation. He had always believed that the job of physics was to penetrate to the core of reality and reveal its essentially lawful nature. To remain satisfied with a merely statistical description of what is observable—as the quantum theorists appeared to be—stuck him as a dereliction of duty, a betrayal of the high purpose that had guided the scientific enterprise since Galileo's day: "I feel that there is a real world corresponding to our sense perceptions. . . . I believe that Minneapolis is a real city and not simply a city of my dreams," Zeleny proclaimed, like a high priest chanting his credo to abash the heretics at the gate.

Among the American physicists who found the new physics uncongenial was Max Mason. Without rejecting the quantum theory out of hand, he adopted what might be called an attitude of hostile indifference; Weaver, under his influence, took a similar stand. By 1930 there could no longer be any doubt that the future in physics belonged to the new ideas; and it is hardly surprising that Weaver saw no place for himself in that future. What is remarkable is that the thirty-seven-year-old university professor was able to pull up roots, shift his field of operations, and, by an act of will, position himself in the midst of the next creative flowering of science—a revolution in biology as intellectually exciting and as philosophically disturbing as that wrought by the quantum theorists of the twenties

Weaver reported for work at the Rockefeller Foundation office on lower Broadway in February 1932. His first day on the job began inauspiciously, to say the least. He got lost. He had already settled his family in what he called a "typical little New York apartment" on upper Fifth Avenue, where his wife was "literally and unpleasantly dizzy most of the time." Boarding a subway train that, according to instructions he had been given, would take him to lower Broadway, he watched the stations go by until he felt a pressure in his ears; this, he realized, meant that the train was passing under a river, which he knew was not a good sign. He got off the train, climbed the stairs, and poked his head above ground to get his bearings. As he later recalled, he must have looked at that moment not unlike a "gopher trying to find out just where in the prairie he was lost." Weaver got to work, hours late, via taxicab—after an expensive and chastening ride through the "wilds of Brooklyn."

The sense of disorientation persisted even after his arrival at the office:

> There was an in-and-out box on my desk, and from time to time people would come and put things in the "in" box. And I had a general intuitive notion that I was supposed to do something prior to their being moved over to the "out" box. But I really had no indoctrination from individuals who had previously been connected with the program at all. . . .

His predecessor Herman Spoehr had resigned the preceding summer, leaving behind an experienced and dedicated secretary and a voluminous file of correspondence. As president of the Foundation, Max Mason had established no formal procedure for explaining to new recruits the ins and outs of what Weaver later called "this strange business of being a philanthropoid."

Weaver learned something about the mechanics of his job from the Foundation's financial officers—the treasurer and the comptroller—and a great deal more from talking to his secretary, from perusal of the correspondence files, and from his first, hesitant meetings with grant recipients and supplicants. From time to time he and Mason got together to refine their ideas about the new program in the biological sciences that they planned to submit to the trustees at the first opportunity.

One thing Weaver soon discovered was that a Foundation officer was expected to travel, to get out and see for himself the people and institutions whose requests for aid kept piling up in his "in" box. Little more than a month after joining the Foundation, he arranged a trip to some academic institutions in the South that had been selected by his predecessors to receive RF funds. He was not impressed by what he saw.

In many southern schools, where scientific research had never been a high priority to begin with, the effects of the Depression were devastating. "The great exodus of able men, combined with a disheartening lack of support, has made the situation precarious," Weaver noted in his diary. At the University of Virginia he found an "indefensibly messy" physics building and a department that was "for the most part playing at research." At the University of North Carolina he was told that "no funds whatsoever [were] available for research in the medical sciences" other than the Rockefeller grant and "what little [could] be stolen from laboratory fees."

But Weaver did not have to go south to see how the economic squeeze was choking off both public and private funding for scientific research. Out on Long Island the Cold Spring Harbor Biological Laboratory was in serious trouble. Although it shared quarters with the Carnegie Institution's Department of Genetics, the Biological Laboratory received no Carnegie funds. Originally a department of the Brooklyn Institute of Arts and Sciences, it had been "adopted" in the 1920s by a private corporation that drew most of its support from wealthy residents of the North Shore. The laboratory, which concentrated on interdisciplinary studies in mammalian physiology (and, after 1928, in biophysics), was said to run on "polo money." But now even the rich were feeling the pinch.

In 1930 Director Reginald Harris had approached the Rockefeller Foundation for help in financing a new "department of experimental biology as applied to medicine." While expressing admiration for the laboratory's work in general, the RF had politely declined this particular request. By the fall of 1931 the stock-market slide—and the coincident death of several of the laboratory's long-standing patrons—had entirely changed the financial picture. When Harris returned to the Foundation to ask for help, it was no longer expansion but the laboratory's survival that was at stake.

Harris's request met with a cold reception from some RF officers, but Mason pushed through a grant of $20,000 to permit the laboratory to pay its outstanding bills. A few months later Harris was back, pleading for more support. This time he was interviewed at length by the newly installed director for natural sciences. Weaver warned him not to expect any increase in aid but encouraged him to apply for an extension of the preceding year's emergency grant. From the diary entry that he wrote after talking to Harris, it is obvious that Weaver was impressed by the laboratory's emphasis on quantitative studies of biological material. And from subsequent developments it is equally obvious that Harris understood the advantages of building a research program around the clearly expressed priorities of the chief science officer of the Rockefeller Foundation.

By his own testimony Weaver's ideas about science patronage had been "tentative and amateurish" when he first assured the officers and trustees of the Foundation that "physics and chemistry were ripe for a fruitful union with biology." His conversations with Harris gave him some confidence that he might be on the right track. But his real education in planning and executing a large-scale program of support for scientific research began when he and his wife sailed, at the end of April 1932, to visit the Foundation's European headquarters in Paris.

# 25

# Riding the Circuit

Although the Paris office of the Rockefeller Foundation was administratively subordinate to New York, it was in many ways the intellectual capital of the Rockefeller philanthropies. The RF's involvement in Europe had begun with the relief work during World War I. Right after the war, Wickliffe Rose decided that his International Health Division needed a permanent European base from which to oversee its rapidly expanding commitments on the Continent. Paris seemed the logical location, and suitable office space was rented on the Right Bank.

Before long, as the IHD representatives were joined by those of the other RF divisions (plus the International Education Board and the Laura Spelman Rockefeller Memorial), the need for larger quarters became apparent. Early in 1927 the Foundation purchased the two top floors of a newly constructed seven-story building at 20, rue de la Baume, a quiet street in the Eighth Arrondissement just off the Boulevard Haussmann. From open terraces on the top floor, one could see the church of Sacre Coeur to the north and the cathedral of Notre Dame to the south. The decor, as planned by the office manager, George W. Bakeman, was quietly impressive but fell short of opulence. Carpeting salvaged from the preceding office was relaid in several rooms; floors were covered with a thick cork linoleum that, Bakeman explained, was "more sanitary and durable than a carpet" and "considerably cheaper."

A minor annoyance to all new arrivals—and a never-ending source of humorous anecdotes to the old hands—was the fact that few Parisian taxi drivers had heard of the rue de la Baume, which ran for only one block and was ignored by most mapmakers. The unsuspecting stranger who gave a driver the address of the RF office (especially a stranger who spoke French with anything less than a perfect Parisian accent) was likely to find himself being rushed to the rue de la Pompe, or the rue de la Bonne, or even the rue de Beaune.

Weaver, who could get along in German at this time but spoke only rudimentary French, had been elaborately briefed on the problem; and he and his wife arrived in Paris on a beautiful spring day armed with a street map and a pencil and paper. With the help of these and much gesticulating, they managed to convince a cabman that there was indeed a "rue de la Baume" in Paris, to which they wished to be taken.

What he found waiting for him there was everything he had missed at the New York office: a warm welcome from his peers, a sense of intellectual camaraderie, an indoctrination into what he later called "the essential intellectual character" of his job. The Paris office was headed by Selskar M. (Mike) Gunn, who had represented the RF in

*Selskar M. (Mike) Gunn*

one capacity or another in Europe for over ten years. Originally trained in bacteriology, he had made a name for himself in public health in New England in the years before World War I, and was recruited in 1917 for the Commission for the Prevention of Tuberculosis in France, sponsored jointly by the Rockefeller Foundation and the French government. He worked for the International Health Division from 1921 to 1927, when he was promoted to RF vice-president and given responsibility for overseeing all Foundation work in Europe. There was nothing dry or aloof about Gunn's intellect; Weaver saw him as "an extremely colorful, extremely romantic, extremely interesting character" with "all of the best charm . . . [and] something of the unpredictability of the Irish."

Gunn ran a lively office. While there was nothing indecisive about him, his management style tended toward the collegial. Once a week all the resident officers got together to discuss current projects and future plans. Debate was informal but vigorous, touching on the underlying philosophy as well as the administrative details of the far-flung European operation. All the divisions except humanities were permanently represented in Paris; and if the European representatives sometimes acted as if theirs was the Foundation's principal office, who could blame them? At the time of Weaver's first visit, in the spring of 1932, both the social sciences and the medical sciences had as many officers in Paris (three and two, respectively) as in New York, while Weaver himself was outnumbered three to one by his European staff. But if the Paris-based officers of the Division of Natural Sciences resented the fact that their new chief was so young and inexperienced, they never let him know it.

The dean of the Paris staff was Lauder W. Jones, a professor of organic chemistry on loan from Princeton. Jones was sixty-three years old, a widower who had lost his only daughter in an accident some years before. Weaver was not alone in observing that the Rockefeller Foundation seemed to be Jones's whole life now. Fluent in French and German, he obviously relished the constant travel that his job entailed. In an office noted for its gastronomes and polymaths, he was, according to Weaver, in a class by himself. His capacity for good food and drink was prodigious; his encyclopedic memory encompassed everything from the specialties of all the fine restaurants of Paris to volumes of German poetry in the original, which he would begin declaiming whenever the mood struck him.

Second in command for the natural sciences in the Paris office was Wilbur E. (Tizzy) Tisdale, who administered the fellowship program.

He was forty-six years old in the spring of 1932. Although he had spent the better part of ten years in Europe, first for the National Research Council, then for the International Education Board and, since 1929, with the Foundation, his command of local languages was virtually nil. His French accent made bellhops wince; when he found that he didn't have the words in French or German to say what he wanted to say, he said whatever he had words for. Perhaps because he never pretended to be a linguist, he charmed his conversational partners into doing most of the talking—a useful attribute in a circuit rider.

Harry M. (Dusty) Miller and his wife arrived in Paris two months after the Weavers. Miller was a year younger than Weaver; he had been doing research in parasitology at Washington University in St. Louis when Tisdale, on a brief leave from his Paris duties, came to the campus looking for an assistant. Miller had spent two years in France with the U.S. Army Medical Corps during and after the First World War, and he already spoke French with a beautiful accent that even Parisian taxi drivers approved of. But since Tisdale himself would have been the last person to appreciate this achievement, Miller always maintained that he had been hired because his first reaction to the prospect of working in Paris had been negative; while most of the people Tisdale spoke to were ready to leave on the next boat, Miller was reluctant to interrupt his research on the parasites of rats and mice. He finally agreed to take the job only when he was assured he could finish writing up his most recent research work in Paris.

The Foundation's Paris staff worked hard and played hard. When Lauder Jones was not on the road (or scouting around Paris for the particular variety of Austrian postage stamps that he collected), he and Daniel P. (Pat) O'Brien of the Division of Medical Sciences would meet at noon in the lobby at 20, rue de la Baume to choose the restaurant where they (and anyone else who wished to accompany them) would dine that day. As Weaver remembered it, this was "a decision of considerable solemnity," not only to the RF staffers but also to the chefs and sommeliers of Paris, who always accorded Jones's party special attention and never failed to congratulate him, when the meal was over, on the exquisite combinations of food and wine he had selected.

The Paris outpost of the Foundation offered homelier occasions for conviviality as well: daily games of hearts and dominoes after lunch, nightly tournaments on the Ping-Pong table in the accounting office. But Weaver's real apprenticeship in the ways of Rockefeller philan-

*RF Paris Office Staff, 1937.* Left to right: *George W. Bakeman, Wilbur E. Tisdale, Robert Letort, Harry M. Miller, Thomas Appleget.*

thropy came on the long trips he took with Jones and Tisdale (and later with Miller) to meet the leading scientists of Europe and get a firsthand look at their work and their working conditions.

In those days the Foundation considered its field of operations to include twenty-six European countries; during his first decade as director of natural sciences, Weaver made regular survey trips to the majority of them. He spent most of his time in England. Scandinavia, Germany, Switzerland, and France, but he also became acquainted with Scotland, Ireland, Holland, and Belgium and paid occasional visits to Finland, Poland, Austria, Czechoslovakia, and the Baltic and Balkan states.

On his first tour with Jones, in the spring of 1932, Weaver's arrival at a major university or research institute was usually the occasion for a formal celebration designed to thank the Foundation for past support and to establish friendly relations with the man who would control future funding decisions. At the University of Munich, for example, Weaver was introduced to most of the famous members of the faculty, including Karl von Frisch, the authority on bees who was in the process of moving to a new Institute of Zoology that had been made possible by a generous Rockefeller grant, and Kasimir Fajans, who had just moved into a new RF-funded Institute of Physical Chemistry.

This was all pretty heady stuff for Weaver, a self-proclaimed small-town boy who, just a few months before, had thought of the table at the Faculty Club in Madison, Wisconsin, as the acme of sophistication. Years later he remembered how he and Jones had started drinking Rhine wine at a ceremonial breakfast in Munich's medieval city hall, gone on to brandy and beer at the home of the rector of the university, continued with cocktails at their hotel (where the barman insisted on making powerful "Jones specials" for one of his favorite customers and for Herr Jones's friend), and ended with a surfeit of wines and liqueurs at a dinner hosted by Professor Fajans. Weaver had always prided himself on his ability to hold his liquor, but he had never seen anything to match Jones's performance on this occasion. His faith in his companion's grip on reality wavered only once, when Jones tried to strike up a conversation in French with the dour cabbie who drove them to the Fajanses' apartment. But at the dinner itself Jones was the life of the party, reciting, in German, highlights from the works of Goethe and Schiller.

If Weaver later treated his first European survey trip as the stuff of legend, it is not hard to see why. Eating and drinking in the company of Europe's intellectual elite, one was able to ignore the rigors of the Depression and the social dislocations that were spreading in the wake of the economic crisis. Ancient seats of learning like the University of Munich had seen many such crises come and go; in these hallowed surroundings the ranting and posturing of Hitler and his National Socialists seemed less frightening. This was, after all, the land not only of Goethe and Schiller but also of Koch and Ehrlich and Planck and Einstein. European science was unquestionably the best in the world, and German science was, arguably, the best in Europe.

Nowhere else did university professors enjoy more prestige, better working conditions, a higher standard of living. A man like Fajans had every reason to feel that he had reached a secure place at the apex of his profession, and Weaver had every reason to believe that, in backing people like Fajans, he would be building the science program of the Foundation on the firmest bedrock. The fact that Professor Fajans was of Polish-Jewish ancestry seemed, for the moment, of no great consequence.

# 26

# Self-Education

When he returned from Europe in the summer of 1932, Weaver's immediate problem was his almost total ignorance of current biological research, the field around which he proposed to organize his division. With characteristic energy, he set out "to minimize [this] difficulty" by a program of self-education even more ambitious than that pursued by his illustrious predecessor Wickliffe Rose. This is how Weaver described it in his autobiography:

> Over my first five years I followed a strict program of individual study in the various relevant areas of the biological sciences. I started with genetics, not because I realized in 1932 the key role this subject was destined to play, but at least in part because it is a field congenial to one trained in mathematics. I went at one after another of the areas in which we were working—cellular physiology, organic chemistry, biochemistry, developmental mechanics, the techniques for studying molecular structure, and so on—and did the best I could, subject to the disadvantage of working alone with no laboratory experience, to familiarize myself with the background material.

Weaver, of course, had to fit this self-education program into the time left over from his other duties. While most of his engergies were directed toward shaping the future policies of the division, he also had decisions to make about new requests for aid and about grants already in place. Through the reorganization of 1928, for example, the Foundation had inherited from John D. Rockefeller, Jr.,'s Bureau of Social Hygiene an obligation to support an extensive program of "research in problems of sex," ranging from studies of reproduction in monkeys to the biochemistry of gonadal hormones. In an arrangement that harked back to the pre-Mason era—when the Foundation tried to insulate itself from the day-to-day business of science patronage—a committee of the National Research Council had been empowered to select the researchers and the research topics. Weaver was simply expected to approve the next annual installment of $75,000, and he did.

The first grant of the Weaver years that looked forward to a more intimate relationship with the scientific community was a $40,000 appropriation to the California Institute of Technology, in Pasadena, voted by the trustees in July 1932. These funds were to be divided between research in the "physics of solids" (which represented a holdover from pre-Weaver priorities) and research in structural chemistry under the direction of an up-and-coming young organic chemist named Linus Pauling, who, in Weaver's words, was "so courageous, so audacious . . . that he was not afraid of the larger molecules." Pauling proposed to extend to complex inorganic and organic substances the powerful analysis of molecular structures that had emerged from the quantum theory. His approach was exactly the sort of thing Weaver was looking for: "closely coordinated work in chemistry, physics, and mathematics."

But Pauling was so outstanding a researcher that (as Weaver himself later noted) it took no particular insight to commit the Foundation to his support. Nor did one or two grants of this kind add up to the coherent program of science patronage that Weaver sought. He had come to the Foundation with the idea of fostering a "friendly invasion" of biology by physics and chemistry. Now, throughout the summer and fall of 1932, he wrestled with the problem of how to present the philanthropic implications of this "invasion" in the most favorable light to a group of trustees who were already worried that the Foundation was frittering away its assets on projects too far removed from the economic and social issues of the moment.

The Board no longer consisted exclusively of men on the Rockefeller payroll or closely associated with the family, as had been true in earlier years—and was still true of the powerful Executive and Finance committees. Alongside such Rockefeller insiders as Raymond Fosdick, Jerome Greene, and Arthur Woods, room had been made for a few broad-minded outsiders from the universities, like James R. Angell, president of Yale, and from industry, Owen D. Young, the chairman of General Electric. There were even two trustees with scientific credentials: Vernon Kellogg, a Stanford University zoologist who had been permanent secretary of the National Research Council in Washington from 1919 to 1931, and David Edsall, an advocate of scientifically oriented medicine who had helped shape the curricula of both the Harvard Medical School and the Harvard School of Public Health during the twenties.

But when considering issues of policy, the trustees were not disposed to defer to "experts," either from their own ranks or from the divisions. The boards of other large philanthropic enterprises might be content to rubber-stamp the decisions of their operating officers, but the trustees of the Rockefeller Foundation took seriously their responsibilities as stewards of the legacy of John D. Rockefeller, Sr.

The most visible symbol of the Board's tradition of active stewardship was, of course, John D. Rockefeller, Jr., who continued as chairman through 1939. Since his eldest son, John D. III, was made a trustee on December 13, 1931, this meant that for most of the decade there were two family members on the Board. Although John III was a mild-mannered, self-effacing young man with a genuine interest in philanthropy, there can be no doubt that his elevation to the Board at the age of twenty-five betrayed a lack of sensitivity to public opinion. Certainly, it undercut the Foundation's repeated protestations that its business was conducted in complete independence from the rest of the Rockefeller empire.

Most knowledgeable observers, including Warren Weaver, found no cause to complain that the Board was acting to advance the interests of the Rockefeller family. In fact, the records show that Junior was outvoted time and again on the boards of his father's foundations. Nevertheless, Junior did set the tone of the Board meetings. And from the perspective of an adventurous young recruit like Weaver, the trustees as a body represented "an orthodox and restraining" influence on the work of the divisions. Nor was Weaver alone in this opinion. During the early 1930s there existed between the officers and trustees of the RF an attitude of mutual suspicion that often erupted into open conflict.

Part of the problem was traceable to Junior's disenchantment with Mason's performance as president. Instead of the passion for detail and documentation that Junior had expected from a scientist, Mason offered extemporaneous brilliance. After an initial period of bedazzlement, the son of the founder make it clear that he was not amused. In Weaver's phrase, Max Mason and John D. Rockefeller, Jr. had "totally immiscible personalities"—especially within the confines of the RF boardroom. For all his courtly manners Junior was as relentless as his father had been in evaluating new departures.

The job of an RF officer proposing a program was to "prove his case"; the job of the trustees was to try to poke holes in the proof. Only proposals that could survive this kind of scrutiny deserved to be implemented. Confrontations were bruising, unsparing. When the Board meetings were held in Williamsburg, Virginia—the site of Junior's ambitious colonial restoration—the directors of the various RF divisions often got boisterously drunk on the train back to New York, like undergraduates celebrating the end of a particularly trying examination period.

Most trying of all were the full-scale inquiries into programs and policies that the Board initiated from time to time. It was through such instruments that Fosdick had effected his overhaul of the Foundation in the 1920s. Another Foundation-wide reappraisal was scheduled for April 1933. The agenda called for a close look at the progress being made by the various divisions in reconciling the RF's new commitment to the "advancement of knowledge" with its obligation to promote "the well-being of mankind throughout the world." As Weaver later recalled, these special meetings gave even the most experienced officers "a perfectly definite sensation of being on the pan."

For Weaver himself this sensation must have been especially acute in the summer of 1932, as he prepared to argue that the RF should invest a major portion of its research-support dollars in investigations along the frontiers of the biological sciences.

By October he was ready to solicit comments and criticisms from his colleagues. The first draft of his proposal, elaborated with the aid of Max Mason, struck a balance between new departures and old programs. On the one hand, Weaver talked about shoring up fiscally undernourished science departments in universities where the tradition of scientific research was already well established: "The policy is frankly one of helping the strong." On the other hand, to avoid spreading the resources of the Foundation too thin, he suggested concentrating research grants in three "major" areas, which he identified as "physics and chemistry of vital processes," "physics of the earth," and "genetic biology," along with three "minor" areas: "fundamental construction problems" (that is, the construction of matter), "physical and colloidal chemistry," and "viruses."

After meeting with Alan Gregg, who was reorganizing the RF's Division of Medical Sciences around a core of psychiatry and neurology, Weaver agreed that closer cooperation between the science divisions would be beneficial. So he added a fourth major area—"quantitative psychology"—and dropped the minor in viruses in favor of "theory of probability and statistics." Viruses had long been a specialty of the International Health Division (whose research laboratory was housed, to the confusion of one and all, at the otherwise independently administered and funded Rockefeller Institute); and it is possible that the director of the IHD, a former Army colonel named Frederick F. Russell, objected to the "new boy" at NS poaching on his territory. Weaver remembered Russell, who retired in 1935, as a "lovely old character . . . who looked and acted just a little bit like Colonel Blimp."

In the fall of 1932 Weaver wrote to Lauder Jones in Paris to ask him what he thought of the revised proposal that had emerged from the interdivisional conferences in New York. Jones's reply was a detailed critique that occasionally took on the tone of a doting father correcting his bright but insufficiently respectful offspring.

Understandably, Jones objected to Weaver's statement that in recent years the Division of Natural Sciences had been squandering much of its budget on grants that enabled scientists "to do only a little more of the sort of research which [would] be widely done whether the Foundation existed or not." Jones was highly skeptical about the wisdom of concentrating RF grants in a few, narrowly defined areas, especially if this meant withholding funds from some of the leading scientists of Europe in order to support men "of less distinction" merely because the latter's work fit into the Foundation's program.

While conceding that there was some truth to Jones's critique, Weaver in the end ignored the caveats of his Paris colleague. His reasons were partly philosophical—he believed that only through concentration could the Foundation make a contribution commensurate with its resources—and partly tactical. With a tightly focused program, it would be easier to convince the trustees that monies allocated to the Division of Natural Sciences would be spent with an eye on social "relevance."

# 27

# To the Barricades

In January 1933 Weaver completed a revised draft of a proposal for a "new program" in science patronage that he intended to submit to the trustees in April. It included an eight-page section entitled "The Benefits from Science." This document summarized just about every argument that had ever been advanced to justify the ways of science to laymen—from the utility of more accurate weather forecasts to the notion that modern science could serve as a model of intellectual honesty and international cooperation in a world torn by sectarian strife. As for traditional spiritual values, what could be more uplifting than a selfless attempt to understand "the heavens, the earth, the atom and the living cell" so as to banish from human society "the curse of mystery, superstition and fear"? In Weaver's unblinking phrase, the never-ending pursuit of scientific truth could be seen as a "clear matter of noblesse oblige."

In bringing this lofty vision down to cases, Weaver sounded a note that would echo throughout his years at the Rockefeller Foundation: "Our understanding and control of inanimate forces has outrun our understanding and control of animate forces." Since the Foundation had no choice but to be selective, the implication was clear: there should be an "increased emphasis" on supporting those sciences that will help us get a grip on the "animate forces," that is, biology and psychology, plus "special developments in mathematics, physics, and chemistry which are themselves fundamental to biology and psychology."

What it all added up to, in the coordinated presentations that Weaver, Gregg, and Mason made to the trustees at the April 1933 meeting, was nothing less than a concerted attack on the "general problem of human behavior, with the aim of control through understanding."

In their assertion that the support of basic scientific research would lead to greater "control" over social forces, the officers of the Foundation were telling the trustees exactly what they wanted to hear. Having proclaimed to the world that they considered the advancement of knowledge to be virtually synonymous with the well-being of mankind, the trustees were in a quandary. The declaration of purpose that had emerged from Fosdick's reorganization was itself a restatement of Rose's frankly elitist strategy of "making the peaks higher." The assumption was that "the benefits from science" would, in time, trickle down to the masses. But the victory of Franklin Delano Roosevelt in the presidential election the preceding fall had signaled, among other things, a rejection of trickle-down strategies in the economic realm. This placed new pressures on the Foundation.

It was a question of timing. In the affluent twenties men like Rose were content to wait for their benefactions to produce social dividends. Science, by its very nature, could not be hurried. All one could do was free the finest minds of the day from financial and administrative worries, and then be patient. But in the spring of 1933, with the tides of anarchy and barbarism rising, patience was a luxury that not even the affluent could afford—or at least could not afford to be seen embracing in public.

On January 30, 1933, Adolf Hitler had become chancellor of Germany. Within a month the Reichstag fire had given the National Socialists an excuse to move against their best-organized adversaries, the German Communists. On March 23 Hitler had assumed dictatorial powers under the so-called Enabling Act. By April 11, the day the trustees and officers of the Rockefeller Foundation sat down to discuss program and policy in a country club–like hotel in Westchester, a comprehensive purge of "non-Aryans" in the German civil service was under way. Since university professors were government employees, the purge reached to the heart of the German scientific establishment. Public book burnings had been announced. During a government-sponsored boycott of Jewish businesses, store windows had been smashed, homes vandalized, and Jews beaten by gangs of brownshirts while the police looked the other way.

Albert Einstein, who was visiting in America when the Nazis came to power, resigned in protest from the Prussian Academy of Sciences, an act that he hoped would arouse the consciences of his German colleagues. But under pressure from the government, the academy issued a statement condemning Einstein for antistate "agitation" and declaring that academy members had "no reason to regret" the resignation of the Jewish Nobel laureate.

In Washington, meanwhile, the newly inaugurated president was assuring his countrymen that they had "nothing to fear but fear itself." Yet on March 5, the day after his inauguration, Roosevelt declared a four-day "bank holiday" to stem a continuing wave of bank failures. Although most of the nation's banks had reopened by the beginning of April, confidence in the stability of all established institutions was severely shaken.

If the years from 1930 to 1933 had demonstrated anything, it was that the pieties of the preceding decade offered no guidance in the new era. Individuals could no longer be trusted to control their antisocial impulses; and the instruments designed to channel the energies of the aggregate toward socially approved goals were obviously inadequate. Faced with this crisis of confidence, the trustees could have repudiated the Rose-Fosdick consensus and sought a new formula for philanthropy in a time of anarchy and barbarism. But they had several strong reasons not to act rashly. As we have noted, changing the direction of a philanthropic enterprise is not easy; having spent three years redefining the purpose of the Foundation, they were reluctant to scrap all their work and start over. Besides, no attractive alternatives were at hand.

A return to purely ameliorative charity—feeding the hungry, sheltering the homeless, succoring the sick—seemed out of the question, especially once the Roosevelt administration made clear that the federal government would not shirk its responsibilities in this area. Nor was there any sentiment within the Foundation for promoting ready-made ideological solutions to the nation's problems. Even if anyone had believed such answers existed, the trustees knew that the public was not about to listen to sermons from the fat-cat stewards of the Rockefeller legacy. The lessons of the industrial-relations fiasco had not been forgotten.

So the message of Weaver, Gregg, and Mason (and of Edmund Day in the social sciences) was welcome indeed. Acknowledging that the times compelled a new sense of urgency, they argued that what was really needed at the moment was more knowledge—knowledge about how things worked, knowledge that people of goodwill could use to restructure society along more rational, and therefore more stable, lines. Of course, no one could predict just how long it would take a given researcher to come up with such useful knowledge. But the trustees were assured that simply by being more selective about the kind of knowledge they chose to advance, they could keep the Foundation on its present course with a clear conscience.

This notion was so appealing that some of the more hardheaded Board members feared it was too good to be true. Since they themselves were not competent to oversee the day-to-day operations of the science divisions, they redoubled their scrutiny of the presiding officers. Given the urgency that everyone felt, it is not surprising that the trustees eventually let themselves be persuaded. Nor is it surprising that the officers were forced to reformulate specific proposals again and again and yet again before the new program received an official stamp of approval.

Throughout this long deliberative process the trend was always toward greater and greater concentration. Lauder Jones's transatlantic doubts about the wisdom of this approach represented a voice from the past. Indeed, the proposal that Weaver presented to the trustees on April 11 was more narrowly focused than the draft that Jones had criticized. It made no mention of chemistry or the theory of probability, and the "physics of the earth" was reduced from a major to a minor interest (before being dropped entirely the following year). What remained was a focus on the "basic problems of biology," which Weaver divided among "four closely correlated and overlapping subfields"—namely, endocrinology, genetics, "biology of reproduction, and psychobiology."

To emphasize that his main interest was "the study and investigation of the phenomena of living things," Weaver chose the descriptive label "Vital Processes" for the entire program. This had the advantage, as he saw it, of getting away from "any ancient limitations adhering to the word biology." It also fit in with the trustees' known aversion to the "artificial" divisions that traditional academic disciplines imposed on the search for knowledge.

Along with concentration, a major theme of the April 11 meeting was coordination. In his policy overview President Mason repeatedly assured the trustees that despite greater specialization within each division, the program as a whole was "pointed toward a structural unity"—the control of human behavior:

> The Social Sciences, for example, will concern themselves with the rationalization of social control. Many procedures will be explicitly co-operative between divisions. The Social Sciences and the International Health Division, for example, may have common interest in the expansion of health control units into the broader services of community centers. The Medical and Natural Sciences will, through psychiatry and psychobiology, have strong common interest in the problems of mental disease.

In practice, coordination between the divisions turned out to be a sometime thing, dependent more on the temporarily converging interests of individual officers than on the expression of some grand strategy. To keep up with changing currents in scientific research, science patronage must remain opportunistic, ready to capitalize on unexpected developments wherever they appear. But too much uncertainty makes financial planners and administrators unhappy. Unquestionably, it was the *appearance* of a coordinated strategy that helped sell the RF trustees on the new program.

In his own presentation at the April 11 meeting, Weaver once again deplored man's lack of progress in "the analysis and control of animate forces." He even managed to tie in the earth sciences, declaring that their purpose was to "furnish information concerning the physical background for the development of man." Before a skeptical listener could ask whether there was *any* scientific investigation that could not be construed as contributing to the welfare of mankind, Weaver went on to explain the criteria he proposed to use in selecting projects.

To qualify for backing, a field of research had to be *"sufficiently developed to merit support, but so imperfectly developed as to need it."* (This ruled out parapsychology at one extreme and classical physics at the other.) The ideal situation, Weaver said, was a field in which Foundation support would stimulate important developments "that otherwise would not occur within a reasonable time." In arguing that a quantitative attack on the basic problems of biology met these criteria, Weaver offered corroborative quotations from "outstanding experimentalists" whose opinions had recently been solicited by a renowned "biological institute." The institute was the Cold Spring Harbor Biological Laboratory. Weaver did not mention that the opinions had been solicited by the laboratory's director as part of a desperate campaign to raise funds from a variety of sources, including the Rockefeller Foundation.

To indicate why he thought that man's age-old problems of mental health were now ready for a quantitative approach, Weaver cited recent experimental work "on electrical phenomena associated with the conduction of nerve impulses." As for genetics, he stipulated that the focus would be on animal rather than plant material, with human genetics targeted for special attention "as rapidly as sound possibilities present themselves." Just in case any of the trustees had not been paying attention to recent developments, Weaver added, "The attack planned, however, is a basic and long range one, and such a subject as eugenics, for example, would not be given support."

The rest of Weaver's presentation was devoted to the administrative implications of the new program—implications that had far-reaching consequences, not only for the Foundation itself but for the international scientific community as a whole.

With RF income shrinking in a depressed economy, new initiatives could be financed only out of funds withdrawn from older programs. Weaver's primary targets were the large capital grants that had figured so prominently in the work of the General Education Board and the International Education Board in the twenties, and which the Foundation had inherited. Of the nearly $12 million in RF aid to the natural sciences between 1929 and 1932, more than half had gone for endowment, for construction and maintenance of buildings, and for equipment. Under Weaver's proposal these capital grants to educational and research institutions in the United States and western Europe would cease entirely. The bulk of Foundation aid in the future would go to support specific research projects in the four subfields of the "vital processes."

Many former RF grantees would still be eligible for support under the new program. But their relationship to the Foundation would be quite different. In the past the primary responsibility for deciding how money was to be allocated *within* an institution lay with the recipients, although Foundation officers were kept informed by means of written reports and periodic conferences with the scientists in charge. Under the new rules the Foundation took a far more active role in deciding which projects got funded. Compared with the Carnegie Institution—with its in-house laboratories and salaried investigators—the RF's control over the research it supported would still be largely "indirect." But simply by entering the marketplace with a more detailed shopping list—for example, by advertising its interest in "animal genetics" rather than the more general "problems of sex"—Weaver and his colleagues were bound to have an increased influence on the pursuit of biological research in American and European laboratories.

This prospect disturbed some of the trustees. To set their minds at rest, they kept asking the officers for more assurances about the relevance of it all; when these were forthcoming, they worried that they were being being sold a bill of goods by overeager specialists. At their April 1933 meeting, the trustees gave a go-ahead to the proposed programs of concentration in the "vital processes" (Division of Natural Sciences) and psychiatry (Division of Medical Sciences), but they also asked Fosdick to organize yet another "Committee on Appraisal and Plan" to reconsider the whole question of "the advancement of knowledge" as a philanthropic goal.

Complicating the normal deliberative process within the RF was the fact that it was being conducted against the background of the Nazis' rise to power in Germany. The swift pace of events on the Continent had the effect of changing the terms of the debate in midstream. Ever since 1929 the proponents of science patronage had had to defend themselves against the charge that their "ivory tower" orientation was callous, even immoral, in a world on the brink of chaos. But Hitler's ascendancy had turned this argument on its head. With the Nazis trying to convert the German universities into bastions of Aryan thought, the very future of science—broadly defined as a disinterested pursuit of the truth—was in doubt. Among the first refugees from Nazi Germany were Jewish scientists who were either forced from their positions or chose to resign rather than submit to humiliating conditions imposed by the new government. Through its long-established association with European science, the Rockefeller Foundation was drawn almost immediately into efforts to help these refugees. (See chapters 28–29.)

Because so much was at stake, it is hardly surprising that the debate over the RF's "new program" occasionally became overheated. In a series of position papers, Weaver speculated about the emergence of a "new science of man" based on "discoverable laws" governing the entire range of human experience: "conceiving, child-bearing, thinking, behaving, growing and, finally dying." At the end of 1933 Weaver and Gregg jointly submitted to the trustees a document that committed their divisions to work together to foster the long-range development of such a science, which they called, for want of a better term, "psychobiology."

In an effort to make the support of basic science sound more socially relevant, the authors referred to certain topics of investigation that are, "as yet, not often found on the firing line, but are even now furnishing communication and supply service without which the front lines could not continue safely to advance."

With the advantage of hindsight, it is apparent that Weaver and Gregg were trying to sell their program to the trustees as a kind of New Deal in scientific research. Weaver returned to the rhetorical barricades in February 1934, with a "progress report" that asked,

> Can we develop so sound and extensive a genetics that we can hope to breed, in the future, superior men? Can we . . . develop, before it is too late, a therapy for the whole hideous range of mental and physical disorders which result from glandular disturbances? . . . Can we release psychology from its present confusion and ineffectiveness and shape it into a tool which every man can use every day? . . . Can we, in short, create a new science of Man?

The contrast between this invitation to a crusade for human betterment (which appeared almost unchanged in the *Annual Report* for 1933) and the rather complacent elitism of Wickliffe Rose's "Make the peaks higher" could hardly be greater or, considering the historical context, more poignant.

# 28

# Reason and Unreason

Weaver's speculations about a biologically based "new science of Man" were born of desperation. In the absence of hard proof that biology would benefit from more quantification, Weaver was asking the trustees to share his vision of a humanity saved from its own destructive tendencies by scientific self-knowledge. When he offered that invitation, he had already seen, at first hand, the forces of destruction at work in the new Germany of Adolf Hitler.

As soon as possible after the trustees' meeting in mid-April, Weaver left on his second fact-finding trip to Europe. By the time he arrived in Göteborg, Sweden, on May 3, there was abundant evidence that the Nazis were serious about implementing, in law, the anti-Semitic and generally anti-intellectual sentiments that Hitler had been articulating since the early 1920s.

The purge of German universities had begun with the promulgation, on April 7, of the Law for the Restoration of the Career Civil Service. It authorized the removal from office of all persons of "non-Aryan" descent, as well as anyone whose loyalty to the new state could not be guaranteed. In support of this law, the German Students Association announced the beginning of a month-long campaign "Against the Un-German Spirit" that would culminate, on May 10, in a public burning of books by such non-Aryan authors as Albert Einstein.

In addition to what he could read in the newspapers about such events, Weaver was privy to information gathered by his Paris-based circuit riders. Hardly a diary entry crossed his desk in the early months of 1933 that did not contain at least one mention of ominous doings in Germany. To clarify the references to otherwise inexplicable dismissals, promotions, and the like, it became necessary to annotate the names of scientific workers with hitherto unheard-of phrases like "K. is a Jew" and "S. is not a Jew."

As early as March 9 Harry Miller was reporting that a young French recipient of a Rockefeller fellowship would not be able to work as planned with James Franck at the University of Göttingen, because Professor Franck, a Jew, was afraid of repercussions if he gave to a Frenchman a position that might otherwise go to a German. A few weeks later Wilbur Tisdale, traveling in the United States, spoke with Paul Weiss, a former IEB fellow who was due to return to Austria after spending two years at Yale. Weiss expressed fear that if he went back to his homeland, he would be returning "to practically certain death." On April 20 Miller was informed by Professor David Keilin in Cambridge that a young researcher named Hans Krebs, who had done excellent work with Otto Warburg, had been driven from his post at the University of Freiburg and was hoping to emigrate to England if he could find the funds.

As news of such outrages spread through the international scientific community, activists on the Continent, in England, and in the United States began mobilizing to aid these early victims of Nazi persecution. The Executive Committee of the Rockefeller Foundation set aside $140,000 to be used "for grants to institutions desiring to provide positions for eminent scholars whose careers have been interrupted by the present disturbed conditions." As the wording implies, this aid was conceived of as a short-term measure that did not violate the Foundation's self-imposed strictures against "relief" programs; only researchers who had a firm commitment for employment outside Germany were eligible.

This "temporary" program was extended by the Board of Trustees on an annual basis through 1939, by which time over $750,000 had been spent to help relocate 197 refugee scholars, nearly half of whom were noted scientific researchers known personally to the circuit riders of the Divisions of Natural and Medical Sciences.

In the spring of 1933, however, it was still possible to believe that the forces of reason would quickly reassert themselves on a continent that many thought of as the true home of rationality. Before he entered

Germany, Warren Weaver's European itinerary took him through Scandinavia, Finland, and the Baltic states. Traveling in the company of Lauder Jones, he paid an especially reassuring call on the Swedish chemist Theodor Svedberg, the inventor of the ultracentrifuge.

Weaver never tired of talking or writing about this device. For one thing, the ultracentrifuge, whose development had been supported by Rockefeller funds from the beginning, provided a clear demonstration of Weaver's claim that biology had much to gain from the careful application of quantitative methods of measurement. Weaver also took pleasure in the fact that the mathematics behind the principle of the centrifuge had been spelled out in a paper that he and Max Mason had jointly written in 1926 and that Professor Svedberg never failed to refer to in his own publications. In addition, as a native of rural Wisconsin, Weaver was plainly tickled by the fact that such a complicated and highly precise scientific instrument had originally been adapted from a machine used on dairy farms to separate cream from milk.

Even aside from the ultracentrifuge, there was something about Svedberg's style of research that deeply appealed to Weaver. After visiting the Swede's newly built institute at the University of Uppsala, he raved, "Not only has every detail been anticipated and provided for, but everything has been done with a real artistic flourish, even the color scheme of the walls and the power and lighting cables being unusual and attractive." As for the work being done there—a systematic study of blood pigments—Weaver summarized it as "an unusual combination of mechanical genius, of patience and experience, and of imagination."

On May 18, 1933, Weaver and Jones crossed by rail from Lithuania into the German town of Königsberg, the onetime seat of the dukes of Prussia and the birthplace of Immanuel Kant. Weaver described his first impressions of Hitler's Germany in detailed diary entries (on which he relied twenty-eight years later when he recorded his reminiscences for Columbia University's Oral History Project).

Having just suffered through the impolite and inefficient railroad service in the eastern Baltic countries, he could not help being impressed by the German workers who took over at the border: "The lights and heat were turned up in the cars; the cars were cleaned; and it was obvious as anything could be that we had entered civilization."

Other signs of the new Germany were less reassuring. Weaver and Jones waited in line at a bank while the entire staff stood at attention to hear a half-hour speech by Chancellor Hitler over a loudspeaker. A picture magazine bought at a newsstand showed a farmer trundling a wheelbarrow full of books by Jewish authors, which were identified as "a fine and proper meal for his sow."

There was no mistaking the "atmosphere of fear and suspicion and secrecy" the Nazi takeover had brought to the German universities and research institutes that Weaver had so admired only a year earlier. One of Weaver's strongest memories was of the "frequently embarrassed" reaction of academic officials when confronted with evidence of discriminatory policies they felt bound to enforce.

At the University of Königsberg, where Kant had lectured for fifty years, Weaver and Jones called on Friedrich Paneth, an Austrian-Jewish chemist whose world-famous chemical institute was a longtime recipient of Rockefeller funds. Outside the institute they found crowds of students making preparations for the upcoming "Purification Day," when Jewish books were to be burned in public. There was no sign of Professor Paneth, but his assistant told the RF circuit riders that the professor had gone to London to attend a scientific meeting and had been unable to return because of illness.

On close questioning, the assistant conceded that Professor Paneth might not be returning from England after all. The dean of the Philosophical Faculty and the rector of the university at first pretended to have no information, then reluctantly admitted that Paneth and several other faculty members had been "given leave," a state of affairs that could turn out to be permanent.*

Weaver, after making clear that a country's "internal politics" were not the Foundation's concern, pointed out that Paneth's RF grant was for his own work and not for the "general support of the Institute"; accordingly, all expenditures had to cease at once, and the unexpended balance had to be returned to the Paris office. Attempts to draw the officers of the university into a discussion of Paneth's fate or of "Purification Day" were politely turned aside. On the way out, Weaver saw an announcement on the rector's door, advising applicants for the summer term that only students "whose parents for two generations [had] been non-Jewish" could matriculate.

By the time Weaver and Jones arrived in Berlin on May 24, the

*Paneth, in fact, remained in England, where he became professor of chemistry at the University of Durham. In 1953 he returned to Germany as director of the Max-Planck Institute for Chemistry, in Mainz.

Nazi assault on "Jewish science" had claimed more notable victims. James Franck, whose courage on the frontlines during the First World War had won him two Iron Crosses and a commisssion as an officer, was known as a man of principle as well as an experimental physicist of genius (he had won a Nobel Prize in 1925 for work establishing the quantized nature of energy transfer). After the first dismissals of non-Aryans from German universities, on April 13, Franck resigned his Göttingen professorship in protest, although as a war veteran he was exempt from the anti-Semitic provisions of the civil-service law. If he had hopes that this action would prompt an outcry against the Nazis by his colleagues, he was mistaken. Forty-two instructors at Göttingen signed a progovernment petition condemning Franck's resignation as tantamount to "an act of sabotage."

Two days later the Prussian Ministry of Education placed on involuntary leave six more Göttingen professors, including the physicists Max Born and Richard Courant, the mathematician Emily Noether, and the statistician Felix Bernstein, who was in the United States on an official fact-finding mission for the (pre-Hitler) German government.

At it happened, Dr. Bernstein was the first victim of Nazi persecution to receive assistance under the Foundation's Special Research Aid Fund for Deposed Scholars. Cut off from his family, unable to withdraw any funds from Germany, he came to the office of Max Mason on May 10 to ask for help in getting through the summer. He had reason to think that Charles Davenport, an admirer of his work on the genetics of blood types, would offer him a temporary position at Cold Spring Harbor (where he had spent part of a sabbatical year in 1929) if the Foundation could help pay his salary and expenses.

Because he had no long-term job lined up, there was some doubt that Bernstein's situation came within the scope of the new program. But Mason, never one to haggle over details, authorized a grant of $2,300, to be paid through Reginald Harris, director of the Biological Laboratory at Cold Spring Harbor. With this money, Bernstein bought a secondhand car and traveled around Long Island, looking for clues to the "heredity of the aging process" in family patterns of farsightedness.

Although he had to hire an assistant to "speak Long Island" to the people he interviewed, Bernstein managed to complete his fieldwork before moving on to Columbia University in the fall. Through some scrimping and the resale of the car, the Cold Spring Harbor Biological Laboratory was able to remit to the Foundation the unexpended portion of the grant, which came to $560.17.

Dr. Bernstein's summer on Long Island was the beginning of a long and painful period of adjustment for him and his family in America. Yet, for all the travail of exile, the Bernsteins were more fortunate than some other members of the German-Jewish intelligentsia on whom the Nazi ax fell with such unexpected swiftness in the early thirties.

# 29

# Despair and Hope

If any Jew in Germany could have expected to be untouched by the anti-Semitic frenzy, it was the world-renowned chemist Fritz Haber. His process for "fixing" atmospheric nitrogen had assured the German army of an uninterrupted supply of nitric acid (necessary for the manufacture of high explosives) during the First World War. A man of intense nationalistic sympathies, Haber had placed his scientific genius at the service of the Kaiser's war machine, not only in the production of munitions but in the manufacture of poison gases as well. After the war he had labored without success to extract gold from seawater, in an effort to pay off the huge indemnities imposed on Germany by the Treaty of Versailles.

In January 1933, when the Nazis came to power, Haber was sixty-four years old. For more than two decades he had been director of the Kaiser Wilhelm Institute for Physical Chemistry, in Berlin. This was one of some thirty institutes organized under the umbrella of the Kaiser-Wilhelm-Gesellschaft (KWG), an agency that channeled private and public funds to the support of scientific research. As a war veteran, Haber was personally exempt from the anti-Semitic laws. But since his institute operated with government funds, his research staff held civil-service status, and in the second week of April he was ordered to dismiss his Jewish assistants. Haber's response was forthright. He informed the authorities that as a German he must obey the law but that as a man he had no alternative but to resign his directorship.

The president of the Kaiser-Wilhelm-Gesellschaft was Max Planck, whose discovery in 1900 of what he called "the elementary quantum of action" had laid the foundations of modern physics. By virtue of his scientific achievements and the force of his personality, he had become the undisputed leader of the German scientific community. Although not Jewish, he had been outspoken in support of the German-Jewish scientists, Albert Einstein among them, who had been vilified as "un-German" by anti-Semites in the 1920s. But like Haber, Planck was a devoted servant of the state in the best Prussian tradition, and he seemed unable to comprehend that the Nazi assault on Jewish scholars and on academic freedom in general was a phenomenon that could not be dealt with in a traditional manner—any more than the behavior of subatomic particles could be dealt with by the formulations of classical physics that he had helped to overthrow.

The seventy-five-year-old Planck was vacationing in Sicily when Haber submitted his resignation to the Prussian Ministry of Education. Efforts (by Haber and others) to persuade Planck to cut short his vacation and return to Germany immediately were unavailing. When he did return, in the middle of May, he paid a call on Chancellor Hitler and tried to put in a good word on behalf of Fritz Haber. According to Planck's account of the interview, the führer reminded him that all Jews were Communists and must be treated as such. When Planck demurred, suggesting that one might make distinctions among Jews, Hitler worked himself into a rage, leaving Planck with no option, as he later told his colleagues, "but to fall silent and take [his] leave."

Weaver and Jones spoke at length with Haber, Planck, and other actors in this tragic drama during the last week of May 1933. It was evident to Weaver that Haber was a broken man, "a pathetic and yet noble figure" who had salvaged from the wreck of his life and work only "his own self-respect." During their conversation Haber told the circuit riders that, despite appearances, the Nazi government *was* concerned about "foreign public opinion." Weaver wrote in his diary, "[Haber] urges us to do all we can in talks with officials, hoping that we will go to the Minister of Education. This is suggested to us many times—also that we go to Hitler himself. We explain always that we cannot enter into questions of internal politics."

In fact, the representatives of the Rockefeller Foundation found it increasingly difficult to separate questions of science from questions of politics in a country whose new rulers explicitly denied any such separation. The most important commitment to German science that

Weaver had inherited from his predecessors was a pledge to build and help equip two major KWG institutes, one for physics and one for cell physiology, in the Berlin suburb of Dahlem. This obligation, which Max Mason had approved in 1930, was contingent on a KWG guarantee to set up an endowment to maintain the buildings.

The Institute of Cell Physiology had just been completed and was operating under the direction of Otto Warburg, who had been awarded a Nobel prize in 1931 for his studies of how cells take in and use oxygen. But construction of the physics institute had been held up by administrative and financial problems within the KWG. The Foundation had earmarked $360,436.75 on its books against the day when the Germans made a formal request for the funds. This commitment was never far from the thoughts of the Rockefeller circuit riders or the representatives of the KWG during their conversations in the spring of 1933.

The general director of the KWG was Friedrich Glum, who, as a nonscientist, oversaw day-to-day administration and financial policy. Weaver was distinctly unimpressed with Herr Glum. When they met over coffee on May 26, Glum mentioned, as a kind of trial balloon, that there was talk of placing the physics institute in the center of Berlin (instead of in suburban Dahlem, where most of the other KWG institutes were located) and of administering it through the University of Berlin (presumably for budgetary reasons). Weaver and Jones seized this opportunity to inform Glum that any such radical change in plans would "completely void the past action." In any case, they said, unless the KWG moved quickly to implement the original plan, the Foundation would be forced to reconsider its commitment.

Inevitably, the talk turned to anti-Semitism. Speaking "with his eyes down on the table," Glum made a halfhearted defense of his government's actions, equating the Nazi animus toward the Jews with "race prejudice in America against the negroes." This prompted a vigorous rebuttal from Weaver, who said that there had never been "general official sanction" of race prejudice in the United States and that most American intellectuals were working to eliminate rather than defend it. Glum made no response.

If Weaver found nothing good to say about the "shallow" and insincere Glum, he was all the more impressed by Planck's "sincerity" and "frankness." Since other accounts of Planck's behavior at this time suggest that the aging physicist repeatedly failed to exert his considerable influence to ameliorate the plight of Jewish researchers, it is possible that Weaver was simply blinded by his admiration for one of the pivotal figures of twentieth-century science.

At this stage in his life, Planck seemed more concerned with defending the institutional integrity of the Kaiser-Wilhelm-Gesellschaft against the attacks of the more radical Nazi leaders than in standing up for the rights of individuals. Apparently, Glum had made a deal with the more "moderate" Nazis in the Ministries of Interior and Education, whereby the KWG would purge itself of political undesirables and continue to run its own affairs. In its annual report, which came out while Weaver and Jones were in Berlin, the KWG leadership urged "the government of the national revival" to permit research laboratories to pursue "their ceaseless scholarly work."

Although the Rockefeller circuit riders heard story after story about researchers dismissed and even physically abused by Nazi stalwarts, it was not clear at the time just how far the government would go in dismantling a scientific establishment that had made formidable contributions to Germany's industrial and military strength during the preceding three decades. A number of Weaver's and Jones's informants spoke approvingly of Planck's cautious policies, on the grounds that the Nazis themselves would soon come to their senses. Otto Warburg, who was half Jewish, changed his viewpoint with each new incident; one moment he was determined to leave the country as soon as possible, and the next he was insisting that if "the professors" simply stood up to the government, the bureaucrats would back down.

At the end of May, Weaver and Jones left Berlin to visit some of the provincial universities, where they found a generally calmer atmosphere. In Leipzig, Peter Debye, a Dutchman who was a leading candidate for director of the proposed KWG physics institute in Berlin, reported that his laboratory was "not seriously affected by the political situation," although a number of Jewish professors, assistants, and advanced students at the university were leaving.

In Munich, where Hitler had staged his beer hall putsch in 1923, Professor Fajans showed the circuit riders his new Institute of Physical Chemistry, toward which the Rockefeller Foundation had contributed $134,000. To Weaver's eye the design and details of the laboratories were on a par with Theodor Svedberg's in Sweden. And while Fajans was anxious about the safety of his friends, he told Weaver and Jones that his own position was secure, partly because of his Rockefeller connection. The moment he had received word that the circuit riders were coming, he had gone to the proper government officials to ask for clarification of his status so that he would know what to tell the Foundation. He had been given a formal letter assuring him of his post, which, as far he knew, made him "the only Jewish professor in Germany" to have obtained this kind of guarantee.

It was with such mixed messages about the situation in Germany that Weaver returned to the United States at the end of June. The former college teacher, who only sixteen months earlier had left his cozy berth at the University of Wisconsin in search of new challenges, certainly comprehended the magnitude of the tragedy he was witnessing. "An irreparable damage has been done. . . . The world renowned intellectual freedom of Germany is a thing of the past," he wrote from Berlin. For the ever-increasing number of deposed scholars, the toll of personal suffering was bound to be high. Yet, for every tale of woe recorded in the diaries of Weaver, Jones, Tisdale, Miller, and Alan Gregg during this period, there were other entries indicating that the worst might soon be over.

In retrospect it is easy to dismiss these hopeful forecasts as mere wishful thinking. Nevertheless, such analyses carried much weight at the time. Some informants who took the long view noted that every successful "revolution" went through a period of extreme irrationality before settling down (more or less) to business as usual. Others were of the opinion that the German army would step in to curb Nazi excesses once the people had been thrown a few Jewish scapegoats. Several times during their travels Weaver and Jones were assured that even Adolf Hitler had been surprised by the zeal of his followers and that as soon as he recognized the damage being done to Germany's reputation abroad, he would take corrective action. Weaver's own tentative conclusion was that "Hitler himself [was] an influence for moderation."

Back in New York, Weaver was hard-pressed to keep up his spirits. Both directly and through RF support for ad hoc groups like London's Academic Assistance Council and the Emergency Committee in Aid of Displaced German (later, Foreign) Scholars, headquartered in New York, he was involved almost daily in efforts to rescue research scientists from Nazi persecution. At the same time, he was responsible for administering the Foundation's long-standing commitments to individual scientists like Theodor Svedberg and Kasimir Fajans and to agencies like the Kaiser-Wilhelm-Gesellschaft and the National Research Council. And in the face of severe challenges to the very existence of an international scientific community, he continued to try to redefine the long-range goals of the Division of Natural Sciences in a way that would be acceptable to the RF Board of Trustees.

Before the end of 1933 seventy-one deposed scholars had found sanctuary and employment with help from the Foundation's Special Research Aid Fund for Deposed Scholars. For those with international reputations, like James Franck and Max Born, there were many job offers. Franck went to Johns Hopkins University, where he stayed until 1938, when he was appointed professor of physical chemistry at the University of Chicago. Both Oxford and Cambridge tried to recruit Born. He chose Cambridge, but there was a problem with money. His funds were tied up in Germany, and the university offered him a salary that came to no more than half his annual income at Göttingen. The Foundation agreed to supplement the Cambridge stipend over a three-year period until an arrangement more suitable for a man of Born's stature could be worked out. Despite his anxiety over the fate of his wife and two children, who had not yet left Germany, Born was grateful for his good fortune. Even before he was settled in Cambridge, he had written to Einstein, "My heart aches when I think of the young ones."

The luckiest of the younger victims of Nazi prejudice were certainly those men who had close ties in one way or another with the Rockefeller Foundation. Hans Krebs, who was so precipitously dismissed from his position at the University of Freiburg that he did not even have time "to pick up his handkerchief," received a Rockefeller grant that allowed him to work with Sir Frederick Gowland Hopkins at Cambridge University in 1933. From there he went on to the University of Sheffield, where he embarked on studies of the metabolic pathway by which living cells utilize foodstuff energy (the citric acid cycle); these studies, which the Rockefeller Foundation helped finance over the course of two decades, won him a Nobel Prize in 1953.

Another young Freiburg émigré was Viktor Hamburger, whose work in the famous embryology laboratory of Professor Hans Spemann had been rewarded with a Rockefeller Fellowship in 1932. After a winter at the University of Chicago and a summer at the Marine Biological Laboratory in Woods Hole, Massachusetts, Hamburger was informed that he had no post at Freiburg to return to. A small grant from the Foundation sustained him at Chicago for another year. In 1934 he joined the zoology department of Washington University in St. Louis; by 1941 he was a full professor and acting head of the department.

While the Foundation worked with others in the scientific community to assist deposed scholars, the circuit riders who remained in Europe continued to report on the devastation of Germany's universities and research institutes, many of which had received substantial sums for construction and equipment from Rockefeller philanthropies dating back to Rose's International Education Board. What must have been a particularly distressing report for Weaver was filed by Harry Miller on December 13, 1933, from Berlin-Dahlem, after a visit to the institute that had been the domain of Fritz Haber.

Haber himself had left Germany in the summer of 1933 for Cambridge, England. His health failing, he would live only a little while longer, dying on January 29, 1934, in Switzerland while en route to Palestine. In charge of Haber's institute Miller found two amiable nonentities. Both professed the by-now-standard ignorance concerning the whereabouts of "missing" assistants, but they insisted that Miller take a look at the liquid-air machine that the Foundation had purchased for Professor Haber and that was just then being installed. They even invited Miller to come back in a month for "a sample of five litres of liquid air."

Although no mention whatsoever was made of Haber's institute in the next two annual reports of the KWG, rumors reaching the Foundation indicated that the institute had been converted to research in chemical warfare. In any case, a glimpse into the Nazi ideal of a research laboratory was provided by the KWG annual report for 1937, which spoke not of scientific investigations but of the creation within the institute of a "work community" *(Arbeitsgemeinschaft)* whose activities included "comradeship evenings" and a weeklong retreat during Pentecost at a castle provided by the minister of education.

Weaver's diary entries describing his experiences in Germany were widely circulated in the New York office. The trustees who read them could hardly fail to make a connection between what was happening under Hitler and the arguments that Weaver and Gregg were advancing in support of their proposed programs in the natural and medical sciences. If there was ever a time for a "new science of Man"—one that would make possible more rational social controls—the time was now. Moreover, the older bricks-and-mortar philanthropy based on Rose's rallying cry to "make the peaks higher" was in a shambles.

The buildings that Rockefeller money had helped to erect in Berlin, in Göttingen, in Munich, were already hostage to the forces of unreason. But the people involved, the Borns, the Francks, the Hamburgers, could carry on their work elsewhere. This was a powerful argument for shifting priorities away from bricks and mortar toward programs—like grants-in-aid and (temporarily) the Special Research Aid Fund for Deposed Scholars—that focused on individual researchers.

Of course, this argument did not address the specifics of Weaver's proposal: his choice of disciplines in which to concentrate the Foundation's buying power. But the real issue that trustees and staff were grappling with during this period of reappraisal was confidence. The trustees did not want to allocate the money themselves; they wanted to be assured that its allocation was in good hands.

Max Mason had asked a group of eminent biologists to assess Weaver's proposal in light of the key question: Which policy was likely to produce a better return on funds invested in research—a concentration in the "vital processes" or a wide-open "scientific opportunism"? After long deliberation the biologists failed to reach any consensus. This left Fosdick free to reaffirm the initiatives taken the preceding year. Under his guidance the trustees' committee on appraisal and plan gave a vote of confidence to the principle of concentration, and

*John D. Rockefeller, Sr., and John D. Rockefeller, Jr., 1935*

to the ability of divisional officers to choose the areas of concentration.

When the full Board ratified these decisions, in December 1934, Weaver had his "sailing instructions" for the rest of the decade. The Board mandated only a minor cosmetic change: the name "vital processes" (which the trustees had never been comfortable with) was dropped in favor of "experimental biology," which no one much liked but which was adopted because no one could think of anything better.

For his part, Weaver had promised to tone down his propaganda for the new program, so as not to raise expectations unreasonably. That was one point the panel of biologists had been able to agree on; the best anyone could hope for in the life sciences was "a slow, painstaking accumulation of knowledge which in a decade or in a few decades will prove perhaps of profound importance."

# 30

# Some Wooden Kitchen Tables

Nineteen thirty-five was the year in which Weaver's program in experimental biology, in his own words, "really got rolling." Beginning in that year awards were made to investigators who did pioneering work in electron microscopy, in the use of radioactive and "heavy" isotopes as tracers to probe biological tissues, in the application of X-ray crystallography and spectrographic analysis to organic material, in the development of more reliable ultracentrifuges and the analytic technique of electrophoresis. These tools and techniques are so taken for granted in today's laboratories that it is hard to appreciate the impact they had when they were introduced in the late thirties and forties.

Although Weaver later characterized his decision to concentrate on experimental biology as a "lucky guess," it was hardly that. To take just one example, neither luck nor prescience was required to foresee the development of the electron microscope in 1931. At the very moment that Weaver, on his first exploratory trip to New York, was discussing the future of biology in Max Mason's office, the first working model of an electron microscope was being tested in Berlin. Physicists had known since 1927 that high-resolution microscopy using focused beams of electrons was theoretically possible. But before a practical instrument could be built, some daunting technical obstacles had to be overcome; Ernst Ruska, who set to work on the problem while still a graduate student in Berlin, belatedly received a Nobel Prize for his contributions, in 1986.

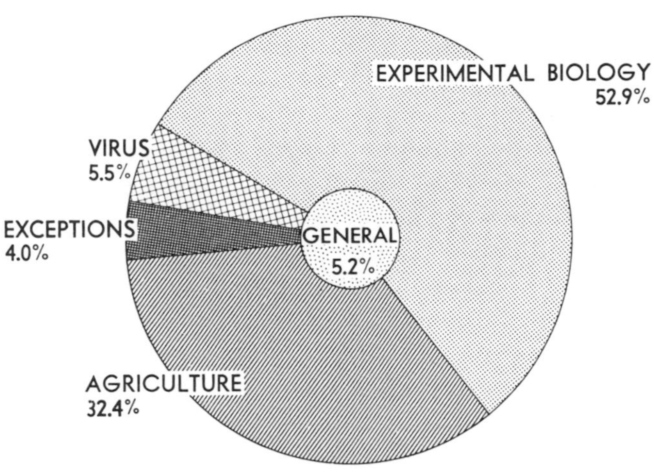


*Chart B*

*How the Rockefeller Foundation divided its support for the nonmedical sciences under Warren Weaver (1932–1959). Total appropriations in the Weaver years: $88,348,093. Appropriations specifically for "experimental biology": $48,947,152.*

In placing his bets on a "friendly invasion" of biology by physics and chemistry, Weaver was simply expressing his confidence in the ability of modern science to keep developing (as physics and chemistry had developed in the nineteenth century) from insight to application, from understanding to control—even when faced with such obstacles as the traditional inviolability of living matter.

By his own calculations, the Foundation's "investments" in biological research under Weaver's direction totaled some $90 million over nearly three decades. Most of the grants that he administered fell into three basic categories: one- or two-year fellowships for young scientists who had already earned their doctorates and who were deemed likely to make important contributions later on; small awards (often under $1,000) to individual investigators for specific research projects or pieces of equipment; and longer-term support for senior scientists working in areas that touched on the Foundation's stated interest in a quantitative approach to the life sciences. Many of those who received grants under Weaver's experimental-biology program were not biologists but chemists and physicists.

The more time Weaver spent in the business of philanthropy, the more cautious he became about attempts to calculate in advance what he called the "leverage" (the return in measurable social benefits) of a specific philanthropic investment. But he had no qualms about making a good retrospective case for the Foundation's acumen in science patronage:

> When the Rockefeller Foundation first gave financial support to the National Research Council fellowships immediately after World War I, this was a pioneering move which, on the basis of all competent estimates that I have ever heard, played a very important role in the development of science in our country.

He was equally proud of what his Division of Natural Sciences accomplished in "encouraging and accelerating and even in initiating the development" of the discipline that he himself first labeled, in 1938, molecular biology.

As one measure of the success of these efforts, Weaver cited the testimony of George W. Beadle, who won a Nobel Prize in 1958 for his contributions to biochemical genetics. Looking back on his career, Beadle identified eighteen recipients of Nobel Prizes, over the period of 1954–65, who had played major roles in the progress of molecular biology; of these eighteen, fifteen had received assistance from the Rockefeller Foundation. Since giving money to internationally renowned scientists takes no special acumen or courage, Weaver stressed that all fifteen had been on the rolls of the Foundation *before* receiving their Nobel Prizes—"on the average over nineteen years" before. Many of these Nobelists, including Max Delbrück, André Lwoff, Jacques Monod, and Severo Ochoa, got their start in research on Rockefeller-funded fellowships.

The Nobelist James Watson, who was not an RF grantee, has described Weaver's coinage of the term "molecular biology" as a "quite significant" contribution, because it got people thinking in a systematic way about what might otherwise have seemed like a bunch of disparate discoveries. Yet Weaver, of all people, knew better than to claim that the development of an entire discipline like molecular biology would have been seriously hampered if a particular source of funds had been unavailable.

The historical concept of an idea whose time has come is especially strong in the history of science. There are many documented cases of two investigators making the same scientific discovery at virtually the

*Howard W. Florey, 1943*

same time. The principle of natural selection came independently to both Charles Darwin and Alfred Wallace. No one, not even James Watson and Francis Crick, has ever suggested that biologists would have remained ignorant of the double-helix structure of DNA if Watson and Crick had taken up another line of work. It would be a rash philanthropist indeed who dared to assert that scientific progress, except in the shortest of terms, depended on his largess.

But there are moments in history when what is crucially important, both for science and humanity, *is* the short term. One of these moments came in the late 1930s; it involved the development of penicillin as the first effective antibiotic—the original "miracle drug" that went into mass production just in time to save thousands, perhaps hundreds of thousands, of lives during the Second World War.

For their roles in the development of penicillin, Alexander Fleming, Howard W. Florey, and Ernst Boris Chain were awarded the Nobel Prize in medicine and physiology in 1945. The Rockefeller Foundation, and especially Weaver's Division of Natural Sciences, provided indispensable support for the work of Florey and Chain. Although the sums of money involved were never large, the penicillin story clearly shows how financial exigencies can influence the outcome of a scientific investigation, and how experienced patrons of science, sure of their goals but flexible in their methods, can play a pivotal role in hastening the emergence of a major discovery.

As Harry Miller remembered it, the RF's involvement in the penicillin story began with some ordinary wooden kitchen tables. The year was 1936. A young professor of pathology at Oxford University named Howard Florey needed the tables to hold his laboratory's ever-expanding collection of bacterial-culture dishes. These are usually called Petri dishes, after Julius Richard Petri, a German bacteriologist (1852–1921) who standardized their shape and dimensions.

A shallow glass receptacle the size of a small ashtray, the Petri dish is as important a tool for the biological scientist as the microscope. In fact, the two go together. In a layer of nutrients and nonnutrient gel on the bottom of a Petri dish, the researcher can grow colonies of bacteria and other microorganisms. The bacteria eat the nutrients and multiply in the gel. By observing changes in the color and clarity of the gel, the researcher can get a rough idea of how the bacteria are doing. For more information he can smear some bacteria-infested gel on a glass slide and view it under a microscope; the bacteria, or their chemical products, can easily be separated from the gel for further analysis. A laboratory interested in the behavior of bacteria will have many cultures growing in Petri dishes at any one time.

Florey, who had been called to the chair at Oxford's Sir William Dunn School of Pathology the preceding year, at the age of thirty-seven, was an ardent experimentalist who was less interested in clinical pathology than in problems of basic science. This emphasis did not endear him to his older colleagues, whose notion of pathology went no further than the traditional dissection and analysis of morbid tissue. Florey did, however, have the backing of two very important people in the British scientific community. One was Charles Sherrington, professor of physiology at Oxford, who won a Nobel Prize in 1932 for his work on the nervous system. The other was Edward Mellanby, the secretary of the British Medical Research Council (MRC), an agency that assisted and administered research projects with funds supplied mostly by the British government but supplemented by grants from private donors, including the Rockefeller Foundation.

Florey's flair for science was noted, and rewarded, early. Born in Adelaide, Australia, on September 24, 1898, he attended the University of Adelaide, where he qualified as a doctor at the head of his class, and won a Rhodes Scholarship to Magdalen College, Oxford. There the young man known as Floss became a protégé of Sherrington, who liked the drive and precision that Florey brought to his laboratory work. But his Rhodes Scholarship was good for only two years, and if Florey was to remain in research after that, another source of income had to be found. He could expect little financial support from home. His father had made a lot of money manufacturing shoes and boots in Australia, then had lost most of it to an embezzling employee just before his death, in 1918.

What Sherrington found for Florey, in the fall of 1924, was an endowed "studentship" at Cambridge, where he would be expected to devote all his energies to research. There was only one catch: the research was to be in a subject, experimental pathology, that Florey had never heard of. Fortunately, the professor of pathology at Cambridge was a Sherringtonian progressive for whom experimental pathology and physiology were virtually identical.

Florey's promise as a researcher blossomed at Cambridge, where he published four well-received papers on cell physiology and blood flow during his first year. When it became clear that he needed advanced training in the techniques of microdissection—the isolation and exploration of small pieces of functioning tissue—Sherrington used his considerable influence to get him what was known as a Rockefeller Travelling Fellowship, administered by the Medical Research Council with funds supplied by the Foundation.

The high point of his trip was a three-month stay in Philadelphia, where he learned how to insert cannula (very thin glass tubes) into the blood vessels of experimental animals. His instructor, who also became his friend, was Professor A. Newton Richards, a pharmacologist who was to play an important role in the penicillin story when his and Florey's paths crossed again fifteen years later.

On his return to England, Florey took up yet another research fellowship, this one at London Hospital. Though the stipend was quite generous for the time—£850 a year—he felt cramped by the emphasis on clinical research to the exclusion of the basic sciences; and he found life in London uncongenial, especially after his marriage, on October 19, 1926, to Mary Ethel Reed, whom he had met when they were both medical students at the University of Adelaide and with whom he had conducted a long and stormy courtship by correspondence. When his influential friends arranged for him to return to Cambridge, as a lecturer in special pathology, he jumped at the chance.

He spent the next five years in Cambridge, doing research, teaching medical students, and playing tennis with a ferocity and unquenchable will to win that scared off even his friends and earned him enough enemies to dash his hopes of getting a coveted Cambridge "blue." "His drive and ambition were manifest almost from the day he arrived," an acquaintance later recalled. "A great fire seemed to burn within him." With Sherrington still advising, prodding, and pulling strings behind the scenes, Florey managed to get away from England for three short trips during these Cambridge years, to hone his laboratory skills under senior scientists in Copenhagen, Strasbourg, and Madrid.

At Cambridge he also met a Hungarian biochemist named Albert Szent-Györgyi, who, aided by a Rockefeller Fellowship, was already on the trail that would lead to a 1937 Nobel Prize for the elucidation of the structure and function of vitamin C. Their conversations left Florey with a new appreciation of quantitative biochemical techniques. The voluble Hungarian assured him that a good biochemist could extract, isolate, and purify "any naturally occurring substance" from any medium, provided there was a quick reliable test for its presence. Florey did not forget what he heard.

His own researches during this period ranged from studies of blood and lymph circulation to studies of the curious properties of mucus secretions in the gut. His interest in the latter had been piqued during his trip to the United States when he suffered from persistent dyspepsia, or "mucus gastritis." Some might have dismissed this as a simple "nervous" stomach, but not Florey. He looked into the medical literature, where he discovered that the mucus-lined wall of the gut is "relatively resistant to penetration and infection by bacteria." No one knew why.

In his search of the literature, Florey also discovered a paper by Dr. Alexander Fleming, who described the antibacterial properties of a substance he had come across, quite by accident, while doing research on the common cold in his laboratory at London's St. Mary's Hospital in 1922. Observing that a bit of his own nasal mucus "dissolved" certain bacteria in a Petri dish, Fleming went on to show that the bacteria killer in the mucus was an enzyme, which he named lysozyme, from the Greek root for "dissolve." He found the same enzyme in human tears, the lining of the gut, and many other animal and plant substances, including egg white. Despite the ubiquity of lysozyme, Fleming's paper had not attracted much attention, since the bacteria that were susceptible to its dissolving action turned out to be quite harmless to man—whereas disease-causing germs (pathogens) were entirely unaffected.

But Florey was intrigued. He reasoned that the action of one natural antibacterial substance might hold some clues to the body's other (and more important) defense mechanisms. Unable to interest Szent-Györgyi in a joint research project, he found another collaborator, applied to the Medical Research Council for support, and in 1929 was awarded the sum of £50 a year for the next three years.

Most of his contemporaries would have considered him a fortunate man. He had a toehold in the Cambridge scientific community, he had found a research problem that seemed important, and he had opened up a "line of credit" with Britain's foremost patron of science. But Florey was not the sort of man to settle into a comfortable niche when the chance for advancement beckoned.

# 31

# A Bushranger of Research

In the fall of 1931 the post of professor of pathology at the University of Sheffield opened up, and Florey applied. Some of his friends were surprised that he would consider moving out of the Cambridge-Oxford-London triangle that most Englishmen thought of as the intellectual capital not only of their country but of the world. Florey's motivation was simple enough; he needed more money. At Cambridge he was earning less than £1,000 a year and could expect only gradual increases. Now that the Floreys had a child (a daughter, born in 1929) they were chronically worried about finances. The Sheffield professorship offered £1,000 a year to start, with more opportunities for improvement.

There were a few problems, however. Florey had no administrative or clinical experience, he had never hired or directed a staff, and, most damning of all to traditional pathologists, he had never conducted a postmortem examination. When informed of Florey's application, one influential clinician stated flatly, "There is no pathologist named Florey." Nevertheless, he got the post, with the powerful support of Edward Mellanby, professor of pharmacology at the Sheffield Medical School, member of the board of the Medical Research Council, and a friend and ally of Sherrington. Fourteen years older than Florey, married but with no children, Mellanby seems to have taken an almost paternal interest in the young Australian.

By the time Florey arrived in Sheffield in March 1932, the university authorities were well aware of his single-minded concentration on research; they had assured him a free hand to continue his laboratory investigations while other pathologists at Sheffield took care of the postmortems. But his expanding research program (which typically included at least one experiment a day) required periodic infusions of outside funds. When Mellanby cautioned him that money was as tight at the Medical Research Council as everywhere else, Florey decided to go over his head and appeal directly to Sir Walter Fletcher.

Fletcher, a physiologist in the Sherrington mold, had headed the MRC from its inception in 1913 as a special subcommittee of the Ministry of Health. When the MRC was reorganized in 1920 as a separate agency charged with raising the standards of biological research in Great Britain, Fletcher became perhaps the single most important figure in the British scientific establishment. As adviser to the trustees of the Sir William Dunn Estate, he had helped spend the fortune left by that merchant banker (who died in 1912 without heirs) on such benefactions as the Sir William Dunn School of Pathology, at Oxford, and the Sir William Dunn School of Biochemistry, at Cambridge.

Any other newly appointed professor at a provincial university might have hesitated to ask the MRC for more money at this particular time. Like the rest of Europe, Great Britain was in an unprecedented economic depression; the preceding September the Bank of England had been forced to go off the gold standard. But Florey had discovered that his Australian brashness could be an asset in the rigidly stratified world of British academe: "I could always . . . do the outrageous thing and still be tolerated. They made allowances for the rough colonials." Two months after arriving at Sheffield, he asked Fletcher for a substantial increase over the £50 a year he was already getting from the MRC. What he really needed, he wrote, was £400 to £600 a year to hire a chemist who could work with him on the study of lysozyme, "a widely distributed enzyme" that presumably had "some universal biological significance."

Fletcher rejected this request, suggesting that Florey look for a collaborator who was already on the Sheffield faculty (and who would not therefore need outside support). But Florey's brashness did not go entirely unrewarded; out of its limited funds, the MRC advanced him another £140 for lab equipment and agreed to support a young lab assistant whom Florey had brought with him from Cambridge on a stipend of fifty shillings a week.

The following year Fletcher died and was succeeded at the MRC by

Edward Mellanby. Now Florey had a friend in the inner sanctum of British science. But if he expected any special favors from Mellanby, he was soon disappointed. In fact, he received no further help from the MRC until 1935, the year he exchanged his professorship at Sheffield for the coveted chair of Pathology at Oxford.

Once again it was Sherrington and Mellanby who cleared the way for him, overcoming objections from traditionalists and personal enemies (one of whom referred to Florey derisively as a "bushranger of research" for his constant grubbing after grants). Florey lobbied hard for the post, which, his wife confided to a friend, had been "his Mecca for many years." Still, it was a close thing. As a member of the board of electors that met in Oxford on January 22, 1935, Mellanby was expected to cast a decisive ballot for Florey. But his train from London was delayed several hours, and the electors were just about to give the nod to a rival candidate when Mellanby made a dramatic entrance and swung the vote back toward Florey.

The professorship brought Florey an annual salary of £1,700 and a magnificent laboratory building, erected with funds provided by the Dunn trustees in 1927 but little used since then. Taking his election as a mandate to "give pathology a good twist away from diagnosis and morbid anatomy," Florey began by putting together a multidisciplinary research team of coworkers whose knowledge and skills complemented his own. Such teams are considered the norm in biological laboratories today. But the concept was unfamiliar to many members of the scientific community in the mid-thirties and constituted another source of friction between Florey and his critics.

As an Oxford professor, Florey had a call on outside funds that he had not had at Sheffield. In the summer of 1935 he persuaded Mellanby to give him £300 for a chemist who would work on the purifying of lysozyme. He found the man he needed (E. H. Roberts) at Oxford's Dyson Perrins Laboratory of Organic Chemistry, a few hundred yards down the road from his own pathology building. Then, with funds provided by the Sir William Dunn trustees and the university, he went shopping for a biochemist who might pinpoint the actual mechanism by which lysozyme "dissolved" bacteria. After several unsuccessful approaches to other men, he heard about a twenty-nine-year-old German-Jewish refugee named Ernst Boris Chain, who was working at Cambridge. The son of a German mother and a Russian émigré father, Chain had left his native Berlin on January 30, 1933, the day Hitler was named chancellor.

Chain had arrived in England with a degree in chemistry, an interest in enzymes, and a talent for music that led him to consider seriously a career as a concert pianist. A recommendation from J. B. S. Haldane, the eminent geneticist, brought him to the attention of Frederick Gowland Hopkins, a Nobel laureate (1929) who had established the importance of vitamins in the diet and who now presided over the Sir William Dunn School of Biochemistry, at Cambridge. Hopkins put the young refugee to work in his lab and was much impressed with Chain's ability, but he had no permanent post to offer. Just when Chain was getting ready to emigrate to Canada or Australia in search of fresh opportunity, he was invited to Oxford by Florey.

At Cambridge, Chain had been investigating snake venoms. It was known that many of these potent neurotoxins (nerve poisons) were proteins. But why did such small amounts have such powerful—often fatal—effects? Chain showed that the active substance was an enzyme that destroyed another enzyme in the victim's body that was essential to the neural control of breathing. He worked not with laboratory animals but with flasks of pulverized tissue from which the structural components of the cells had been removed, leaving only the physiologically relevant chemicals. By adding small amounts of neurotoxins, he could conduct controlled experiments that took the mystery out of an obscure, even exotic, biological phenomenon. "For the first time," Chain wrote, looking back on this research with pride, "the mode of action of natural toxins of protein could be explained in biochemical terms."

Chain put the finishing touches to the snake-venom work in his first months at Oxford. Then he turned his attention to lysozyme. Florey's interest in this substance had been based on his conviction that it was the first line in the body's antibacterial defenses and that an understanding of how it worked might provide an entry into the entire problem of natural immunity. For Chain the fact that it was an enzyme was sufficient reason to study it.

The investigation of lysozyme was by no means the only project under way at the Oxford pathology lab, and finding money to support an expanding research program once again became Florey's major concern. To ensure Chain's long-term collaboration, he arranged a three-year grant (£300 in salary, plus £100 for expenses) from the

British Empire Cancer Campaign. And at Chain's request he recruited the twenty-five-year-old Norman Heatley from Cambridge. Dr. Heatley, who was adept at microdissection and micromanipulation, was supported by an MRC stipend.

But for every successful approach to a funding agency, there were many rebuffs; and the entire process of assembling and maintaining a multidisciplinary laboratory on a patchwork of small grants—most no longer than a year in duration and some as short as three months—was exhausting. During his first two years at Oxford, Florey completed only two scientific papers, far below his rate of productivity both before and after this period.

Despite the Depression he had had good reason to expect an easier time getting established in Oxford. Aside from his rising reputation as a researcher and the prestige of his new position, his arrival at Oxford coincided with the announcement of a major gift to the university—a gift of £2 million expressly targeted for the support of medical research.

The donor was William R. Morris (1877–1966), the son of a farm laborer, who started manufacturing motorcars in 1913 and ended up as Great Britain's counterpart to Henry Ford. His company's products ranged from the humble Morris Minor to the sporty MG. By 1934, when he took the title Lord Nuffield, he was devoting less and less of his time to making money, and more and more to giving it away. He had dreamed as a youth of becoming a surgeon, and when he was presented with a plan to expand and improve the medical school at Oxford, he agreed to finance it.

The £2 million would have been impressive at any time. But coming as it did in the middle of a depression, such largess was bound to provoke an especially fierce struggle over who would get the juiciest slices of pie. When the faculty in-fighting was over, the clear victors were the advocates of clinical research, who got new professorships, new buildings, and new equipment, while the "pre-clinical" departments (that is, the basic scientific disciplines like physiology and biochemistry) had to settle for a few crumbs or nothing at all.

The disappointment must have been particularly galling for Florey, since the Nuffield benefaction had been set in motion by Hugh Cairns, a surgeon who had preceded Florey as Rhodes scholar from South Australia, who succeeded him by one year as Rockefeller traveling fellow, and who had since become a good friend. But Lord Nuffield had a clear bias toward the clinical side, and his personal style of philanthropy was closer to Andrew Carnegie's than to John D. Rockefeller's. One of his special desires was to establish a chair of anesthetics at Oxford; when informed that this subject was not considered strictly academic by medical educators, his counter argument was simple and effective: no chair of anesthetics, no two million pounds.

Meanwhile, Florey, having assembled a research team without a major grant of any kind, was finding his progress hampered by a shortage of the most basic equipment, including those wooden tables. As a former Rockefeller fellow he was aware of the Foundation's interest in medical research. He knew that the officers of the Foundation periodically visited Oxford to keep in touch with research there, although no one had yet visited his new team. Without waiting for the RF to come to him, he decided to introduce himself. On May 9, 1936, he wrote a letter to Daniel P. (Pat) O'Brien, the associate director of the Division of Medical Sciences of the Rockefeller Foundation, who worked out of the Paris office.

After a brief paragraph expressing regret at having "missed" O'Brien on his previous trips to Britain, Florey came to the point. As part of his laboratory's effort to develop "the chemical aspect of Pathology" he had hired two biochemists. "The difficulty I find myself in at present," he wrote, "is that the funds of the Department are inadequate for the equipment for these people." If the Foundation could consider a grant of £250 (which at the current rate of exchange came to about $1,280), he continued, it could "rest assured that the money would be put to good use by the Biochemists working here." For further information about the "aims of the Department," Florey suggested that O'Brien get in touch with "Professor E. Mellanby."

A few weeks later Florey had his reply—a politely worded letter of rejection. "The field in which you are working might have been of interest to the Division of Medical Sciences a few years ago," O'Brien wrote. "Unfortunately, under our present program it is not one in which we would be in a position to consider a recommendation for aid."

There was no way for Florey to know, from O'Brien's vague wording, that the Division of Medical Sciences had recently begun to concentrate its resources on psychiatry and related disciplines and that projects like Florey's now came under the purview of the RF's Division of Natural Sciences. On the same day that he turned down Florey's request, O'Brien passed along copies of the correspondence to Wilbur Tisdale, who had succeeded Lauder Jones as the top-ranking Division of Natural Sciences officer in the Paris office.

Tizzy, as he was known to his colleagues, had been on the fellowship staff of the National Research Council before taking charge of the International Education Board's European fellowship program in 1926. He had assumed a similar role for the Foundation after the reorganization of 1928, and was now one of the RF's most traveled circuit riders.

In his memo to Tisdale, O'Brien seemed torn between a strict interpretation of Foundation policy and his desire to help Florey: "I do not think that his scheme is likely to fall within present MS [Medical Sciences] program. I feel, however, it should be carefully considered under any circumstances, because of Florey's importance and the significance of the development at Oxford."

Two weeks later Tisdale was in Oxford, accompanying Warren Weaver, who had come over from America on one of his periodic tours of European laboratories. While Weaver spoke with Professor Robert Robinson of the Dyson Perrins Laboratory, Tizzy dropped in on Florey.

Each of the circuit riders had a distinctive style of diary keeping. Weaver's entries were long and detailed, with a strong narrative structure. Harry Miller's were more chatty and discursive, even gossipy at times. But Tisdale's were strictly business, at times almost telegraphic.

In recording his first meeting with Florey, on June 15, 1936, he wasted no words summarizing the interview:

> Prof. Florey needs balances, micro balances, vacuum distillation apparatus etc. to a value of about £250. The MRC does not give equipment, the Royal Society is swamped with applications, and the University is not willing to help him further. F. is an exceptionally good man and he is working on interesting problems, but they are not directly in either MS or NS programs. I told F. I would discuss the matter, and that if it was possible for the Foundation to help him I would communicate with him further.

The laconic Tisdale was not a man to waste time either. Eleven days later the RF's Paris office authorized payment of up to £250 ($1,280) to "the Department of Pathology of the University of Oxford directed by Prof. H. W. Florey to provide equipment necessary for his researches in the fields of enzymes and cell growth."

In line with RF policy the money was not released to an individual directly but to an institutional entity, with a clear indication of the person who would be responsible for spending it. This policy had a dual purpose. On the one hand, the Foundation wanted to affirm its ties to the institutions that made up the permanent core of the scientific community. On the other hand, by naming the individual in charge of a project, the Foundation was protecting itself if, for reasons ranging from personal disability to political pressures, control of the work passed into less competent hands—as had happened in so many cases after the Nazi takeover in Germany.

But the distinction was a subtle one, and often ignored in the routine administration of grants. For example, in his letter to Florey announcing the good news, Tisdale suggested that the Oxford workers could avoid red tape by simply ordering what they needed from their usual suppliers and then informing the Foundation when the material arrived in "acceptable condition," whereupon the RF's comptroller would "pay directly to the suppliers."

No mention of the much-needed kitchen tables ever appeared in official correspondence between Oxford and Paris, perhaps because such a detail would have sounded undignified. Presumably, that request came under the "etc." noted in Tisdale's original diary entry. All the invoices and paid bills have disappeared from the Foundation's files. But Dusty Miller was quite sure that Florey used part of this grant, his first from the RF, to buy wooden kitchen tables to hold the ever-increasing number of Petri dishes in the biochemical section of his laboratory, where attention would soon turn to the antibacterial action of a mold product called penicillin.

# 32

# Keeping the Team Together

Why did Tisdale and his colleagues take such an interest in Florey despite the fact (duly recorded in Tizzy's diary) that the work at Oxford fell outside the Foundation's newly defined fields of concentration? An entry from O'Brien's diary gives a clue.

As part of their continuing effort to thrash out precisely where the boundary lay between their two divisions, O'Brien and Miller sat down to discuss the Oxford situation in the fall of 1936. It was Miller's opinion that, with a few exceptions like Professor Robert Robinson, the science faculty at Oxford was "not strong." Through their contacts in the British universities, the circuit riders were aware of the fierce opposition that Florey's appointment had roused. In supporting him, however modestly, they were demonstrating not only their willingness to bolster the natural sciences at Oxford but also their close working partnership with the progressive wing of the British scientific community represented by Sherrington and Mellanby.

In all, the Foundation gave away just over $11 million during 1936. Of this total, the Division of Medical Sciences accounted for $1,623,750 and the Division of Natural Sciences for $1,370,350. Among the larger commitments was a five-year program of support (at a projected cost of £8,000, or $40,800) for Professor Robinson, Florey's neighbor at Oxford and one of Europe's foremost organic chemists. Robinson had sixty-five people working with him in the Dyson Perrins Lab, including a team of X-ray crystallographers headed by a young woman named

Dorothy Crowfoot who would later (under her married name of Hodgkin) win a Nobel Prize for determining the three-dimensional structure of several important molecules, among them penicillin. Robinson himself would receive a Nobel in 1947 for his work on plant alkaloids.

Unlike the large grant to Robinson, Florey's $1,280 did not rate a special mention in the Foundation's *Annual Report* for 1936. It figures only as part of the sum of $80,000 set aside for the fifty-nine grants-in-aid approved by the Paris office during that year. Even as a grant-in-aid, it was small potatoes; some went as high as $6,000. Nevertheless, Florey had been bucking heavy odds; 270 applications to the Foundation for research aid were rejected outright in 1936.

The value of very small grants, such as the one given to Florey, had been debated at the Rockefeller philanthropies ever since the days of Frederick Gates, who warned against a "retail" approach to philanthropy that tried to do a little of everything and ended up without a measurable impact in any one area. Rose and Buttrick were also opposed on principle to what the latter referred to as "chicken feed" grants. Not until 1929, when all three were gone from the RF Board and Max Mason had assumed the presidency, was the small grant-in-aid explicitly recognized as an instrument of science patronage at the Foundation. In 1935, the year before Mason resigned, his protégé Warren Weaver still found it necessary to justify grants-in-aid in these words: "While such grants are always relatively small in amount, it frequently happens that the results accomplished through the opportunities they afford are of far-reaching significance."

There is no question that Florey and his collaborators got results in the months following their receipt of the Foundation's grant. By the end of 1937 they had a pure extract of lysozyme, and Chain had demonstrated that the enzyme worked by breaking down a particular polysaccharide, a complex carbohydrate molecule. Only microorganisms that incorporated this molecule in their cell walls were susceptible to lysozyme's dissolving action; unfortunately, as Fleming's crude tests had already indicated, the polysaccharide attacked by lysozyme was not found in the microbes that caused serious diseases in humans.

Still, it was a sound piece of research, and Chain wrote up his findings for publication. Following standard procedure, he searched through the scientific literature for accounts of similar work that he might want to refer to in his paper. He found more than he had bargained for. As he put it later, he had "stumbled, more or less accidentally, across the well-known phenomenon of microbial antagonism." There were many reports of substances produced by bacteria, molds, yeasts, and other microorganisms that inhibited the growth of bacteria—among the most remarkable being a paper by the same Alexander Fleming who had discovered lysozyme.

Unlikely as it sounds, Dr. Fleming had had another fortuitous encounter with a contaminated Petri dish. Writing in the June 1929 issue of the *British Journal of Experimental Pathology,* he reported that he had been studying some variants of staphylococcus when he noticed a strange mold growing in a dish whose cover had been removed and replaced many times. He could see at a glance that wherever the uninvited mold grew, the colonies of staphylococcus did not thrive; in fact, there was a bacteria-free zone directly adjacent to the mold. Recognizing that he had something "interesting," Fleming removed a piece of mold from the dish with a scalpel and put it into a tube of nutrient broth. In a series of experiments, he showed that this mold (which belonged to the genus *Penicillium)* would inhibit not only staphylococcus but other pathogens as well, including the microorganisms responsible for pneumonia, diphtheria, gonorrhea, and meningitis. Furthermore, concentrations of "mould broth" strong enough to stop bacteria from growing were not toxic to mice or rabbits and did not appear to interfere with the body's normal defense mechanisms, such as white blood cells.

One might think that, with such encouraging results, Fleming's paper would have caused an uproar in medical circles. In fact, nothing of the sort occurred, perhaps because the very concept of a substance capable of destroying a broad range of harmful bacteria without harming their animal hosts ran counter to the most advanced medical opinion of the time. Fleming himself had helped create this skepticism about germ-killing drugs by his demonstration, during the First World War, that the antiseptics then available were worse than useless in combating infections in open wounds, since they were more deadly to the body's own white blood cells than to pathogenic microorganisms.

No one, including Fleming, seemed to realize that penicillin was different. Fleming did ask some of his young associates at St. Mary's Hospital in London to try to concentrate the active ingredient from the mold broth (dubbed "penicillin") so that it could be subjected to more elaborate tests. When these efforts were thwarted by the extraordinary instability of the stuff, Fleming apparently lost interest. A handful of other experimenters, stimulated by Fleming's paper, also tried and failed to isolate the antibacterial substance from the culture medium. Whatever this penicillin was, it was hard to produce in quantity, hard to purify, and hard to preserve.

By the time Chain came across Fleming's paper, interest in penicillin among British medical researchers was almost nil. Even those who were receptive to the controversial concept of chemotherapy had their hands full exploring the new sulfa drugs, which had been introduced to the world by German scientists in 1935 and which had already proved effective against the often fatal puerperal fever that struck birthing women and their newborns.

Fleming himself had become interested in the sulfa drugs; his papers in the late thirties describing his work with these chemical germ killers contain no references to penicillin. Yet throughout the decade he maintained in his laboratory a culture of *Penicillium* mold that he used as a kind of bacteriological weed-killer. Since the "mould broth" wiped out certain species of bacteria and had no effect on others, he dosed his Petri dishes with it whenever he wanted to eliminate a susceptible species from a culture during the course of an experiment. In other words, ten years after its discovery as a contaminant in a Petri dish, penicillin had been forgotten by virtually everyone except its discoverer—who kept a small supply handy to decontaminate Petri dishes.

Chain was originally attracted by Fleming's 1929 paper on penicillin because, in Chain's words, it represented "one of the most impressive and best described" of all the examples of bacterial antagonism he had come across. He thought that the active substance might be an enzyme—"a sort of mould lysozyme"—in which case there was a chance it would yield up its secrets to the same techniques he had already used successfully on lysozyme and snake venom.

In the summer of 1938 he and Florey decided to look into the whole phenomenon of bacterial antagonism; Chain would concentrate on the isolation of antibacterial substances and the study of their chemical and biochemical properties, while Florey explored their biological properties. In an article they wrote jointly in 1944, Chain and Florey attributed at least some of their success in this endeavor to good luck: "It was hoped that a systematic study might lead to the preparation of new compounds of biological interest. By great good fortune one of the first to be investigated was penicillin. . . ." This makes it sound like a happy accident. Although luck *was* involved, their decision to begin with penicillin actually had more to due with the precarious state of their finances.

In a laboratory so strapped for cash that the purchase of standard off-the-shelf chemicals seriously strained the budget, the launching of a major investigation involved an act of faith as well as intellectual judgment. The Oxford researchers had to believe that bacterial antagonism was not only a theoretically interesting problem but also one that could be profitably attacked with the tools at hand. By a curious twist of fate, there just happened to be a free supply of penicillin available at the Sir William Dunn School of Pathology.

Florey's predecessor, Professor Georges Dreyer, had obtained a sample of Fleming's original "mould broth" (presumably from Fleming himself) because he wanted to test a theory that the active ingredient was a bacteriophage, a virus that attacks bacteria. Dreyer played with the culture long enough to establish that it contained no bacteriophage and then set it aside. But Chain knew that it still existed because he remembered meeting a woman researcher in the corridor shortly after his arrival at Oxford and asking her about a tray of Petri dishes she was carrying; she had told him that they contained a substance she was using to eliminate unwanted bacteria from experimental cultures. Now Chain sought her out, and when she confirmed that her "weed-killer" was penicillin, he got a sample from her and began growing it himself. For a biochemist who had been warned by his boss only a few months earlier not to order any more supplies (not so much as a glass rod!) this was a significant development indeed. "I was astonished at my luck," Chain wrote. "I could hardly believe it was true."

In their later public statements about the sequence of events that led them to penicillin, both Chain and Florey insisted that their only motive in launching the investigation was pure scientific curiosity. They went so far as to deny having had any interest in the life-saving potential of the substances they chose to study. The possibility of therapeutic applications, Chain wrote in 1971, "did not enter our minds when we started our work on penicillin." Florey was only a little less certain about what drove them: "I don't think that the idea of helping suffering humanity ever entered our minds." There is no reason to doubt that the main impetus behind their decision to look into bacterial antagonism was what Florey called "the desire to know and the desire to know first." But there is evidence that other, more mundane factors—notably, a desire to keep the grant money flowing—also played a role in their decision.

In the fall of 1937 the Oxford Department of pathology had been overdrawn at the bank by £500, an impressive sum considering that it represented Chain's salary for the entire year. To Mellanby, Florey complained, "I have the utmost difficulty in finding money for even relatively small essentials in my work." Mellanby, who was facing difficulties of his own at the MRC, had nothing but sympathy to offer in response, and even that he offered grudgingly. Indeed, he seemed annoyed that his Oxford protégé, instead of settling down to solid research, was spending so much time playing at what Mellanby described to the RF's Pat O'Brien as "Oxford medical politics."

Of all people, the hard-pressed secretary of the Medical Research Council might have appreciated Florey's determination to scrounge up a few extra pounds for research when he could have simply sat back and enjoyed the lifetime sinecure that an Oxford professorship provided. The MRC itself was lurching from crisis to crisis. Its funds came from a block grant that the British Parliament voted every five years. The most recent appropriation, in 1934, had been for £1 million. But by the beginning of 1938 disbursements were running far ahead of projections, and Mellanby was desperately searching for additional funds to fulfill the MRC's obligations.

He had repeatedly sounded out the RF circuit riders about increased Rockefeller support for the MRC, without success. In March 1938 he confided to O'Brien that he was withdrawing £10,000 from a £30,000 MRC fund earmarked for nutritional studies (which happened to be his own scientific specialty) in order to meet other, more pressing commitments. In April he revealed to Miller that political pressures had sometimes forced him to "spend relatively large sums for what he considered not to be real research." Miller noted the example that Mellanby gave:

> The miners in Western England have a lung disease which is not silicosis, and the Member of Parliament from that region raises such a fuss in the House of Commons that the Home Secretary runs to [Mellanby] with an urgent appeal to do something, which he cannot well refuse.

Undaunted by Mellanby's admonition to make do with what they had, Florey and his colleagues now turned their attention to the same source of funds that Mellanby was eyeing—the Rockefeller Foundation. In August, Florey ran into Tisdale in Zurich, where both were attending an international congress of physiologists. Listening to a speaker eulogize prominent physiologists who had died during the preceding year, Tizzy was startled to hear the name of John D. Rockefeller, Sr., included in the honor roll. The explanation offered was that Rockefeller (who died on May 23, 1937) had done more for the study of physiology, through his benefactions, than any other single individual. Perhaps this emboldened Florey, who took the occasion to detail his own money problems to the RF circuit rider.

Florey explained that key members of his research team were living on MRC stipends that were due to run out the following spring. He wondered if the Foundation could give him £600 to £700 a year for the next three years. Tisdale was anything but encouraging. His characteristically terse diary entry ends with this sentence: "I was unable to give him any definite answer and held out no hopes that this could be done."

It was shortly after Florey returned from Zurich that he and Chain decided to launch the investigation that led to the development of penicillin. The two men lived near each other, and after work they often walked home together across the grassy parklands behind the Dunn building. As Chain remembered it, their conversations on these walks touched not only on scientific matters but also on their mutual determination to make the laboratory "as independent as possible of both University and government financial support and to look for private funds." He said, "I thought that a long-term project . . . would suit us best so that we would not have to go through the agonizing experience of fund-raising every year." To Chain a study of microbial antagonism "seemed ideal for the purpose."

Florey has left no such definitive account of his state of mind at the time. But a good friend of his, an Australian physiologist, recalled a late-night conversation he had with Florey after his return from Zurich. Florey doubted that the MRC would support the lysozyme investigation much longer, since after four years of hard work it had "produced nothing." He spoke of penicillin as a possible lifesaver—in a financial rather than a clinical sense. He described it as a "tough project" that might lead to "another blind alley" (the way lysozyme had), but he added, "There is no question we will now have to go for penicillin." The very existence of the research team he had assembled was in danger. "My worry," he said, "is that I've got the bacteriologists and the biologists, and I've got my team together. If the money doesn't come along, I might not be able to hold them together and it would all be finished."

# 33

# Lebensraum

Florey's daughter remembers her father telling her that the decision to "go for penicillin" came one night toward the end of 1938, while he was standing (presumably with Chain) beneath a "big old chestnut tree" in the park adjoining the Dunn building. What is certain is that Chain told Florey of his fortuitous discovery of a batch of Fleming's penicillin sometime late in 1938; by the new year Chain and an assistant were already hard at work trying to isolate and purify the antibacterial ingredient.

This project, which ran into some unexpected difficulties, went on side by side with the final stages of the lysozyme investigation. Meanwhile, Florey continued to juggle grants in an effort to keep his team intact. The British Empire Cancer Campaign came through with some more money; and Harry Miller agreed to arrange a Rockefeller Travelling Fellowship to allow the young Norman Heatley to study advanced techniques of micromanipulation in Copenhagen at the Carlsberg Laboratory, whose RF-supported biochemical research dovetailed nicely with Heatley's own investigations of cell metabolism.

If by this time Florey was feeling a little like Lewis Carroll's Red Queen, who had to run as fast as she could just to stay in the same place, who could blame him? He hated to lose Heatley for a year, but the alternative was to lose him entirely. Venting his frustration in a conversation with Miller, he outlined a scheme to put the entire structure of British medical research on a sound financial footing by borrowing a leaf from Britain's dentists. Each dentist was committed to contributing one pound a year to a permanent fund for dental education and research. Why, Florey wondered, couldn't practicing physicians do the same, with the accumulated funds "to be used entirely for research"? If Miller had a personal opinion about the feasibility of this scheme, he did not commit it to writing. And Florey, having gotten the matter off his chest, apparently never mentioned it again.

Early in 1939 Florey gingerly broached the subject of a systematic study of microbial antagonism, including penicillin, in a letter to Mellanby. Chain was trying to produce pure extracts of penicillin and other bacterial antagonists so that controlled studies of their germ-killing properties could get under way. While the Oxford researchers waited for Mellanby's reply, an international crisis far graver than the Great Depression was gathering momentum on the Continent. The war would, of course, have an enormous impact on the work of Florey, Chain, and their collaboraters. What seems hard to credit, even in retrospect, is the enormous impact that the research at the Sir William Dunn School of Pathology in Oxford would have on the war.

In the words of Britain's Prime Minister Neville Chamberlain, the Munich Agreement of September 1938 assured "peace in our time." Just a few months earlier, Germany had annexed Austria with only weak protests from France and England. At Munich the British and French acquiesced in a transfer to German sovereignty of 10,000 square miles of Czech territory, whose inhabitants, mostly German speaking, seemed to welcome their new status. In return for control of this "Sudetenland," Hitler promised to respect the boundaries of the rump state of Czechoslovakia. In fact, the respite from German demands for lebensraum lasted only a few months.

By the end of 1938, Fascist agitators in what was left of Czechoslovakia were calling for a linkup with the Third Reich. Since there was no significant German-speaking population in the remaining territory, Hitler's imperial ambitions could no longer be cloaked in the politically acceptable rhetoric of "self-determination." Warily, the British government moved to strengthen its armed forces, opening a voluntary "national register" for war service. The French, whose army and navy had been partially mobilized since September, took further steps to increase military preparedness. But when the crisis came to a head in March 1939, neither of the Allied Powers intervened, and Czechoslovakia was dissected into several so-called autonomous provinces under German "protection."

For the British these events marked the end of Chamberlain's policy of appeasement. On March 31 Britain joined with France in a pledge to resist any threat to Poland's independence; this joint declaration was quickly transformed into a mutual defense pact, which the British hoped to make the cornerstone of a grand alliance against German expansionism. While British diplomats tried to woo the Soviet Union into this "peace front," the military budget was increased again; and on April 27 limited conscription was introduced.

In Oxford this escalation of military preparedness touched Florey's lab but did not seriously disturb the work going on there. One spring afternoon the entire staff turned out to construct an air-raid shelter in the backyard. But the English had no intention of repeating their tragic error of 1914, when many young scientists marched off to war along with others of their generation and were lost in the trenches, decimating the ranks of researchers for years to come. This time, rather than draft scientists as cannon fodder, the government planned to make use of their skills in war-related work in their own laboratories. For example, Florey agreed to give a third of his time, if necessary, to "poison-gas research" for the War Office. And at the request of the Ministry of Health, a blood-transfusion service was established in Oxford, linked to local hospitals and headquartered in the School of Pathology under Chain. To facilitate travel between hospitals, the government provided Chain's group with a van.

Florey was pleased by such tangible recognition of his laboratory's importance to the nation. However, these developments did nothing to solve his money problems. His unceasing requests for aid had already earned him a reputation as an "academic highway robber." But as he saw it, he had no choice. Instead of increasing its support as his lab grew, the university had recently warned him to expect a reduction in departmental funds. When he asked the Nuffield Research Grants Committee to support Dr. Margaret Jennings (one of his assistants, who, twenty-eight years later, would become the second Mrs. Florey), the committee not only turned him down but did so in a manner that Florey found insulting. He demanded and got an apology. But he got no money, and he had to return, hat in hand, to Mellanby, who came up with £200.

Through the summer of 1939 (while Hitler's pressure on Poland increased almost daily), Chain continued to make slow progress in purifying penicillin, and he and Florey continued to refine their strategy for financing their investigation of antibacterial substances. They decided to go through all the usual channels before approaching the Rockefeller Foundation again. In April the RF had announced yet another big grant to Professor Robinson—£23,000 to build and equip a three-story research wing adjacent to the Dyson Perrins Lab. Since Robinson was a friend, it is likely that Florey knew something about the talks that led up to this commitment. Although he might have wished that some of the money had come his way, he was undoubtedly relieved to learn what the Rockefeller circuit riders had told Robinson—that British war planners considered Oxford a relatively safe area.

What Florey certainly did not know was how much the Foundation, in which he rested his principal hopes for support, had itself been affected by the crises of the past year. Much of the energy of the circuit riders had been devoted to aiding displaced scholars from Germany and other countries where civil and academic liberties had been curtailed. Initiated in May 1933 with an appropriation of $140,000, this "emergency" program was renewed and extended each year through 1939; all told, 122 scholars were relocated with RF assistance, at a cost of nearly $750,000.

Meanwhile, at the request of Raymond Fosdick (who had succeeded Mason as president in 1936), the head of the RF's Paris office, Mike Gunn, was drawing up a detailed contingency plan "in the event of a crisis arising similar to that of September last." Gunn had reported to Fosdick right after the Munich Agreement, "Europe is still very jittery and the general outlook is distinctly unfavorable." But Europeans were adept at keeping up their spirits in the face of disaster, and the Foundation officials in Paris had adopted a similar attitude. Whenever possible, Gunn wrote, he and his wife managed to get away for weekends to Fontainebleau, where they had "become members of the Fontainebleau Golf Club."

All in all, the mood in Paris was less gloomy than the mood in the Foundation's New York office. Writing to Gunn on March 6, Fosdick passed along some depressing financial news. The treasurer had just estimated that income from the RF's portfolio in 1939 would amount to less than $6 million, "the lowest figure the income has ever reached in the history of the Foundation." As it turned out, actual income was

$6,627,450. But even when the trustees added nearly $2 million from principal, total appropriations for 1939 were down substantially from previous years.

The final dismemberment of Czechoslovakia in March 1939 prompted Gunn to put his contingency plan into action. A thirteen-room furnished villa in the Brittany seaside town of La Baule was rented for a year. The idea was to move the entire office to La Baule in case Paris came under attack. Several of the circuit riders, including Harry Miller, knew La Baule as a pleasant if not particularly elegant family resort, about an eight-hour drive from Paris. (Years later, Miller offered this laconic summary of his vacations there: "Same pension, same room, same fleas.")

Since no one could be sure when or where Hitler would strike next, Gunn had received authority to transfer RF personnel from endangered areas to safer locations in Europe without prior consultation with the New York office. Meanwhile, the families of officers stationed in Europe were given permission to sail home, whenever they wished, at Foundation expense. There was no rush to the boats. All but a handful of dependents chose to stay put for the moment, as did the fifty-eight fellows whom the Foundation was supporting in various European countries (including two in Germany).

There was only one important change in personnel. Tisdale left for New York in early April for what turned out to be permanent sick leave. He never returned to Paris, and Miller, who had been running the fellowship program, assumed responsibility for all NS projects in Europe. Miller and his wife and their son and daughter had been residents of France since 1934; they had settled in a large villa in the pleasant Paris suburb of St. Cloud. In the spring of 1939 this villa was designated a way station on the Paris–La Baule escape route; the Millers' large garage was stocked with hundreds of liter cans of gasoline.

On April 28 Hitler repudiated the nonaggression pact with Poland that he had signed in 1934, and the French government redoubled efforts to put the country on a war footing. At the beginning of May the French army was authorized to requisition civilian automobiles for its own use. When Mike Gunn requested an exemption for the six cars belonging to the Foundation's American employees, his request was granted, in specific recognition of the RF's longtime support of French science. But there was no exemption from the mounting fear of German attacks. Gas masks were issued to the entire staff, with detailed instructions on how to use them. It was agreed that officers should refrain from extended trips for a while, that no major new projects should be considered, and that all fellowship and grant-in-aid activities should be maintained at the present level.

Having completed all preparations that seemed advisable for the moment, the Paris office of the Rockefeller Foundation, like the rest of western Europe, waited for Hitler's next move. When the blow did not fall immediately and the approach of summer found Hitler spending more time in his Berchtesgaden retreat than in his Berlin chancellery, there was a natural tendency to believe that perhaps the worst was over. By June, Gunn had relaxed his restrictions against international travel, and O'Brien paid a brief visit to London, where Mellanby bent his ear about the politicians' latest insult to medical research: while rejecting his plea for a modest £2,000 supplement to the general research fund (on the grounds that all available money was needed for rearmament), Parliament had authorized £1 million for radium treatment of cancer—a politically popular but scientifically unproved innovation—without so much as consulting the MRC.

Still, the mere existence of politics-as-usual at Westminster could be taken as a sign that war was not imminent. On June 12 Gunn wrote Fosdick that the feeling in London and Paris was that "war, if it should break out, would not come until [September] after the crops were all in." He added, "In Paris, where the sun has been shining the last few days and the 'season' is on, there seems to be a general feeling of optimism."

By the last week of August, not even the most determined optimist could find anything encouraging in the news. The Soviet Union and Germany announced a nonaggression pact on August 21. This development, so at variance with the ideological conflict between the two countries, caught the Allies by surprise. But the consequences for Central Europe were clear enough: the Germans now had a free hand to deal with Poland as they had with Czechoslovakia. On August 22 the British reaffirmed their pledge of solidarity with Poland. With German armies already massing on the Polish border, a military response by Britain and France seemed inevitable.

Ernst Chain, who had been vacationing in Belgium, cut short his trip and hurried back to Oxford. In Paris the American embassy urged Americans who had no compelling business abroad to return to the United States. Prof. Henri Laugier, head of the Caisse Nationale de la Recherche Scientifique (CNRS)—the French equivalent of America's National Research Council and the British Medical Research

Council—advised the Rockefeller Foundation to shift its operations to La Baule without delay.

Gunn resisted the suggestion, but he took every precaution short of actually closing the Paris office. Two secondhand trucks were purchased, and more cans of gasoline were stored at La Baule. All staff members were given the opportunity of transferring immediately to the villa there, which could sleep twenty (counting sofas). On August 26 a truckload of files and two secretaries made the trip to La Baule. That same day Miller's wife and their two children sailed for New York on the liner *Roosevelt.*

By August 29 the French were no longer permitting long-distance telephone calls in foreign languages; even American embassy staffers were cut off if they spoke in English. Most of the vital RF records were already safe in La Baule; the remainder were piled in the hall outside the Paris office, waiting to be taken down by truck or private car. Although public transportation was becoming capricious, several staff members managed to get to La Baule by train. In Berlin, Hitler gave the Poles twenty-four hours to submit to his demands.

On Thursday, August 31, O'Brien drove to La Baule in his own car. The next day, while news reports told of the German invasion of Poland, the remaining RF officers, including Gunn and Miller, led a four-car caravan out of Paris. They spent the night at Miller's St. Cloud villa, and arrived in La Baule on Saturday morning. Sunday at 11 A.M. everyone gathered around a radio to listen to the announcement that France and Britain had declared war on Germany. The Americans stood in stunned silence; the French secretarial and service staff wept openly.

# 34

# Embattled

In the first days of the war, the British government evacuated more than a million civilians from "danger zones" throughout the country in what the press referred to as "the greatest exodus since Moses." There were also contingency plans aimed at safeguarding the nation's scientific and industrial strength. Most laboratories in London were moved to locations considered less vulnerable to air raids or invasion. In Oxford the Sir William Dunn School of Pathology became host to a Health Ministry unit that was supposed to do diagnostic work if epidemics broke out in the south of England.

Whether Florey's personal timetable was affected in any way by the onset of the war is not clear. He said repeatedly in later years that he never for a moment considered the penicillin effort mere "war work." So it may have been only coincidence that the formal request for MRC support of his laboratory's "antibacterial" investigation, which he and Chain had been polishing for weeks, was in the mail on September 6, three days after Britain entered the war.

In this application Florey reminded Mellanby that there existed "no really effective substance against bacteria in vivo"—that is, in a living animal as opposed to a test tube or a Petri dish: "I have long had the feeling that something might be done along these lines." He proposed to inject animals with pure extracts of substances like penicillin, which Dr. Chain was now preparing, in order to "study their action as antiseptics." What he and his coworkers needed immediately was £100 for equipment and supplies, and some assurance that the project met with official approval.

Mellanby replied within the week: "It seems to me the new line of work will be interesting." But money was tighter than ever. The MRC was prepared to provide no more than £25 for expenses in the current year; future requests would be considered sympathetically as the project matured.

This response seemed calculated to confirm Chain's view that government bureaucrats could not be trusted. Of course there was a war on. But it was government policy—vigorously espoused by Mellanby himself—to keep the nation's most qualified researchers at work in their own labs. For a wartime government to withhold a paltry £75 from a major laboratory about to embark on a major study of antisepsis seemed the height of folly to Chain. Yet, as Florey knew from his earlier experiences with the Nuffield committees, private philanthropy was not necessarily more receptive to needy scientists. As if to prove this point, Florey applied for a grant to the same Nuffield committee that had turned an unsympathetic ear to his request for Margaret Jennings's stipend. Once more, he was turned down, apparently because Lord Nuffield's scientific advisers concluded that the work on penicillin had little chance of success.

With Chain's three-year British Empire Cancer Campaign grant coming to an end, it was imperative to assure his salary for the year, since without Chain there could be no investigation. At Florey's urging, Chain applied to the MRC directly—the first time he had approached that agency in his own name—and was awarded a stipend of £300 for the year, plus £100 in expenses, for his work on "naturally occurring antibacterial substances."

But no sooner had one weak spot in the laboratory's financial ramparts been plugged than others began to give way. Although Oxford was filled with evacuees from other parts of England (especially London), undergraduate enrollment for the academic year 1939–40 was expected to drop. This meant a shortfall in all departmental budgets that depended on student fees. In the reshuffling of priorities that was bound to come, Florey had no confidence that his lab would even hold its own.

Of more immediate concern, however, was the plight of Norman Heatley, whose one-year fellowship (awarded by the MRC, funded by the RF) for study in Denmark had been postponed. This left him stranded in more ways than one, as Florey explained to Miller in a typically wry turn of phrase: "It is very unfortunate for Heatley that he cannot carry on his programme as he has been left through no fault of his own without any source of income for the last six weeks, which he finds something difficult."

Heatley's unfortunate situation was hardly unique. In fact, Miller and the other circuit riders had spent most of September trying to untangle the mess that the war had made of the fellowship program. The key decision—to cancel all new appointments of fellows and all sailings of previously appointed fellows from America to Europe or from Europe to America—had been made in New York on September 5 and conveyed to La Baule via cable that same day. But the cable did not arrive until September 7. This was typical of the near-breakdown in communications in the days just after France entered the war.

Transatlantic telephone service was suspended. Although Gunn kept up a steady stream of telegrams and letters to New York, advising Fosdick of his actions and asking for guidance, he had no way of knowing which, if any, of his messages were getting through. Even when cables began to arrive in La Baule, they contained no reference to his letters. This was hardly surprising, since the letters Gunn wrote in early September were not delivered in New York until early October. Obviously, the first order of business was to reestablish contact between La Baule and the rest of the world—starting with Paris, where Henri Laugier of the CNRS had kept the RF offices from being requisitioned by the military through the simple expedient of having them assigned to his own organization.

A skeleton staff of four RF clerks and secretaries continued at 20, rue de la Baume, under the supervision of Alexander Makinsky, a White Russian émigré who had worked for the Division of Medical Sciences since 1926. Telephoning in France was no longer a matter of picking up the receiver and giving the operator a number. Permission for all calls had to be secured in advance from the proper authorities. With the help of Laugier and local officials, Gunn arranged for at least two phone calls a day between La Baule and Paris. As a gesture of appreciation, he presented the mayor of La Baule with 5,000 francs to help the town meet the expenses of caring for the large number of refugees who, like the RF itself, had fled to Brittany from cities considered to be prime bombing targets.*

Once the La Baule office was running with a semblance of effi-

*Later, Gunn contributed another 5,000 francs to the town relief fund.

ciency, the officers turned their attention to the many RF fellows whose plans had been stymied, in one way or another, by the war. Of the 302 fellows under appointment on September 3, 1939, roughly a third were just completing their year abroad; the rest either had been just about to embark, like Dr. Heatley, or had just arrived only to find their sponsoring laboratories closed or converted to war work.

Even before they received Fosdick's cable suspending the fellowship program, Miller and his colleagues had been trying to get in touch with all the far-flung fellows, if only to assure them that the Foundation was aware of their plight. Now all fellows in foreign countries, especially the fifteen Americans in Europe, were urged to go home. Those who chose this option needed help in booking passage on ships that were already crowded with their compatriots. By the end of October eleven of the American fellows had returned to the States to take up newly arranged posts at American institutions; similar opportunities were created for the seven American fellows who had not yet sailed when war broke out.

Other cases were not so easily resolved. A Czech physician whose homeland no longer existed was stranded in Switzerland, along with an Englishman who had waited too long to return home. The latter had to be supported by the RF until the end of the war. Two of the four German fellows in the United States decided to accept academic appointments there rather than go back to Germany.

In New York, meanwhile, Fosdick and his colleagues were trying to get a clear picture of the war's impact on the Foundation's balance sheet. Even before the first batch of wartime mail arrived from La Baule, the comptroller's office had prepared a report showing that, as of August 1, the unpaid balance of appropriations in Europe amounted to more than $4 million. Determining what part of these obligations would have to be postponed or canceled outright was the melancholy task assigned to the European officers.

In a letter to Gunn dated September 7, Fosdick summed up his thoughts about the RF's role in the crisis: "first, in a dark world to keep burning the candle of scientific research in the alleviation of human misery." For example, he envisioned the International Health Division and the Division of Medical Sciences combining forces to combat the epidemics of typhus and influenza that would inevitably come in the aftermath of war.

But, Fosdick went on, such programs could and should be administered from New York:

> I think of La Baule merely as a temporary retreat, and not as a permanent location during the war. . . . As soon as the immediate questions of fellowships and running grants are settled, I would imagine that we would of necessity have to face the drastic step of disbanding the French personnel and bringing the American personnel back to New York for reassignment.

To the officers at La Baule, these were fighting words. Although Poland lay prostrate before the German blitzkrieg, no one in western Europe was in immediate danger. Despite the press of urgent business—dispatching emergency funds to stranded fellows, securing the numerous government permits that were now necessary to use a phone or operate a car or open a bank account—the circuit riders' thoughts about their role in wartime Europe were pointing in an entirely different direction from Fosdick's. Miller assured Warren Weaver in a letter dated September 8, "On one point, at least, we are all in accord—the desirability of the RF carrying on in Europe some sort of a program. . . . The RF plays a far more important role in Europe than is appreciated by non-Europeans."

The same theme—that the Foundation should remain on the scene, shoulder to shoulder, as it were, with the people it had assisted through the years—was sounded again and again in the reports that Miller, O'Brien, and Gunn filed during the first few weeks of the war. Not only would the continuing presence of the RF bolster the morale of the European scholars, but also, by preserving the network of contacts built up over the last quarter of a century, the officers could keep abreast of developments and be ready to act quickly when fresh opportunities arose. Florey's penicillin proposal (still in the drafting stage at Oxford) would soon give Miller a chance to show the validity of this argument.

Even in France the war seemed like someone else's bad dream. Although a few air-raid alerts had been sounded in Paris, no bombs had dropped; and the rue de la Baume office was functioning smoothly, relaying incoming communications to La Baule (by phone or mail), and executing Gunn's instructions. In La Baule, the circuit riders were eager to hit the road again. Despite the influx of refugees and the recent arrival of British medical officers to set up military hospitals, the mood of the moment was best captured by one of the secretaries, an American woman, who wrote to a friend in New York, "It is so beautiful and peaceful at La Baule, which heightens one's everpresent sense of unreality."

The relative tranquillity of La Baule did little to allay the anxiety of the circuit riders, who felt frustrated by the policy of retrenchment that was clearly gaining support in the home office. Nor did the reopening of communications with Europe shake the conviction of Fosdick and his advisers that a full-scale office in wartime France, and especially in La Baule, was more trouble than it was worth. As Fosdick explained to Gunn in a letter in early October, there was no thought of the RF's backing out of Europe; the intention was to "center responsibility" in New York, with one or two officers stationed in England or some neutral country and "some sort of token or symbolic office in Paris under Makinsky." The International Health Division had just ordered its overseas representatives home. A decision affecting the other divisions could be expected soon.

Gunn, however, had already taken steps to show how wrong Fosdick was in his assumption that the officers in New York had a "clearer picture of the military and political situation . . . than anyone in a belligerent country can possibly have." The first step was to allow Miller and O'Brien to travel to Paris and London on fact-finding expeditions. The traditional division of labor within the European office was temporarily suspended, and each circuit rider was authorized to talk to any RF grantees he could find, no matter what their field.

Gunn was understandably cautious at first. Before Miller was permitted to drive from La Baule to Paris on September 13, he had to promise that he would sleep outside the city limits, just in case the Germans began their long-awaited air raids. Yet Miller's trip was most notable for the heavy traffic he encountered on the roads leading to Paris. Parking was as difficult as ever in the rue de la Baume. Public transportation—buses and the Metro—had recovered from earlier disruptions. Miller was pleased to report that life in Paris appeared quite normal, with the exception of a few closed shops, "people walking about with gas masks," and the blackouts at night—the last an especially depressing phenomenon in the City of Light.

O'Brien, meanwhile, had gone on to England, where he learned that construction of Professor Robinson's new building at Oxford would have to be postponed because of shortages of steel and labor. There is a brief mention of Florey in O'Brien's report, but the reference is so garbled that only someone already familiar with the British scientific community could have puzzled it out. Writing about the current state of "NS activities" in England, the circuit rider who had represented the Division of Medical Sciences in Europe since 1926 noted, "The activities of the M. R. C. have been transferred to Cambridge in considerable part and are going nicely, many of them in the laboratory of Prof. Florey, Pathology."

If O'Brien confused the Sir William Dunn Schools at Cambridge (Biochemistry) and Oxford (Pathology) he would not have been the first visitor to do so. But this was precisely the kind of mistake that no circuit rider on his *accustomed* rounds would have made. Read in this light, O'Brien's report could only reinforce doubts about the ability of the RF to conduct its business in wartime Europe without an experienced circuit-riding staff.

Not surprisingly, both Miller and O'Brien tended to underplay the obstacles they encountered on these wartime journeys. Even the long delays in obtaining permits and visas were cited as arguments for maintaining an RF office in Europe, since only representatives familiar with the local scene would know how to cut through the red tape with requisite speed. There is no question that the obstacles were formidable. On O'Brien's first trip to London, he crossed the Channel not on the familiar Boulogne–Dover steamer but on a small ferry of the type usually reserved for transporting cars; the reason for the switch, apparently, was to offer a smaller target to German submarines. On O'Brien's return trip, the English authorities confiscated virtually everything in his briefcase, including fellowship records, routine correspondence, even blank notepaper and personal photographs. He was assured that all the items that passed inspection would eventually be forwarded to the Foundation's Paris office.

During the last week in September, while the Russian and German armies raced to divide up Poland between them, the French government announced a new set of regulations governing travel within France. Once again, foreign nationals had to obtain new permits for themselves and for their cars; and once again, Henri Laugier intervened in the Foundation's favor. As a result of his efforts, the RF was granted privileges exceeding those granted even to the American embassy.

O'Brien did not miss this opportunity to point out, in words he knew would get back to New York, that the Rockefeller Foundation's "unique" position in France was based on more than its longtime role as a dispenser of funds. Others who handed out dollars, like the Carnegie Foundation, were not extended the same courtesies. It was the

RF's reputation for objectivity, its genuinely international character, that counted most in these hard times. If the RF were to withdraw its presence, he wrote, the French would consider it both "a national catastrophe . . . [and] an international disaster."

# 35

# Pure Science

After the fall of Poland, on September 27, western Europe drifted into a period of uneasy peace that became known as the sitzkrieg or "phoney war." Gunn began allowing clerical staff as well as officers to travel. For many it was a time to see family and friends who had been left behind in Paris. When Gunn and his wife visited Paris in early October to catch up on Foundation business and keep an appointment with their dentist, they found the city "very calm," although the streets were less crowded than usual and the nightly blackouts horribly depressing.

The French, like the English, had acted to insulate their scientists, and the academic community in general, from frontline duties.* But French scientists were not allowed to continue with their prewar research. In early October, Miller spoke with Boris Ephrussi, one of Europe's leading geneticists, a man who had received no fewer than three Rockefeller fellowships (from the IEB and the RF). He had been excused from active military duty to do "research work for national defense," the nature of which he was not at liberty to discuss; even the whereabouts of his lab was a secret.

*Between 1914 and 1918 half of the students at the Ecole Normale Supérieure, in Paris—the institution that traditionally supplies French universities with their most eminent professors—were killed in action. In addition, 560 graduates of this school who were already professors were mobilized; 119 of these were killed.

Since he obviously could not continue the work on which his most recent RF grant was based, he asked Miller to stop all payments until the war was over. Finances at the moment were the least of Ephrussi's worries. His main concern was the preservation of the living material that he had collected and bred painstakingly over the last few years: fruit flies (of the genus *Drosophila)* and maggots (of the genus *Calliphora).* If something happened to his collection, he would be forced to start again from scratch, and the results of his recent labors would be lost to science.

Miller assured Ephrussi that he could use the money remaining in his RF account, some 6,000 francs, to pay an assistant to tend the most important strains of *Drosophila.* As for the maggot larvae, the only man to whom Ephrussi felt he could entrust this valuable material with confidence was George Beadle of Stanford University, in whose lab Ephrussi had worked during his most recent fellowship. Miller agreed to arrange for shipment of a thirty-kilo box of dried maggots to Stanford, the cost to be charged to the unexpended portion of Ephrussi's grant.

Meanwhile, at the rue de la Baume office, functionaries of the Caisse Nationale de la Recherche Scientifique had taken up desk space alongside Makinsky and his staff. A deal was worked out with the government; the CNRS would share the services of the two charwomen and pay half their salaries. Daily telephone calls between Paris and La Baule remained the rule, but there were fewer and fewer urgent matters to discuss; toward the end of the month, Gunn stopped passing along to New York a carbon copy of the daily phone log.

In La Baule the most pressing question was how to heat the Foundation's villa, which had five fireplaces but no central furnace. Although the weather was still pleasant, the house had been equipped only as a summer residence; and despite Fosdick's misgivings Gunn and his associates had every intention of wintering in La Baule, a point they made abundantly clear in their letters and memos to New York.

While all the circuit riders agreed that a full Foundation presence in Europe was necessary, each officer tended to see the issue from his own perspective. Gunn, who had been in China on RF business when the Japanese invaded Manchuria in the summer of 1931, kept stressing the relative calm of the situation in western Europe. O'Brien had visions of a vast program to combat wartime epidemics among French civilians. Robert Letort, the French national who served as comptroller of the European office, argued that even with the cancellation of some programs, this was the worst possible time to shift financial control to New York. With all researchers subject to mobilization, cooptation, or worse, close oversight by knowledgeable officers was more important than ever "to be reasonably sure" that RF funds were "used for the purposes indicated."

Miller reminded Weaver that the circuit riders were "playing ball in each division with only the most prominent and distinguished investigators." He saw no way that a New York–based envoy without benefit of a seasoned European backup staff could possibly handle "the business of all divisions."

It was not until October 26 that Miller was able to fly to England to look into NS projects there. The first bombs of what would come to be known as the Battle of Britain had already fallen, on Edinburgh, and fear of air raids was widespread. Yet, despite the rigidly enforced blackouts, the "extensively sandbagged" buildings, and people everywhere carrying gas masks, he reported that the spirit of the British seemed unimpaired.

At the same time, even after a series of interviews with highly placed informants, he found it impossible to come up with the specific project-by-project forecasts that the New York office had requested. Some laboratories, like the Cavendish in Cambridge, had lost the bulk of their staffs to war work; others, especially those devoted to the biological sciences, were maintaining a semblance of prewar order. Even in those labs, however, priorities had changed. For example, instead of synthesizing scientifically interesting proteins, as he had proposed to do, Professor Robinson now wanted to use his most recent RF grant to synthesize "medically important" hormones.

On November 1 Miller dropped in on Florey at the Sir William Dunn School of Pathology at Oxford. He found the laboratory crowded with evacuees from London institutions. In a brief conference Miller and Florey resolved Norman Heatley's financial problems.In lieu of the young man's aborted traveling fellowship, the RF would pay him a stipend equivalent to his Oxford salary for one year. When the twelve months were up, Professor Florey would "take care of him" as before.

But Florey could hardly take care of Heatley unless the future of his laboratory was assured. He confided to Miller that he had been "just on the point" of asking the Foundation to support a major new investigation of "naturally produced bacteriological inhibitors . . . [that] stop in a striking way the growth of staphylococci." Florey was not yet able to say exactly how much money he would need, but he estimated that £1,500 a year for three years would be sufficient.

In a detailed memo dispatched to the RF's New York office a few days later, Miller paraphrased Florey's words as follows:

> In spite of the uncertainties of the war situation he felt that the work could go on unless Oxford were almost totally destroyed by bombing; and felt that he would have to have assurances over a three-year period as it would be practically impossible to secure the necessary trained personnel on a year to year basis.

Miller concluded with this qualified endorsement: "F. is a distinguished young investigator who has the full confidence of Mellanby, and HMM feels that his request if and when received should be given serious study."

RF officers customarily muted their enthusiasm in official correspondence. But from the special attention Miller gave this report, Weaver must have guessed that his Paris-based circuit rider, a trained bacteriologist, was more than a little excited over what he called "this bacteriological inhibitor business."

It was standard procedure during this period to send one copy of each important memo to New York via the regularly scheduled Pan American Airways Clipper, and a backup copy by boat. Clipper mail generally made it across the Atlantic in a week. But even before Miller returned to Paris, he summarized the high points of his trip in a cable that he wired to Weaver from London on November 3. He knew that Weaver was preparing for the annual trustees meeting, set for December 5–7 in Williamsburg, Virginia, and he knew that the sooner he alerted Weaver to a proposal like Florey's, the better would be the chances of getting it funded. In a 141-word cable otherwise devoted to news of far more eminent members of the British scientific community (such as Robinson, Mellanby, and Sir Henry Dale, president of the Royal Society) Miller used up twenty words to say, "FLOREY OXFORD WILL SUBMIT TO YOU ME THREE YEAR PROPOSAL ON ENZYMES AS BACTERIA INHIBITORS ANNUAL MAXIMUM POUNDS FIFTEEN HUNDRED."

It is significant that in his long follow-up memo, Miller mentions the war only as an inconvenience to the Oxford project. The United States was technically a neutral country, and in their conversation at

Oxford, Florey had apparently gone out of his way to play down the wartime potential of his research. He had assured Miller that the work he contemplated would not be "secret" in any sense and that the results could be freely published—which was the sine qua non of pure science.

Of course, from his earliest dealings with O'Brien, Florey had reason to suspect that a "medical" pitch for experimental pathology would fall on deaf ears anyway. By casting his proposal in terms of pure science, he was simply telling the decision makers in the Division of Natural Sciences what he assumed they wanted to hear.

In fact, there is evidence that Miller helped Florey shape his proposal along these lines. In an oral reminiscence that Florey taped during a visit to Australia in 1967, he recalled his meeting with "the Rockefeller man" in the fall of 1939. What he especially remembered was how Miller had "raised his eyebrows meaningfully" when the subject of bacterial inhibition came up; and how he had indicated that while the Foundation was swamped with medical projects at the moment, a request framed in "biochemical" terms might have a better chance. Chain remembers Florey telling him much the same story.

On November 20 Florey and Chain mailed copies of their formal "Application to the Rockefeller Foundation for a Research Grant" to both Paris and New York. In a covering letter Florey described the proposal as growing naturally out of the "biochemical work" that his lab had been pursuing over the past three years with some success, despite a chronic lack of funds. "With the outbreak of war," he added, "these material difficulties have naturally not been lessened."

But he was quick to point out that the department was intact and ready to carry on its long-range mission of applying "Biochemistry to Pathology and Bacteriology." After praising the abilities of Dr. Chain, "the guiding chemist" of the lab, and noting that the proposed investigation was "in no sense a 'forced' plant" (a reference to war-related research), Florey tacked on this sentence as if it represented a minor, though possibly not unwelcome, consideration: "It may also be pointed out that the work proposed, in addition to its theoretical importance, may have practical value for therapeutic purposes."

# 36

# Foundation at War

Having resolved to try their luck with the Rockefeller Foundation, Florey and Chain figured they had nothing to lose by aiming high. As Florey later put it, they decided to "ring the bell" for assistance rather than go for the dribs and drabs they had been accustomed to.

The first section of the grant application, written by Florey, recapitulated in glowing terms the biochemical work (snake venom and lysozyme) that the lab had carried out since 1936, the date of the last RF grant.

The second section, entitled "A chemical study of the phenomenon of bacterial antagonism," was drafted by Chain. After a brief rundown of microorganisms (such as lysozyme and "the mould penicillium notatum") known to produce antibacterial substances, he drew attention to the fact that these substances had "been hitherto very little studied." Nevertheless, evidence suggested that they were nontoxic to mammals yet highly destructive to such pathogens as staphylococci, pneumococci, and streptococci: "They seem therefore to possess great potentialities for therapeutic application." In defense of this statement, Chain cited recently published work by René Dubos of the Rockefeller Institute in New York, who had isolated a substance from soil bacteria that protected animals against pneumococcal infections.*

*Unfortunately, later investigations by Dubos and others showed that this germ-killing substance was too toxic for internal administration, although it was used with some success in the topical treatment of infected wounds.

The goal of the entire investigation, according to Chain, was to "study systematically the chemical fundamentals of the phenomenon with the aim of obtaining in purified state, and suitable for intravenous injection, substances against various kinds of pathogenic microorganisms."

In the third and last section of the application, Florey got down to numbers. He asked the Foundation for £5,000, to cover three years of research. He proposed to hire another "fully qualified biochemist," two more technical assistants, and a mechanic. Total RF-supported wages: £1,170 annually. To keep his researchers supplied with chemicals, glassware, and similar necessities, he would need another £500 per year.

In addition, because of the "inadequacy of present technical assistance," his lab had pressing needs for basic equipment that he hoped the Foundation would supply. Under the heading of "non-recurring equipment grant," he appended a shopping list for sixteen items with "approximate" price tags. These ranged from a "fast centrifuge" (£150) to an "ice-crushing machine" (£10). Inflation might increase the total, he said, but he was sure he could make do with an equipment fund "not exceeding £1,000."

By the time this proposal reached La Baule, Miller was getting nervous about the upcoming trustees' meeting. If the officers in New York were leaning toward a reduction of the RF presence in Europe, the trustees might prove even more "isolationist." This was, after all, a period of intense agitation in the United States against American involvement in "Europe's war." As if to allay his own anxiety, Miller sent Weaver several reminders during the last weeks of November that Florey's proposal was expected "soon."

On December 8 Miller got his first inkling of Weaver's reaction. In a long letter that touched on a broad range of Foundation business, Weaver included this paragraph: "The application of Florey appeals to me, but I seriously question whether a three-year grant is justified under present circumstances. Could this not be handled on a year to year basis?"

From the context it is clear that Weaver was responding not to Florey's formal application (which was probably still en route to New York) but to Miller's memo recounting the discussion in Oxford on November 1. Almost certainly, then, Weaver did not see the full proposal before the trustees met in early December. In normal times the board would not have discussed such a small grant anyway, but these were hardly normal times, and if Weaver had had Florey's application in hand, he might very well have alluded to it in his remarks, entitled "NS and the European War," which he presented at Williamsburg.

In any case, despite his plea for patience on the part of the New York office, Miller himself was impatient to get the Florey grant rolling. The day after Weaver's letter arrived, he dashed off a "first reply" in which he raised no objections to handling Florey on a year-to-year basis so long as "F. had informal assurance that the second (and third) year support would be forthcoming if conditions at Oxford were not radically changed."

Miller even tried to get over to England for a quick meeting with Florey. When that proved impossible, he wrote to Oxford acknowledging receipt of the grant application, asking for clarification of some minor points (Had the university withdrawn funds from Florey's department recently? What was Dr. Chain's previous stipend?) and apologizing for not being able to discuss these matters in person. As for the possibility of a three-year-grant, Miller tried to let Florey down easily. While the Foundation was not prepared to offer "binding assurances," Miller spoke about "reasonable assurances for continued support over the desired period of time"—contingent on an annual review of his progress and of conditions in Oxford.

Miller sent a copy of this letter to New York, along with a memo in which he suggested a departure from standard RF procedure. Could the initial grant to Florey start on January 1 and run nine months instead of the usual twelve? This would bring future annual grants (assuming they were approved) in line with the British academic calendar, which began in the fall. Miller explained, somewhat wistfully, "I suppose I am trying to look too far ahead to the time when the war will be over, and most appointments will be again made for a year from the beginning of the academic term."

Then Miller added a personal note. Admitting that he felt in "need of a little holiday," he said that he hoped to get away for "10 days—2 weeks around Christmas." But he assured Weaver that he would keep in touch with La Baule by phone and that the moment he was informed that New York had approved the Florey action, he would telegraph Oxford so that Florey "could go ahead with the appointment of personnel."

Finally, Miller reminded Weaver that he and his colleagues were "looking forward eagerly" to news of the trustees' decision concerning a strong RF presence in wartime Europe.

The trustees had, in fact, approved the basic outline of a "war program" at their Williamsburg meeting, but it fell far short of the large-scale commitment to the European intellectual community that the Paris staffers had been arguing for. The Board cited two examples of the kind of emergency work that the Foundation hoped to tackle in wartime Europe. One was a grant of $51,250 to keep the London School of Economics in business, despite its evacuation from Cambridge and its loss of substantial income from international students. The second was Miller's assistance to Boris Ephrussi in shipping his maggots to Stanford University.

The Board went on to acknowledge the importance of "intimate and continuing contacts with European scientists, universities and laboratories." And the circuit riders were given greater freedom to commit grant-in-aid funds (up to $5,000 per project) without regard to divisional boundaries and without prior approval from New York. At the same time, the Board adopted a statement that noted, "necessarily, this will have to be a modest program. No large sums are contemplated." Indeed, most of the money released for new grants-in-aid came from funds earmarked for the moribund fellowship program. As if to underscore the difference in attitude between headquarters and field staff, the La Baule office was informed that, until further notice, all grants in Europe were to be strictly limited to one year.

In truth, despite attempts to patch it over, the rift between the European and the New York offices of the Foundation was growing steadily wider. Although Fosdick and Gunn corresponded on a first-name basis ("Ray" and "Mike"), each found it hard to understand the other's reaction to the terrible events that were overtaking them. In October, Gunn had written,

> It is not easy for the Foundation to be strictly neutral, and it is, indeed, impossible for the individual members of the staff to be neutral. Personal convictions are so strong when England and France are fighting for ideals which are necessarily ours. This of course does not mean that we would engage in any activities which are definitely related to the conduct of war. But it is another matter to say that we should not be involved in aiding activities which are related to a successful peace, if such a thing is possible.

How to draw this fine line was precisely the issue that worried Fosdick. In a letter to Gunn dated November 30, he expressed gratitude for the reports from Europe, which were now "coming in thick and fast." But, after thinking over what he had read, he felt the need to sound a note of caution. If the Foundation hoped to remain "an international force for the future," it was absolutely necessary to maintain "detachment and objectivity" during the war:

> We have our own personal sympathies as officers, and I assume that they are on the side of the Allies, but the causes and ideals which the Rockefeller Foundation is trying to serve transcend the present conflicts. I say this because we shall be under pressure from time to time to support enterprises whose real aim, if carefully and shrewdly analyzed, is to further a war program . . . and we ought to be quick to detect such disguises.

Two weeks later he thought he had detected just such a Trojan horse. Pat O'Brien was busy planning an ambitious program to improve "infant and maternal health" in France. As he had outlined in a letter to his divisional chief Alan Gregg, the program would involve the cooperation not only of civilian public-health officials but of the French military authorities as well. Fosdick's frank assessment:

> This seems to me too much like propaganda. It would be notice to the world that the Rockefeller Foundation is fighting on the side of the Allies. . . . Isn't this an attempt to secure from us a dramatic public expression of good will and backing? We are not interested in such expressions. We are interested in maintaining scientific standards and in preserving values that are apt to be snuffed out during the war.

Fosdick advised Gunn to rein in O'Brien before the French jumped to conclusions about the chances for trustee approval of such a program—which he hinted were exceedingly slim.

When Fosdick spoke of the importance of maintaining the Foundation's integrity as an "international force," he was only quoting the circuit riders' arguments back to them. He devoted some eloquent paragraphs to the same theme in his *President's Review* for the 1939 *Annual Report.* Not even the most casual reader could have mistaken the target of Fosdick's outrage when he gave examples of the "intellectual blackout" that was descending on Europe: "The University of Warsaw has ceased to exist . . . the University of Prague has been shut by the German government . . . the 20,000 student population of the University of Paris has shrunk to 5,000." And yet Fosdick did not explicitly blame the Nazis (as he would in the 1940 *Annual Report*). For the moment, War was the prime enemy.

Speaking directly to an American readership (which presumably included some influential people who were trying to keep the Untied States out of "Europe's war"), Fosdick took pains to explain why the Foundation had become involved with European scholars in the first place. He alluded to the preeminence in mathematics of the English, the Russians, the French. He pointed out that the recent "realization of the old dream of the alchemists"—the transmutation of elements through "splitting the atom"—was based on results obtained by "the Italian physicist Fermi" and "by Hahn and Strassmann of Berlin." And in case anyone wanted further proof that good science depended on international collaboration without regard for "flags and boundaries," Fosdick cited "the amazing development and application of sulfanilamide—that beneficent gift to Mankind."

He noted that the first, crude preparation of this powerful germ killer, known originally as prontosil, emerged from work in Germany, "oddly enough in connection with the commercial dye industry." Fosdick went on to tell the story in brief:

> In 1935 a German scientist—Dr. Gerhard Domagk—published the results of his experiments with mice under carefully controlled laboratory conditions, showing the extraordinary effect of prontosil on streptococcus. The Pasteur Institute in Paris then picked the matter up, and subjecting prontosil to organic analysis discovered that its activity was localized in one distinctive part of its molecular structure. This potent factor in prontosil, separated from the rest of the molecule, is what we now know as sulfanilamide. At this point Queen Charlotte's Hospital in London, with a grant from The Rockefeller Foundation, tried the drug on women suffering from streptococcal infection associated with puerperal or childbirth fever, immediately reducing the death rate from such infections by 25 per cent. The Johns Hopkins School of Medicine was the next institution to carry forward the exper-

iments, and in the last three years research on this drug has been developed, with brilliant results, in laboratories and hospitals on both sides of the Atlantic.

When he wrote that paragraph, Fosdick was probably unaware that the RF's Division of Natural Sciences was considering a grant proposal from a scientist at Oxford who wanted to investigate an entirely different group of antibacterial substances. On December 18 Florey had responded in detail to Miller's questions about the funding of the Sir William Dunn School of Pathology. About half his income, he said, came from the university; the "Dunn Endowment Fund" provided another quarter, and the rest was made up of student fees and and small grants that Florey said he "found by shaking a hat in all possible directions." The most important source of outside funds was, of course, the Medical Research Council.

By the time this reply reached Miller, he was vacationing in the Pyrenees, in the small ski resort of Alpe d'Huez. On the last day of 1939 Miller sent a memo to Weaver, expressing his satisfaction with Florey's clarifications. Although he conceded, "There is no 'fire', and a little delay will not ruin anything," he went on, "For some reason I am anxious to get him started as of January 1." To this end, he again suggested bending Foundation rules. If Weaver would cable his approval at once, Miller would immediately inform Florey that the grant was operative, and the paperwork could be completed later. He was aware that this meant Weaver "would be granting money without having seen the action, which is unorthodox, or impossible from [the comptroller's] standpoint. However, there *is* a war on."

In case this plan was unacceptable, Miller said, the moment he returned to La Baule, he would send Weaver draft copies of several alternative "actions" covering still other possibilities, such as a nine-month or a full-year grant commencing February 1. Gunn, he noted, had already given his blessing to this unorthodox procedure. If the Pan Am Clipper functioned as usual, all these papers should arrive in New York before the end of January, and Gunn—who had been called to the home office for consultation—could add his signature there.

Gunn's trip had come up quite suddenly. On December 15 he had received a cable from Fosdick that began, "NEED OF CONFERENCE BETWEEN NEWYORK AND PARIS ACUTE STOP SUGGEST YOU RETURN TO NEWYORK AT ONCE ARRIVING HERE NOT LATER THAN FIRST WEEK IN JANUARY." The wording had been worked out at an "emergency conference" in Fosdick's office attended by all the senior staff in New York. The consensus was that something had to be done to bridge the growing "chasm" between the viewpoints of New York and La Baule. If the heavy traffic of letters, cables, and memos across the Atlantic had not done the trick, the only alternative was face-to-face consultation. Fosdick had hoped to visit Europe early in 1940, but Chairman John D. Rockefeller, Jr., had vetoed that plan, suggesting that this was a poor time for "the pilot to leave the pilot-house." If Fosdick couldn't go to Gunn, Gunn would have to come to New York, to defend as best as he could the position of the circuit riders.

# 37

# Legends and Luck

Gunn and his wife were uneasy about traveling together by air, and Fosdick did not insist that they try for a Clipper reservation. The first boat they could get passage on was the Italian liner *Rex,* leaving Genoa on January 2, 1940. At Fosdick's request Gunn assembled, with the help of the circuit riders, a comprehensive report on the "dislocation of universities and laboratories" in Europe. From his ski resort in the Alps, Miller assured Gunn that Weaver already knew everything that he (Miller) did about NS projects in Europe.

Miller's attempts to expedite the grant to Florey (by bending the Foundation's rules a bit) proved unsuccessful. When it became clear that neither the money nor word of a firm commitment could be in Florey's hands by the new year, Miller sent an apologetic note to Oxford, asking Florey to be patient and assuring him that his application was being looked on with much favor.

On his return to La Boule, Miller was chagrined to learn that *all* the options he had proposed to Weaver exceeded the $5,000 limit that divisional officers were allowed to authorize without formal approval by the Board of Trustees or the Executive Committee. On January 19 Weaver cabled Miller that the Executive Committee could not possibly act on the Florey proposal until late March. "THEREFORE AND ALSO BECAUSE FREQUENTLY ADVANTAGEOUS FIRST TRY OUT SUCH PROJECT ON MODEST LEVEL SUGGEST ADJUSTING AMOUNT TO COME WITHIN GRANTINAID LIMIT." At the current rate of exchange—about four dollars to the pound—this meant a grant of no more than £1,250.

Informed of this turn of events by Miller, Florey could not resist noting in reply that his lab "could have utilized with profit the sum we asked for." Nevertheless, he proceeded to pare his proposal down to size, chopping £700 from the "capital expenditures" list, asking only £300 for "running expenses such as chemicals, etc.," and dropping his request for another trained biochemist (£600). He assured Miller that even with this bare-bones budget, the research program could proceed as outlined, although not "at quite the same rate as with the original scheme."

On the morning of February 19, 1940, the officers in La Baule—including Gunn, who had returned from New York a few days before—approved Florey's grant for £1,250 or $5,000. The Florey action was the largest of the eight approved at this meeting. Miller had every reason to be pleased, but he didn't have much time to savor it. That same afternoon he left La Baule, together with O'Brien, on the first leg of a trip to look over RF interests in Holland. It would be his last fact-finding tour in Europe. With Tisdale unable to return to France, and with Fosdick unwilling to let Weaver leave the New York office even for a brief time, it had been decided to call Miller home for a few weeks of conferences. He sailed from Lisbon on March 15.

Before he left, however, he got off a letter to Florey to let him know that the money would be available to the Oxford laboratory as of March 1. Miller warned that the grant might have to be terminated before the year was up if "for any reason the work [was] seriously interrupted" because of the war. But in the action itself (which Florey did not see) Miller emphasized that the NS officers were prepared to give "sympathetic consideration" to extending the grant for another two years if results warranted and conditions in England permitted.

As might be expected, the arrival of Miller's letter was greeted with jubilation in Oxford. As Chain remembered it, this was the decisive moment in the sequence of events that led to the development of penicillin. "There was no race, not a race to produce penicillin for the war, not a race against anyone else," he told a journalist many years later. "If anyone else had been working on the subject I would not have been interested, it was getting the whole field opened up that interested me, and the Rockefeller grant was the most important thing for that."

No one questions the accuracy of Chain's memory of his own feelings at that crucial moment. But in a memoir published in 1971 in the *Proceedings of the Royal Society,* he went on to record his impression that the Foundation promised the Oxford laboratory $5,000 a year for five years, "which at that time seemed royal generosity to us." This is a curious misstatement.

Even $5,000 for one one year—which in fact is what Miller's letter offered—would have seemed generous to Chain and Florey, since this sum came to one-fourth of the lab's income for the preceding year. And it is true that the grant *was* renewed in subsequent years, just as everyone had hoped. Was Chain, with the advantage of hindsight, simply conflating all these grants into one delicious moment of triumph for the sake of a good story? The possibility cannot be discounted. Yet, in the introduction to his memoir, Chain goes out of his way to decry the buildup of "romantic legends" surrounding the discovery of penicillin, and he declares his intention of replacing legends with the truth—"a sequence of simple, sober and logical events."

In a biography of Florey published a year after Chain's memoir and five years after Florey's death, the RF grant seems to have grown even larger. This account (which is clearly based on Chain's recollections) makes no mention of the long process of negotiations with Miller. We are told that once the application was on its way to the Foundation, "Florey had no means of knowing whether or not the report had made the journey safely, and he tried to forget the attempt in the busy weeks ahead. It had been his habit for a long time not to give rise to hopes that could easily bring rejection and more despair." Then, weeks later,

> fortune suddenly touched Florey and lifted an enormous load from his mind, casting off the worry that had been a part of his life for years. It came in a letter from New York, and he could hardly believe it was true. . . . The Rockefeller Foundation had approved their application and would meet their request for aid—in full. . . . A magnificent annual grant of £1670—for the next five years. And with this, another £1000 to spend on initial gearing up for the push into penicillin. It was munificent and, as Chain observed, it was a "princely sum" for the time.

In this retelling the "five-year" grant has grown to the equivalent of $6,680 a year—"an assured income of more than £9000 . . . in one envelope," a windfall that gave the Oxford workers "new heart, and added impetus to their work."

This exaggeration of the size and duration of the initial RF grant runs through almost all accounts of the penicillin story, including Chain's obituary in *Nature.* Chain himself is almost certainly the source of the confusion. And the more closely one examines the error, the more curious it becomes. To begin with, Chain was a man who prized accuracy and attention to detail; indeed, a biochemist who is sloppy about details will have trouble getting replicable, much less epochal, results. Furthermore, Chain was known for his unswerving concern for financial security, a concern that eventually led to an acrimonious break with Florey and the rest of the Oxford group. Of all the errors of memory for Chain to make, confusing the first of a series of one-year grants with an "assured income" for five years is the least likely—unless he was intentionally kept in the dark about the terms of the RF grant.

We have seen how worried Florey was that his failure to obtain sufficient funding would lead to the breakup of his carefully assembled team. Unquestionably, Chain was the key member of this team. Ever since Florey's 1929 conversation with Szent-Györgyi, he had recognized the importance to his own career of a first-class biochemist as a collaborator. How far could he have gotten if Chain had left the lab before the penicillin work started?

Although the lysozyme investigation had been disappointing to Florey (because it led to no fresh insights into the body's immune system), Chain's elucidation of the enzyme chemistry had been a complete success. With lysozyme, and the earlier snake venom work, Chain was rapidly building a reputation of his own; and Florey, who always kept his eye on the main chance himself, had good reason to worry that his biochemist might defect to some other lab with better access to the dwindling research funds available in England.

Neither man was known for putting sentiment above ambition; and we have Chain's own testimony that their relationship, while intellectually stimulating, was never easy or especially warm. Even in their early days together, their discussions often ended, according to Chain, in shouting matches in which "the very walls of Florey's office would shudder" with the force of their arguments. Later on, they were often reduced to communicating by means of written memos. After the war Florey confided to another RF circuit rider that he considered Chain to be a troublemaker in his department. Not long after that, Chain left Oxford in search of the recognition and the financial rewards that he believed his accomplishments had earned him. But in 1940 Florey needed Chain, and both men knew it. How much would Florey have done to ensure his collaborator's loyalty?

This was, after all, a man already known as a "bushranger of research" and an "academic highway robber." Even his friends knew he could be "ruthless and selfish" when his plans were thwarted. Miller had previously been enthusiastic about supporting a three-year project. Could Florey, an ambitious and dedicated scientist about to begin a major investigation, have fudged the truth a bit to boost morale in his laboratory during those extraordinarily trying times? Did he inflate the Foundation's one-year grant into a multiyear commitment in order to tell his colleagues what they wanted to hear? The evidence is all circumstantial. But the suspicion that Florey was not always candid about lab finances can be confirmed from a later incident, also involving Chain and the Foundation.

In the summer of 1945 Chain was in New York on a government-sponsored mission having to do with declassification of wartime reports on penicillin research. He was still a member of the Department of Pathology at Oxford. During a conversation at the Foundation office, he and Miller discussed the status of the RF's most recent grant to the Oxford team—£1,200 for the year March 1945 to March 1946. Miller was surprised to learn that, according to Chain, the entire sum had already been spent. He immediately wrote to Florey, offering to arrange emergency assistance to tide the laboratory over until the next formal grant request, which, he suggested, should be less "modest" than the last.

Florey's reply was swift and strangely defensive:

> I am very distressed that Chain should have conveyed to you the idea that the grant which we have obtained from you was inadequate. . . . *Chain does not know how the finances of this department are run,* and as far as I am aware he has never at any time been held up because of our inability to find money for what he has required. There have naturally been difficulties in obtaining materials but that has had nothing whatever to do with the supply of money. . . . We are particularly indebted to you for having furnished us with materials and apparatus which are unprocurable at any price in this coun-

try. To say that the grant runs out before the end of February has conveyed to you a totally erroneous impression which I hope I have corrected. (Emphasis added)

When Florey composed this letter to Miller, he and Chain had been collaborating for ten years; their joint efforts, which had given the world the first true antibiotic, were about to be crowned with joint Nobel Prizes (announced one month later, on October 25). If at this time Chain was still in the dark about the finances of the Sir William Dunn School of Pathology, it is not beyond the realm of possibility that five years earlier Florey had stretched the facts when reporting the good news about the RF grant to his colleagues—even at the risk of being exposed as a liar at the end of the year.

Whatever tactics Florey adopted to keep his team intact, the precise wording of the RF grant was soon to be academic; within a few months Florey, Chain, and the others obtained exciting results that virtually guaranteed continuing support from the Foundation as well as other funding agencies.

With Heatley growing the *Penicillium* mold and testing the extracted "juice" for potency against microbes in Petri dishes, Chain was able to concentrate his energies on what he did best—separating the active substance from the rest of the juice and determining its chemical properties. He was prepared for technical difficulties, but his very first experiment produced a shock that might have ended the investigation then and there.

Working with a new assistant—Edward P. Abraham, who had moved down the road from Professor Robinson's laboratory—Chain discovered that penicillin was *not* an enzyme. As an enzyme chemist, he might have lost interest at this point; instead, he felt challenged to explain the strange results he had obtained. Although the material behaved in some tests like an acid of low molecular weight, it displayed the same instability in solution that had baffled all previous investigators—a combination of properties that was not consonant with the known laws of chemistry.

There was a more positive side to this puzzling finding. Despite his confident assertion in the RF grant application that antibacterial enzymes were "probably non-toxic to mammals," there was always a chance that the injecting of proteins into an animal would provoke a fatal allergic reaction. With a smaller molecule, however, this risk was much reduced—in which case the bacteria-inhibiting power that Fleming had observed years ago in a Petri dish took on much greater clinical significance.

Over the next few weeks Chain, Abraham, and Heatley redoubled their efforts to get a purified extract of penicillin. They played with temperature and pH levels; they moved the active substance back and forth from one solvent to another; they adopted the new method of freeze-drying that had just been introduced on a large scale in England to preserve blood serum. By the middle of March they had succeeded where all their predecessors had failed: they had a small quantity of brown powder that was relatively stable (both dry and in solution) and twenty times more active against microbes in vitro than the most active sulfa drugs known.

The production of this powder and the determination of its antibacterial strength were greatly facilitated by Heatley's invention of a "simple and quick quantitative test for the substance," thereby fulfilling Szent-Györgyi's prophecy that, with such a test, any naturally occurring substance could be isolated and purified. As it turned out, the brown powder that emerged from Chain's lab was far from pure. It contained about 1 percent penicillin and 99 percent impurities. But it was so effective as a bacterial inhibitor that the investigators were understandably fooled into thinking they had the unadulterated substance. The brown powder stopped the growth of staphylococci and other bacteria even in a solution of one part "penicillin" to *500,000* parts water.

Chain was so elated by these findings that entirely on his own (Florey was away from the lab for a day) he arranged for a toxicity test in live animals. This was essentially a replication of Fleming's tests on rabbits and mice more than a decade earlier, but the results were far more impressive. A healthy mouse weighing twenty grams showed no ill effects when injected with ten milligrams of the brown powder dissolved in distilled water. This was an enormous dose of biologically active material for a mouse. When Florey heard what Chain had done in his absence, he was furious at first. Chain was not supposed to butt into the biological side of the investigation. Then Florey repeated the experiment, and got the same remarkable results.

Subsequent tests on rats and cats confirmed that the brown powder, so deadly to bacteria in a dish, was apparently harmless to healthy mammals. Not until weeks later did the investigators learn how lucky they had been in this initial step in the development of penicillin as a chemotherapeutic agent. It turned out that their supposedly "pure" powder actually contained over thirty different impurities, any one of which might have proved fatal to an animal. As Chain put it, "It borders on the miraculous that none of the impurities was toxic enough to mask the non-toxicity" of the real penicillin.*

Oblivious to any danger, the Oxford investigators pressed on to the next logical step—the crucial step that Fleming had failed to take a decade early with his crude preparation of "mould broth." They set up an experiment to determine if their penicillin could cure sick animals.

*The allergic reactions sometimes provoked by penicillin itself, and other potentially harmful side effects of the pure drug, were not identified until years later.

# 38

# Evacuation

As Florey's team closed in on a major discovery, the war news suddenly took a grave turn. After months of "phoney war," German troops and naval forces massed along the Baltic coast within striking range of Norway and Denmark. Winston Churchill, the first lord of the admiralty, persuaded the British cabinet to let the Royal Navy mine Norwegian waters in an attempt to forestall a German invasion. But it was too late. By April 7, the German invasion fleet was already at sea; two days later, German troops began landing in Norwegian ports. On the same day, Hitler's forces occupied Denmark. By the beginning of May, despite the support of British naval units and an Anglo-French expeditionary force, Norwegian resistance was broken.

In La Baule the first reaction of the circuit riders was to try to sit out this crisis as they had the preceding ones. Miller was expected back in France in two weeks; no one saw any reason for him to change his plans. Having survived a bitter cold winter in a villa with unreliable plumbing and inadequate heat, Gunn and his office manager George Bakeman had spent a lot of time looking around for more suitable quarters and had settled on another villa in La Baule, not far from the old one. As Gunn explained to Fosdick, all the other choices (such as returning to Paris or relocating to the south of France) were either too expensive, too isolating, or too dangerous. In giving his reluctant approval, Fosdick betrayed a continuing uneasiness with the

European situation, which Gunn's "emergency" visit to New York had failed to dispel.

Underlying the transatlantic disagreements about specific matters (Where should the European office be located? How large should it be? What decisions should it be authorized to make without consulting New York?) was a deep divergence of opinion about the war itself. To many Americans at home, it was not at all clear that their country should become involved in the Allied struggle against Hitler's Germany. In the spring of 1939 Congress had rebuffed the Roosevelt administration's efforts to modify the two-year old Neutrality Act, which prohibited the sale of arms to belligerent nations outside the Western Hemisphere. And immediately after the start of hostilities in September, the United States had pointedly reaffirmed its neutrality.

Under continuing pressure from the administration, there was a slow erosion of the isolationist position, but debate continued over what role, if any, the United States should play in the armed conflict. It was not until November that Congress acted to permit arms shipments to European belligerents. As late as January 23, 1940, O'Brien was still trying to answer, for his boss Alan Gregg in New York, the "doubt in many American minds as what the Allies are fighting for." He was aware that his answer might seem simplistic to Americans who were looking for "hidden motives," but he gave it anyway: to the Allies the war was a struggle for "existence itself as interpreted by them before the outbreak of hostilities."

The "chasm" between the viewpoints of the New York and the European offices proved too wide and deep to be closed in two weeks of conversations. In all his meetings with Fosdick and Thomas Appleget, the Foundation's administrative vice-president, Gunn had refused to concede that the La Baule office was a bad idea. He *had* agreed to begin reducing the European staff, principally through early retirements. Once back in Europe, however, he apparently realized that letting go a large number of trained and loyal employees (mostly of French nationality) could be seen as a first step toward curtailment of RF activities in Europe; and he began to drag his feet.

Meanwhile, Fosdick's own doubts about the value of the La Baule operation had been reinforced by Alan Gregg, whose fact-finding trip to Europe (from mid-January to mid-March) overlapped Gunn's visit to New York. Gregg, whose Division of Medical Sciences focused on aiding research in psychiatry and neurology, was shocked by the "depression and tension" that he observed in the La Baule office. He concluded that the events of late summer and fall had had a far more severe effect on the morale of the European staff than anyone in New York could have guessed from reading the correspondence. This alone might account for the inability of the two offices to resolve their differences. His conversations with the circuit riders themselves led him to say, "Nobody but [Gunn] wants to stay at La Baule (for continuing which I see no sense whatever)."

Gunn's letters present an entirely different picture. With the unanimous approval of his colleagues, he had signed a one-year lease on the new villa. On May 6 he assured Fosdick that the new quarters were "a very great improvement" over the old; he also reported that "the morale of the French people" was holding up well and that there was "a little less tenseness at the moment" about the international situation. Looking ahead to the summer, he noted, "July would probably be taken mostly for vacation, and Carroll and I are planning to go to Vichy again if circumstances permit."

Four days later German armies invaded Holland, Belgium, and Luxembourg. In England, Winston Churchill replaced Neville Chamberlain as prime minister. The next day a French and British expeditionary force was dispatched to Belgium, where the Belgian army was crumbling before the German blitzkrieg. On May 12, German mechanized divisions crossed the Meuse River into France. Within a week the British and French forces fighting in Belgium were cut off from the rest of the Allied armies by a German thrust to the Channel coast. Their only hope of escape was by sea—from the beaches and harbor of Dunkerque, the last sizable port in their control.

The English were accustomed to falling back to their island fortress, outwaiting and outwitting would-be conquerors. But what chance would they have if the Allies' hastily improvised evacuation plan failed, and thousands of their best-trained soldiers were killed or captured before the defense of the homeland even began? It was in this atmosphere of almost unbearable tension that Florey and his coworkers readied their first test of penicillin's curative powers.

On May 25, a Saturday, eight white mice were injected with a strain of streptococci that causes puerperal fever. Four of the mice received shots of varying concentrations of Chain's brown powder dissolved in distilled water; the other four animals were left to fight the infection on their own. The next morning, when Florey returned to the lab, the four untreated mice were dead; the four who had received "penicillin" were alive (although two of them, who got smaller doses, died a few days later).

Norman Heatley later recalled that Florey's only comment on this occasion was "It looks very promising." Edward Abraham remembered Florey looking elated as he went over the results. But all those present agreed that Florey immediately ordered Chain to produce another batch of brown powder, and plans were laid for a comprehensive series of tests on some three hundred mice infected with streptococci, staphylococci, and one of the family of microbes responsible for gas gangrene, the most dreaded of all infections that breed in dirty battlefield wounds.

While Florey and his collaborators rushed to complete this series of experiments, the British and French managed to evacuate some 350,000 troops from the beaches and piers of Dunkerque, despite constant bombardment from German land and air forces. The Germans captured Dunkerque, and 40,000 remaining French troops, on June 4. The next day the main German army, led by tank divisions, opened its drive toward Paris. Ten days later the French capital fell.

From the moment the Germans attacked the Low Countries, the correspondence between Gunn and Fosdick took on a somber, fatalistic tone. Although Gunn kept writing that the French in general, and his staff in La Baule in particular, were "behaving splendidly," and although Fosdick kept assuring Gunn that New York had full confidence in his judgment, both men began alluding to the inevitable closing of the Foundation's offices in France and the establishment of a small "token" presence somewhere, perhaps in Lisbon, with Alexander Makinsky in charge.

Yet even now Gunn found reasons to postpone a decision. Among other things, he was waiting for Miller to return from the States (his Italian liner was due in Genoa on May 14) so that the matter could be aired at a full staff conference. On May 25, the day Florey and his coworkers first pitted their brown powder against a bacterial infection in a live animal, Gunn and his coworkers (including Makinsky, who had come down from Paris for the day) decided unanimously to keep the offices in France open.

Perhaps O'Brien, Miller, and the other circuit riders were simply unwilling to desert their friends while the fate of the Dunkerque evacuation hung in the balance. But while hoping for the best, the circuit riders prepared for the worst. As a "precaution," reservations for the remaining American personnel were secured on two passenger ships that were due to sail for New York in June.

By the end of the week, even Gunn had to admit that the worst was at hand. The successful evacuation from Dunkerque, while of great strategic importance in the long run, was hardly a victory for the Allies. Everywhere on the Continent Hitler was triumphant. What he coveted, he took; and what he wanted next was obviously Paris. On May 29 Gunn decided to empty the Paris office, leave Makinsky in charge of a skeleton staff at La Baule, and order all the Americans home. "It no longer seems reasonable to keep going as we have," he wrote to Fosdick. "I do not believe that my presence in France is needed much longer. . . . My difficulty is to sleep enough. How one's brain whirls in the night." The ever-ebullient O'Brien, Gunn also noted, "seems to think he will be back in Europe in a couple of months. I wonder."

O'Brien sailed on June 1 from Genoa. The Gunns and Miller left seven days later on the S.S. *Washington* from Bordeaux. Miller had been back on European soil less than a month. The voyage home was a long one, since the ship stopped at Lisbon and Cobh, Ireland, to pick up refugees; just outside Portuguese waters there was a delay while the captain tried to convince a shadowing U-boat that the *Washington* was an American ship, and therefore protected by the rules of neutrality.

The staff members who stayed behind in France had an even more harrowing time. On June 3 Makinsky motored from La Baule to Paris to help evacuate the remaining employees at the rue de la Baume office. Driving without lights through roadblocks and aerial bombardments and past long lines of refugees, he reached Paris the morning of the tenth, and left that afternoon with several passengers for the return trip to La Baule. One former secretary who was staying in Paris promised Makinsky that she would look in periodically at 20, rue de la Baume to check the mail; and the two loyal charwomen agreed to take care of the place until "further notice."

In the face of the German blitzkrieg, even La Baule no longer seemed like a safe haven, and Makinsky prepared for a further evacuation if necessary. Since gasoline was virtually unobtainable in La Baule, he drove to Robert Letort's summer home in Alençon to reclaim some 230 liters in tin cans that had been hidden under a woodpile in the garden.

Paris fell on June 13. When the French government announced four days later that it was suing for peace, Makinsky made his move. He declared the La Baule office closed, left a trusted employee of French nationality in charge of the villa, and headed south with his family and several RF staffers. In Bordeaux, a city crowded with refugees and people seeking exit permits to Spain and Portugal, the RF party survived a "heavy bombardment" that claimed two hundred lives. Through a combination of luck and his extraordinarily wide network of influential friends, Makinsky managed to get all the necessary exit permits by the evening of the twentieth.

Traveling without lights, Makinsky's car collided head-on with another automobile on the road out of Bordeaux, but the RF staffers were able to press on to Bayonne, where they had the damage repaired. In Biarritz, where they spent the night of the twenty-first, Makinsky found "elegant crowds" enjoying "cocktails and bridge parties," while at Hendaye, near the Spanish border, a line of five hundred cars waited to get across the International Bridge to safety. Makinsky and his party crossed into Spain on the twenty-second and two days later arrived in Lisbon, a city filled not only with French refugees but also with Portuguese tourists celebrating the eight-hundredth anniversary of their country's independence.

After being assured that there were no vacant rooms anywhere in the city, Makinsky ran into Professor Celestine da Costa of the University of Lisbon, a former Rockefeller fellow who had studied in England in 1934 under the auspices of the Division of Medical Sciences. Da Costa had been pulling strings to assure the smooth entry of Makinsky's party into the country. With his help, Makinsky found a hotel room for the night. The next day, June 25, he began making arrangements to set up the Foundation's wartime office.

# 39

# Almost Miraculous

In Oxford there was no longer any talk about the penicillin project's being an exercise in "pure" science. No one could predict what role penicillin might play in battlefield medicine until its antibacterial power was demonstrated in human subjects, but Florey had high hopes. On June 11 he wrote to Mellanby, "We have been working on a substance that shows the greatest promise of being a chemotherapeutic agent."

The large-scale animal trials that began in June amply confirmed this promise. In one trial, twenty-four of twenty-five mice receiving penicillin survived a normally fatal gas gangrene infection. By August the Oxford researchers were taking steps to safeguard their discovery in the event of a German invasion. Florey, Heatley, and several others smeared the inner linings of their coats and jackets with some of the *Penicillium* mold. If they were forced into exile (presumably in the United States), some live spores could be recovered from the cloth, and the work could begin again. If any of them fell into German hands, their captors would have no reason to suspect that the brown stain left by the mold juice was anything but ordinary mildew.

Florey's new awareness of the possible military value of penicillin did not lead him—at least not yet—to consider suppressing the news of his laboratory's stunning success. He and his collaborators immediately wrote up the results of the animal trials and submitted the paper to the *Lancet,* the leading British medical journal. The censors raised no objections, and the full report, "Penicillin as a Chemotherapeutic Agent," appeared in the August 24, 1940, issue, with the authors listed in alphabetical order: Chain first, followed by Florey, A. D. Gardner, N. G. Heatley, M. A. Jennings, J. Orr-Ewing, and A. G. Sanders.

The paper announced the isolation from the *Penicillium* mold of a stable brown powder and then gave details of in vitro and in vivo tests demonstrating that the antibacterial activity of this powder was "very great." The next-to-last paragraph described the results as "clear-cut," one of Florey's favorite terms. And in case anyone plowing through the details had missed the significance of the figures, attention was drawn to the efficacy of penicillin against the "anaerobic organisms associated with gas gangrene." Florey had certainly kept his word to Miller about not letting the war interfere with publication of the work supported by the RF grant.

In fact, his sense of gratitude to the Foundation got him into trouble with his old friend Edward Mellanby. A few days after the paper appeared, Mellanby sent an indignant letter to Florey. What had roused the ire of the secretary of the British Medical Research Council was not the possibility that publication in the *Lancet* might alert the Germans to a potentially useful battlefield medicine. Instead, Mellanby took offense at what he felt was Florey's inadequate recognition of MRC support for the penicillin research. In the last paragraph of the *Lancet* paper, Florey had written, "In addition to facilities provided by the university we have had financial assistance from the Rockefeller Foundation, the Medical Research Council and the Nuffield Trust. N. G. Heatley has held a Rockefeller fellowship. To all of these we wish to express our thanks."

Mellanby objected to both the form and the tone of this acknowledgment: "You gave a great boost to the Rockefeller people and it seemed the M.R.C. had played a minor role. . . . I doubt whether this accords with the facts." To back up this criticism, he enclosed his own accounting of MRC support for the penicillin work: £50 pounds to Florey for expenses, £300 plus expenses to Chain, £200 to Dr. Jennings, and over £500 to Dr. Orr-Ewing. He went on,

> I doubt whether the Rockefeller is supporting you to that extent. It seems to me that your method of dealing with this is wrong tactics, partly because these Rockefeller grants in this country follow discussion between O'Brien and myself, and

partly because if you have a good thing in your country you might as well give it proper credit.

In reply, Florey noted, somewhat apologetically, that the *Lancet* editors had cut down his paragraph of acknowledgments to save space, and he implied that Mellanby would have been happier with the fuller version. The rest of the letter, however, was anything but apologetic. He informed Mellanby that the negotiations for the Rockefeller grant were conducted "not through O'Brien at all but with their chemical people." He divulged the size of the RF grant ($5,000, or £1,250) and then contrasted this figure with his own "reckoning" of MRC support for the penicillin project, which, during the preceding year, added up to a "maximum of £800." That gave the RF a clear three-to-two advantage, even without counting Heatley's emergency fellowship, a matter of some £300.

Denying any intention of slighting the MRC (Hadn't he listed the MRC ahead of the Nuffield Trust in his acknowledgments?) Florey held Mellanby to a strict standard of accounting: "By the way, the £50 you credit to me was for the tail-end work on lysozyme"—and so did not belong in any discussion of the penicillin ledger.

Freed from the pressures of fund-raising, Florey and his collaborators could now devote their full time and energy to obtaining sufficient supplies of penicillin for the most important experiment of all: a test of the drug's ability to cure infections in human patients.

At the beginning of the summer, Florey had sounded out several pharmaceutical companies to see if they were interested in manufacturing penicillin in large quantities. For various reasons, these overtures came to nothing. So Florey decided that his laboratory would undertake the wholesale farming and harvesting of penicillin from the *Penicillium* mold. This would have been a formidable task at any time. But it was especially daunting now, as the representatives of several drug companies had made only too clear to Florey. How could he expect to secure the proper equipment or find and train sufficient workers in a country mobilized against invasion and already reeling under wave after wave of German bombing raids?

There had been scattered raids on Channel shipping and industrial targets earlier in the year. But the first major clash in what came to be known as the Battle of Britain, between the Royal Air Force and the German Luftwaffe, took place on July 10. During the next month the defenders more than held their own; by the middle of August the Germans had lost twice as many aircraft as the British. But the Luftwaffe kept coming, in obedience to Hitler's orders to destroy the RAF and crush Britain's will to resist. Through August and September, bombing sorties numbered over a thousand a day; good weather came to mean bombing weather to the British, who kept one ear cocked for sirens and one eye on the nearest shelter. Increasingly, attacks were directed at cities; civilian casualties, especially in night raids, mounted. When the sun went down, Londoners took shelter in the city's subway stations.

Despite the rigors of the London blitz and the destruction of industrial, rail, and port facilities, the British did not crack. A German invasion flotilla had been assembled across the Channel, but Hitler kept putting off the date for the assault while he waited for signs that the RAF had been eliminated as a fighting force. At the end of October the invasion was postponed indefinitely. Heavy bombing continued into the following spring (when the bulk of the Luftwaffe was shifted to the east for the invasion of the Soviet Union), but the British count October 31, 1940, as the final day of the Battle of Britain.

It was a modern siege that ended in a victory for the defenders. Yet the cost was enormous. Bombs had fallen on London, Coventry, Liverpool, Bristol, Portsmouth, Plymouth, Birmingham, Manchester, Glasgow, and Cardiff. One home in every five had been damaged or destroyed. In London alone, more than fourteen thousand civilians were killed.

Oxford was spared direct bombing raids. But it was hardly immune from the depression and anxiety that pervaded England during the summer and autumn of 1940. Like many other children in the British Isles, the Floreys' daughter and son were shipped off to America, to spend the rest of the war at the home of the Yale professor John Fulton, a former Rhodes scholar whom Florey had known since his first years at Oxford. The Florey children left in July 1940. By this time the Oxford team had begun putting together a makeshift penicillin factory in the Dunn building.

Hoping to save space and increase their yield of active material, Florey, Chain, and Heatley experimented with new ways to grow the mold. In place of the standard laboratory flasks, they tried a variety of utensils, including milk bottles, biscuit tins, a bronze letter box, and some old-fashioned metal bedpans. The latter worked the best. But when Florey tried to buy six hundred new bedpans, he found that such production capacity was yet another casualty of the war. He was further informed that designing and manufacturing similarly shaped vessels in glass would take too long and cost too much.

Fortunately, Florey knew a physician who practiced in Stoke-on-Trent, in the center of England's Potteries district. His friend located a ceramics firm that could turn out the required pots within two months. A now enthusiastic Mellanby agreed to foot the bill. Borrowing the van that had been authorized for Chain's blood-transfusion service, Heatley drove up to the factory and brought back the first batch of glazed mold pots on December 23, 1940. By Christmas Day the mold spores were beginning to multiply in the new vessels.

Since it turned out that penicillin grew best when the vessels were kept at near-freezing temperatures and shaken periodically, Florey had to find technicians willing to spend many hours a day in the laboratory's cold room, agitating the mold. Eventually, six "penicillin girls" were hired for this work. Their wages of £2 a week were covered by an MRC grant.

By the end of 1940 Florey had enough penicillin to begin thinking about clinical trials. Refinements in the extraction process had revealed that the brownish tinge of the original powder was due to an impurity; the purer powder now being produced was yellow. But when a solution of this powder was injected into a human patient—a woman dying of cancer at Oxford's Radcliffe Infirmary—the result was a sudden high fever, a sure sign that some impurities remained. Using a technique known as column chromatography, Abraham separated out the fever-producing interloper. The way was now clear for the first real test of penicillin's curative powers.

Another patient at the infirmary, a forty-three-year-old Oxford constable named Albert Alexander, had scratched his face on a rosebush. The wound became infected with staph and strep germs. Sulfa drugs failed to halt the infection, and the man was dying when Ethel Florey suggested him as a possible test case. On February 12, 1941, Constable Alexander began receiving intravenous injections of penicillin.

Almost immediately his condition improved. Within four days his fever was gone. But the infection had not been eliminated, only weakened. On the fifth day the lab's entire supply of penicillin was exhausted—even though at least half of each dose was laboriously recovered from the patient's urine, reprocessed, and injected back into his body. For a few days after the penicillin ran out, Constable Alexander's natural defenses dueled with the remaining bacterial invaders; the latter proved too strong, and he died on March 15.

The outcome was disheartening, but the experience served as a powerful stimulus to Florey and his colleagues; they felt sure that with enough penicillin they could have saved the life of a patient who was beyond the reach of any other treatment in the medical arsenal. Over the next three months they put this faith to the test. With five more patients—including four small children, who were chosen because less penicillin was needed to treat them—the Oxford team achieved results that the quiet-spoken Heatley described as "almost miraculous." All five patients were cured of serious infections (although one died subsequently of an infection-related complication).

Once again the collaborators wrote up their results and submitted the finished paper to the *Lancet.* Once again no one tried to stop publication on the grounds that penicillin might someday play a decisive role in battlefield medicine and should therefore be treated as a war secret. In August 1941 the *Lancet* published the Oxford team's second paper on penicillin. After describing the first six human cases, the authors stressed the need for increased production of the drug to make possible even more extensive clinical trials.

The British censors' lack of concern about what Florey and his coworkers said in print probably reflected Mellanby's own attitude toward the work at Oxford. Although he was now convinced it had merit, he saw no great urgency in its development. Indeed, the first person to sound a warning about possible enemy interest in penicillin was Florey himself. Early in 1941 he had learned, through private sources, that the Germans were trying to obtain samples of the mold through Swiss front men. Someone in Hitler's Germany was apparently reading the *Lancet.* Florey wrote immediately to Mellanby, urging a ban on the release of any *Penicillium* mold from the National Type Collection Laboratories in London. Mellanby saw no cause for alarm. "After all," he replied, "you are so far ahead it seems no one else can compete with you."*

A different kind of competition was on Chain's mind when he raised the question of who should benefit financially from penicillin's devel-

*The Germans never succeeded in producing penicillin during the war; the sulfa drugs, on which they relied to treat war wounds, had little or no effect on many of the worst infective agents, including the microbes that cause gas gangrene.

opment. On the Continent it was standard practice for large chemical companies to take out patents on discoveries made in their research laboratories. If the Oxford workers were allowed to patent the processes they had pioneered, Chain argued, they could finance future research out of royalties once the drug became commercially available. The leaders of the British scientific community rejected this argument. Mellanby and Sir Henry Dale of the Royal Society decided that since "the people" had paid for the work, the benefits should be shared equally by all. In other words, no patents for Chain, Florey, or anyone else.

Chain was furious, but Florey accepted the ruling with a minimum of fuss; perhaps the prospect of an indefinite sum of money at some future date did not seem sufficient reason to provoke a controversy that could only divert him from the task at hand—organizing the production of penicillin on a really large scale.

# 40

# Good as His Word

While Edward Mellanby seemed perfectly satisfied with a "cottage industry" approach to penicillin production, Florey was not. He needed at least a kilogram of penicillin (at the purity then achievable) to conduct truly convincing tests on large numbers of patients. And he knew that his makeshift factory at the Sir William Dunn School of Pathology could never turn out that much drug quickly enough to meet his own timetable.

Even before the first six clinical trials were completed, Florey had decided that wartime England was no place to scale up penicillin production. That left only one alternative: the United States, whose massive lend-lease shipments were helping Britain replace the arms and supplies lost at Dunkerque and in the blitz. In the spring of 1941 Florey resolved to go to the United States "to ginger up penicillin production over there" (as he later put it). He asked for, and got, Mellanby's blessing for this initiative. But when it came to financing and arranging the trip, Florey turned once again to the Rockefeller Foundation.

The RF had already extended the original penicillin grant for a year. In a letter to Miller describing his laboratory's accomplishments during 1940, Florey had written, "I don't think I am too optimistic in thinking that this is a very promising line." As for the war, he noted that he was having more and more trouble obtaining certain supplies and equipment, but he added, "The only thing which seems likely to stop us in the coming year is the actual bombing of the building."

These words must have been enormously reassuring to Miller, who, as a junior member of the Division of Natural Sciences, had gone out on a limb for Florey, and who, along with the other circuit riders, had resisted until the very end the decision to close down the European office. The great fear of the circuit riders was that they would have to abandon the scientists and scholars with whom they had worked so closely for so long. But in Great Britain only six major projects (out of the thirty on the books in 1939) had been canceled. Although the entire program was now being administered from New York, the Foundation's involvement with the British scientific community had narrowed hardly at all.

The Foundation also remained deeply involved in efforts to assist scholars fleeing from Nazi-occupied countries. In his *President's Review* for 1940, Fosdick finally dropped all pretense of evenhandedness in his references to the war. The German conquerors, he noted, did not even pay lip service to the "ideal of objectivity" in scholarship; students and professors who resisted the imposition of a Nazi "cultural program" were imprisoned or killed.

The Foundation's original program for displaced scholars, which aided nearly two hundred individuals from 1933 to 1939, was clearly inadequate to deal with the new wave of refugees after September 1939. So the trustees authorized a stepped-up rescue effort. Working with other agencies—including the Carnegie Corporation, the New School for Social Research, and the Emergency Committee in Aid of Displaced Foreign Scholars—the Foundation appropriated more than $250,000 to help bring European scholars to safety, and to find them at least temporary appointments at the New School and other institutions.

The principal escape route for refugees leaving western Europe went through Lisbon, where Makinsky had established an office in the Portuguese Ministry of Health building. Wartime Lisbon was an extraordinary city, where representatives of the belligerent countries rubbed shoulders with each other, with diplomats from neutral countries, and with refugees, spies, and adventurers of all sorts. In such an atmosphere Makinsky's skills at negotiation (and his mastery of the genteel art of cutting red tape with an appropriate "gift") proved indispensable.

He knew at least eight languages—Polish, German, Russian, French, Portuguese, Italian, English, and Persian—and had friends in all the right places and in a number of extremely useful "wrong" places as well. (The German minister in Lisbon was pleased to help Makinsky on more than one occasion.) When the need arose to improve communications between Lisbon and Paris, Makinsky found friends willing to let him use the diplomatic pouches of Portugal and Chile. At the last moment the RF rejected both these offers as being potentially compromising, whereupon another friend of Makinsky's, a French businessman who traveled between Marseilles and Paris every week, volunteered to transmit verbal messages in both directions.

Clearly, Fosdick had been right when he assured Gunn that closing the La Baule office would not mean abandoning European science and scholarship. In fact, pressures for an expanded RF presence on the Continent were once again building. Gunn had scoffed at O'Brien's prediction that he would be back on French soil in two months. As it turned out, O'Brien was off on his timing, but not by much. At the beginning of July 1940, the Executive Committee appropriated $500,000 for a new "Rockefeller Foundation Health Commission," whose mandate was to work closely with the Red Cross and other relief agencies in combating "war-engendered epidemics, such as typhus fever, influenza, malaria, etc." This was not considered relief work per se, since it would be up to the other agencies to supply food, shelter, and clothing while the Health Commission provided the technical advice and trained personnel needed to bolster the local "health machinery."

When a three-man survey team caught a Pan Am Clipper from New York, to begin a fact-finding tour of England, France, Spain, and Portugal, O'Brien was on board. By the end of August he was in La Baule again, arranging for transport of the RF's European files back to Paris, where the Health Commission temporarily made use of the old rue de la Baume office.

Once the commission was in operation, O'Brien moved on to London, where he set up shop as a representative of the entire Foundation in early 1941. This was about the time when Florey began thinking about a trip to the States. He would probably have talked to O'Brien, or written to Miller in New York, about financing the trip, if he had not learned that Warren Weaver was coming to England. As the director of the division that had so generously funded the penicillin work, Weaver was clearly the man to see. There was only one problem. Weaver was coming to England not as an officer of the Foundation but under the auspices of the National Defense Research Council

(NDRC), where he had been working for nearly a year on the development of more accurate antiaircraft devices.*

Weaver's boss at the NDRC, Vannevar Bush, had asked Weaver and two colleagues to see what progress the English were making with antiaircraft weapons. Leaving New York on March 3, Weaver sailed to Lisbon on a slow and barely seaworthy ship that had been pressed back into service (presumably from a scrap heap) when the war began. From Lisbon, Weaver's party flew to England on a plane whose windows had been blacked out with heavy cardboard and plywood panels to escape detection by the Luftwaffe.

During his stay in London and on inspection trips to coastal antiaircraft installations, Weaver learned to duck German bombs in the best English "carry on" manner. In early April he was injured not by a bomb but in an automobile accident on a blacked-out highway. His car blew a front tire, veered off the road (narrowly missing an oncoming convoy of military vehicles), crashed through a hedge, and turned over in a meadow.

Weaver had to spend several days recuperating in his hotel bed in London. On one of these days, April 14, Florey showed up for an interview. Weaver recorded their conversation in a typically detailed diary entry. He explained to Florey that even though he was on leave from the Foundation, he would pass along to the New York office any proposal that Florey wished him to.

The Oxford pathologist briefly recounted the story of the extraction, purification, and testing of penicillin, up to the first trials on human subjects, which were just getting under way. Then he announced that he was sure the antibacterial substance could play an important role in the war itself—if he could talk one or more of the big American drug companies into beginning large-scale production immediately. Weaver assured Florey that if the British authorities gave him permission to go abroad, the Foundation would pay his way. At the end of the report that he forwarded to New York, Weaver emphasized that Florey's work was "of the very highest potential importance." He added, "We certainly ought to do all that we can to accelerate its progress."

When he got back to the States, Weaver set in motion the machinery to do just that. The key was a grant-in-aid of $6,000 (approved on June 19 by Weaver's wartime surrogate at the Foundation, Frank B.

*The devices that emerged from his office played a vital role in England's defense against the German V-1 "buzz bomb" attacks later in the war.

Hanson) "to provide travel funds and living expenses" for Florey and Heatley. It had been decided that Heatley would also make the trip, to share with American scientists the tricks he had developed to keep the *Penicillium* mold alive and growing. But money alone could not get the two men across the Atlantic. They needed exit permits, visas, tickets on planes from London to Lisbon and Lisbon to New York—and each of these matters required intervention at the highest government levels.

It took a coordinated effort by Mellanby, officials at Oxford University, and RF representatives in New York, London, and Lisbon to get all the pieces in place. O'Brien persuaded the director of hygiene of the British Air Ministry to expedite clearances for Florey and Heatley on a London–Lisbon flight. When Pan Am officials rebuffed Miller's request for seats on the Lisbon–New York Clipper, Hanson telephoned the U.S. surgeon-general, who agreed to ask the State Department to look into the matter; three days later Florey and Heatley were booked on a Pan Am Clipper leaving Lisbon June 30. Their London–Lisbon flight was already set for June 27; O'Brien told them that they would be guests of the Foundation while they waited for the plane to New York.

On the morning of June 26, Florey and Heatley slipped out of Oxford carrying a briefcase full of glass vials wrapped in cotton wool. Inside the vials were freeze-dried samples of *Penicillium* mold with which they hoped to "seed" an American penicillin industry. They went by train to Bristol, where they boarded a small plane that took them to a blacked-out field, somewhere in England, where they transferred to a larger plane for the trip to Lisbon. They were greeted a few hours later by Makinsky, who gave them each $50 in escudos, confirmed their Clipper reservations for the thirtieth, and escorted them to their hotel.

After a year and a half of wartime austerity, Florey and Heatley were dazzled by the sunny, prosperous appearance of Lisbon, a city "untouched by war." They passed the next three days walking in the sun, eating in the well-stocked restaurants, and worrying about the condition of their mold samples, which they had left locked in the hotel safe. Their transatlantic flight was delayed one day, but they arrived safely in New York on July 2. As far as they could tell from a cursory inspection, their mold samples had survived the trip.

Florey spent July 4 with his children at the home of Dr. John Fulton in New Haven; then he and Heatley embarked on a round of meetings that the Foundation had set up with drug company executives, scientists, and government officials. One contact led to another until, two weeks later, they found themselves in Peoria, Illinois, talking with researchers at a recently opened Department of Agriculture laboratory that specialized in the study of fermentation. These talks were so successful that Heatley decided to stay on in Peoria, supported by another RF grant, to help USDA scientists work out improved methods of penicillin culture.

Florey's efforts to enlist private industry in penicillin manufacture were less successful at first. A major stumbling block was the possibility that penicillin would turn out to be easily synthesized; no one wanted to invest large sums of money in a full-scale "penicillin farm" only to be undercut by a cheaper manufacturing process. Chain had assured Florey that synthesis of penicillin would be an enormously difficult technical undertaking. But the drug company executives remained wary, and Florey's mission appeared to be foundering when on August 7 he sat down to dinner in a Philadelphia club with his one-time fellowship adviser, Dr. A. Newton Richards.

By a coincidence that a fiction writer would hesitate to invent, Richards had just been named chairman of the Committee on Medical Research of the Office of Scientific Research and Development, a new government agency headed by Vannevar Bush, who was now President Roosevelt's "science czar." Over the dinner table Florey described the laboratory work, the animal and human trials, and the odds against successful synthesis in the near future. He must have made a good case. By the time he returned to Oxford, at the end of September, he had secured Richards's promise to throw the full weight of the Committee on Medical Research behind the production of penicillin.

This committee wielded great financial and political power. It could allocate government funds for high-priority projects and "request" the cooperation of private business. With Bush's prodding, friendly competition between the major drug companies and the USDA in Peoria became the order of the day; advances made in one lab were shared with other researchers on a need-to-know basis, and the yield of penicillin kept increasing.

But all of Florey's enthusiasm could not hide the fact that the new drug had been tested on only six human subjects, two of whom had died. Not only that; the *Penicillium* mold was a living organism, given to capricious behavior, and coaxing it to produce more penicillin on demand was a tricky business. Seen in this light, Richards's decision to back Florey with all the resources of his new committee seems like a great gamble. What led him to take the risk? Both Heatley and Abraham were convinced that the answer lay in Richards's personal knowledge of Florey, gained fifteen years earlier when the young Australian spent several fruitful months in Richards's laboratory as a Rockefeller fellow. After weighing all the scientific evidence, Richards, it appears, took a chance not so much on a new, largely untested drug but on the word of a man he knew and trusted.

The results exceeded everyone's expectations. Clinical trials initiated by the American government with the first American-made penicillin not only confirmed all of Florey's claims but revealed that the drug was the first safe, effective treatment for syphilis and gonorrhea. By October 1943 the U.S. Army had contracted for all the penicillin the drug companies could turn out, and production was assigned a higher priority than any other item on the military's shopping list, with the exception of the still-secret atom bomb.

# 41

## Miracles

As it happened, the success of Florey's mission to America did not mean that he got the "kilo" of penicillin that had been the original impetus for the trip. Just as the American drug industry was gearing up to produce penicillin in quantity, the Japanese intervened—at Pearl Harbor. With the entry of the United States into the war, all penicillin made in this country was reserved for use by Americans.

Back in England, Florey had to start all over again. This time he managed to persuade some British firms to manufacture penicillin for him. Although these operations were on a far smaller scale than the American effort, the penicillin they supplied, together with the output of his own lab, enabled Florey to launch a definitive series of clinical trials in the first months of 1942. On August 2 Florey wrote to Sherrington, describing the results of those trials as "almost miraculous." Yet even now his stock of penicillin was so limited that only 15 of the 187 cases could be treated systematically with intravenous injections. The rest received local applications of the drug to combat surface infections, a far less satisfactory procedure. Florey told Sherrington that the principal obstacles to widespread availability of penicillin were no longer scientific. The problems now were lack of money and lack of support from the government. "If, say, the price of two bombers and the same energy was sunk into the [penicillin] program, we could really get enough to do a considerable amount."

Four days later Florey received an urgent phone call from Alexander Fleming. No communication had passed between the two men since the summer of 1940, when Fleming, having seen the first *Lancet* paper, showed up at Oxford one day to look around the lab. As members of the Oxford team remembered that visit, Fleming had displayed little enthusiasm for what he saw; it was not even clear how much he understood. To Chain he was a figure out of the past in every way: "I frankly had assumed he was dead."

What rekindled Fleming's interest in penicillin in the summer of 1942 was a personal matter. A good friend of his had come down with streptococcal meningitis that would not yield to sulfa drugs. Tests conducted in Fleming's own laboratory indicated that only penicillin could help; and Florey had the only penicillin suitable for clinical use in England.

In response to Fleming's plea Florey hurried up to London by train, carrying all the penicillin he had at the moment. Treatment of Fleming's friend began that night, August 6, and continued for thirteen days, with Florey rushing more penicillin to Fleming as it came off the lab's production line, and Fleming himself injecting it into the patient. Several doses were introduced directly into the cerebrospinal cavity, the first time the drug had been administered that way. By the beginning of September, the man, who had been given up for dead, was discharged from the hospital; Fleming himself characterized the cure as "miraculous."

One result of this unplanned collaboration between the two men most responsible (in entirely different ways) for the development of penicillin was that the British government finally awakened to the potential of the new drug. Fleming telephoned another friend of his, a fellow Scot who headed the Ministry of Supply, to tell him about the miracle he had just performed with penicillin. The word spread, and following a series of high-level meetings the General Penicillin Committee was organized; this committee played much the same role in expediting British production of the drug during the war as Dr. Richards's Committee on Medical Research did in the United States.

But hearsay evidence about a new "miracle" drug was not enough to persuade doctors—especially hard-pressed military surgeons—to risk its use in place of medications like the sulfa drugs that they were more familiar with. So Florey took it on himself to demonstrate penicillin's efficacy under battlefield conditions. In the spring of 1943 he and his fellow Australian Hugh Cairns led a team of ten surgeons to North Africa, where they spent three months treating wounded soldiers. To stretch their meager supplies as far as possible, Florey devised a revolutionary method of treatment for open wounds: he applied penicillin and sulfonamides directly to the raw wound, inserted rubber tubes to allow for drainage (and facilitate the administration of more drugs later on), and then sewed up the broken skin.

This procedure ran directly counter to the experience of military surgeons, who knew that closing up a dirty wound was to invite the deadly anaerobic infection of gas gangrene. One surgeon who watched Florey at work was heard to mutter: "It's murder, bloody murder." But Florey's innovation, while hardly recommended under normal conditions, served its purpose in North Africa—to spread the benefits of penicillin over the largest possible number of patients. Operating in tent hospitals where the temperature hovered above one hundred degrees day after day, treating wounds that in most cases were already infected, Florey and Cairns were able to report that of 171 patients who received penicillin in this manner, 104 recovered completely; and most of these were able to return to their military units in record time.

By the time of the joint D-Day landings, in June 1944, the British and American armies had enough penicillin on hand to treat their battlefield casualties with intravenous and intramuscular injections. The results were as satisfying as Florey and his collaborators could have hoped. Ninety-five percent of all wounded soldiers treated with penicillin on the European front recovered. This included cases of open fractures and severe burns, as well as the dreaded gas gangrene. In the 1914–18 war, fifteen out of every one hundred cases of gas gangrene resulted in death; during the Second World War the rate of recovery from such infections was close to 100 percent.

With the flow of penicillin from private pharmaceutical firms steadily increasing, the makeshift Oxford "factory" was no longer needed; and in June 1944 Florey gave the order to dismantle it. This freed staff, space, and funds for research into the basic structure and mode of action of penicillin. In 1945, after two years of calculations on the best mechanical calculators then available, Dorothy Crowfoot Hodgkin of the Dyson Perrins Laboratory managed to tease out the correct structure of the penicillin molecule from X-ray pictures of penicillin crystals. It was a heroic effort, which won her a Nobel Prize in 1967; with today's electronic computers, the calculations would probably take less than a month.

As government and industry assumed the major responsibility for the development of penicillin, the Rockefeller Foundation's relationship to Florey's lab necessarily changed. He continued to ask for RF aid, and his requests were invariably approved. These ranged from smaller grants-in-aid—$6,500 in 1942, $4,860 in 1943—to more substantial appropriations, such as $18,000 in 1945. But whereas the earlier, smaller sums constituted a substantial proportion of Florey's entire budget, the later grants were important primarily because they provided Florey with a dollar account in New York, out of which he could buy American-made equipment that was either unavailable in Europe or that could not be purchased with British pounds, because of currency and import restrictions.

The postwar files of the Foundation are filled with correspondence between Florey and the New York office concerning the purchase of even the most basic lab supplies—including Petri dishes. During this period Florey dealt mostly with Gerard Pomerat, a Harvard biologist who had succeeded to the circuit-riding tradition of Tisdale, O'Brien, and Miller. The Foundation's last grant to Florey, which ran from 1953 to 1955, brought the total of his RF support through the years to just over $76,000.

Of course, once Florey's scientific reputation was secure (and especially after he, Fleming, and Chain won the Nobel Prize in 1945), Florey found himself more and more often on the other side of the grant-giving business. He served as adviser to both the MRC and the Nuffield Medical Research Committee; and in 1948 he was asked to replace the retiring Mellanby as secretary of the MRC. He declined—since the job entailed a great deal of administration, which he hated—but he seized other opportunities that came his way to advance the cause of basic scientific research. In 1947 Lord Nuffield offered him £50,000, with no strings attached, to spend as he wished on his own work. Instead of taking the cash, Florey suggested that Lord Nuffield use it to endow three research fellowships at Lincoln College, Oxford. The first three fellows, named in 1948, were Edward Abraham, Norman Heatley, and Gordon Sanders, all longtime associates of Florey's.

Although he remained active in antibiotic research until 1952 and continued to perform experiments until the last year of his life, Florey accepted his role as an elder statesman of science with considerable grace, and without any diminution of his passion for hard work or his knack for raising money.

He had been a fellow of the Royal Society for nearly two decades when he was elected president in 1960. The president of the Royal

Society is the nominal leader of the British scientific community, although some incumbents treat the office as largely ceremonial. Not Florey. At his insistence the government raised the number of endowed "research professorships" in British universities from two to twelve and provided more fellowships for promising young researchers. At Queens College, Oxford, where he became provost in 1962, he secured funds from the Ford Foundation and other sources for the construction of a building expressly designed to house fellowship students from Western Europe. The first of its kind at Oxford, it is now known as the "Florey building."

But most of his energies during his five-year term as president of the Royal Society went into an even larger and more challenging fund-raising venture. For several years before he assumed office, it had been clear that the venerable society had outgrown its home in London's Burlington House. Larger and more suitable quarters were available in an equally fine building, Carlton House on Pall Mall, but the society did not have the £500,000 needed to refurbish the new rooms. By the time Florey stepped down as president, in 1965, the money was on hand—half provided by the Nuffield trustees and the rest pledged by other private foundations and public agencies.

When the Royal Society moved into its new home, on November 21, 1967, Florey—now Lord Florey—attended the opening ceremonies with his new wife, the former Margaret Jennings. (Ethel had died the preceding year, and Florey and Dr. Jennings were married in June 1967.) Of all the honors that came his way, he must have taken special pleasure in the tribute paid to him by the physicist P. M. S. Blackett, who had followed him as president of the Royal Society. Blackett gave Florey full credit for the elegant new quarters in Carlton House, summing up his accomplishment in the blunt statement "He got the money."

Three months later, on February 21, 1968, Florey's heart, which had given him trouble for more than fifteen years, gave out. Although he did not become rich from the development of penicillin—his entire estate was valued at £30,554—there is no question that he received in abundance what he craved most: recognition by his peers in the scientific community.

The desire for recognition is no less powerful a motivation for patrons of science. Public acknowledgment of a job well done is, in fact, the only reward that a philanthropist can reasonably expect. So it is not surprising that the Rockefeller Foundation should have made efforts to publicize its pivotal role in the penicillin story. What is curious is that these efforts at institutional self-promotion should have aroused such ill feeling and that even those in a position to know better should have consistently gotten important details of the story wrong.

We have seen how Florey's acknowledgment of Foundation support in the first *Lancet* paper provoked an angry outburst from Mellanby. But this quarrel between the secretary of the MRC and his unruly protégé was settled in private. The first signs of public rancor concerning the pedigree of penicillin surfaced in the summer of 1942, just before Fleming and Florey collaborated to save the life of Fleming's friend. An editorial in the *Lancet* urged the government to do more to stimulate the production of penicillin in Great Britain. The *Times* of London seconded the *Lancet*'s plea, in a "leader" that described the new antibacterial drug as more powerful than the sulfonamides. This was the first reference to penicillin outside the medical press; while no names were mentioned, the research was ascribed to a group at Oxford. Four days later a letter appeared in the *Times* correspondence column, suggesting that credit for the discovery of penicillin should go to one man: "Professor Alexander Fleming." The letter was signed by Dr. Almoth E. Wright, Fleming's original mentor at St. Mary's Hospital.

The omission of Florey's name from the first public discussion of penicillin enraged those insiders who knew the true story. The next day the *Times* ran another letter, this one from Professor Robert Robinson. While acknowledging Fleming's perspicacity in observing the antibacterial action of penicillin and in preserving the original mold, Robinson lauded "Professor H. W. Florey" and "his team of collaborators, assisted by the Medical Research Council," for their labors in demonstrating "that penicillin is a practical proposition."

Scenting a good story, members of the press descended on Fleming's lab in London and on the Sir William Dunn School of Pathology in Oxford. Florey was never comfortable talking to reporters, and in this case he had reason to believe that too much publicity could only spur the Germans to develop their own penicillin. On the advice of both Mellanby and Sir Henry Dale, he refused to speak to the press. But Fleming and his friends were under no such constraints. The impression left by the resulting stories was that penicillin had been given to the world solely through the efforts of a keen-eyed observer of contaminated Petri dishes at St. Mary's Hospital. So widespread was this impression that the Nobel committee, according to one rumor, almost awarded the 1945 prize to Fleming alone.

In failing to stake a public claim to credit, Florey also missed a chance to correct the impression, left by Robinson's letter, that it was the MRC alone that had financed the development of penicillin. In 1944, when the need for strict secrecy was past, the Rockefeller Foundation decided to toot its own horn. In his *President's Review* for that year, Fosdick recounted the RF's record of support for the Oxford lab going back to the original $1,280 grant in 1936. He could not resist commenting, "Seldom has so small a contribution led to such momentous results."

This comment had repercussions that Fosdick could not have anticipated. The editorial page of a Republican newspaper in Massachusetts, lauding the success of this "investment . . . by a privately endowed institution," took the occasion to lambaste the Democratic administration in Washington for its tilt toward "government-controlled charities, education and benevolences, including scientific research." The editorial received national and even international attention when an excerpt ran in Ed Sullivan's syndicated column on April 2, 1944.

A columnist for the *London Evening News* picked up the item and appended a comment of his own: "Now I cannot help thinking that research in this country must be shamefully starved if an Oxford professor, for a paltry sum of less than £500 for sensational research work, has to go to the United States with a request for aid." Mellanby, of course, was furious; he labeled the Rockefeller Foundation's attempt to hog the credit for the support of penicillin as "simply grotesque." Before the dust settled, the controversy reached the floor of the House of Commons, where a government spokesman offered figures showing that the MRC had contributed in excess of £7,000 to Florey's researches since 1927.

More to the point was Florey's own accounting, which appeared in the definitive two-volume *Antibiotics* in 1946. Florey took this opportunity to express his gratitude to Oxford University for providing support in the form of laboratory space, materials and salaries—a "fact sometimes forgotten." As for specific grants for the penicillin work, his breakdown looked like this: MRC (1939 to 1945) £8,287; the Rockefeller Foundation (1940 to 1945) £6,140; the Nuffield Provincial Hospitals Trust (1943 to 1945) £5,646. He concluded with a phrase that could have been appreciated only by someone familiar with the situation and with Florey's gift for ironic understatement. "At no time was the work held up for lack of funds," he wrote, but "the tenure of appointment and stipends of the academic research workers . . . left something to be desired."

Confusion over the Rockefeller Foundation's role in the penicillin story extended even to the controversy about patents that erupted after the war. What happened, in brief, was that the American companies whose interest in penicillin had been aroused by Florey's 1941 visit went on to reap handsome profits from the manufacture of penicillin. Part of these profits came from license fees paid by British firms, which discovered, to their chagrin, that a number of industrial processes indispensable to the large-scale production of penicillin had been patented by American drug and chemical companies.

The propriety of these patents can be argued, but one error that has crept into most histories can be cleared up here. The Rockefeller Foundation did *not* share in any royalties paid to American firms or individuals in connection with the penicillin patents. The rewards that the Foundation derived from its association with Florey were strictly intangible: public goodwill (on which the officers gratefully drew when Foundation policies came under attack in Congress in the 1950s) and the satisfaction of demonstrating that a carefully conceived and vigorously executed program of science patronage can both advance knowledge and promote "well-being throughout the world."

But while the Foundation managed to make its corporate role in the penicillin story known to a wider public, the individuals responsible for the actual decisions were left to enjoy their success in anonymity. This is not surprising. In most cases public acclaim for such labors is neither expected nor welcomed. Typically, the work is conducted in confidence, and the behind-the-scene details cannot be revealed until years later. As a result even well-meaning attempts to give credit where credit is due often fall wide of the mark.

I do not know whether Dusty Miller ever saw a letter that Dr. John Fulton of Yale wrote to Alan Gregg on October 4, 1944. But if he did, Miller no doubt had a good laugh over it with his boss Warren Weaver. Fulton, a friend of Florey's since their Oxford days together in the early twenties, was one of the few people in a position to know the truth about the Foundation's involvement in the development of penicillin. This is what he wrote to Alan Gregg, the director of the Foundation's Division of Medical Sciences, five months after D-day: "One of the brightest things you and Dan O'Brien ever did was to support Florey in the early days and to bring him over in 1941 after he had treated six cases. Few others than you would have had the imagination. . . ."

# VISIONS

# 42

# A Puzzlement

Historians of science take a generally favorable view of Weaver's years at the Rockefeller Foundation. Robert Olby had nothing but good things to say about Weaver in *The Path to the Double Helix,* an influential history of the birth of molecular biology. A more recent book on the same subject, *The Eighth Day of Creation,* by Horace Freeland Judson, cites James Watson's testimony that early RF support for "the first work in molecular biology . . . had a great influence on science." In an often cited article published in 1976, Robert E. Kohler painted Weaver as the pioneer "manager of science," a man who had both the daring to formulate an innovative policy of science patronage and the administrative skills to carry out that policy on an international scale. According to Kohler, Weaver's activities at the Rockefeller Foundation not only "deeply influenced several disciplines, notably biochemistry," but "in some ways his mode of management anticipated the subsequent major governmental patrons of science, such as the National Science Foundation and the National Institutes of Health."

Scholarly opinion on Weaver is not all laudatory. A sharp dissent was voiced a few years ago by Pnina Abir-Am, a young historian at the University of Montreal. Her analysis owed its ideological edge to the writings of the French historian Michel Foucault, who has argued that scholarly disciplines, like all communal structures, are best understood as expressions of underlying "power relationships." In other words, even among supposedly truth-seeking professionals, the most important decisions always reflect more basic considerations, such as, Who is on top? Who is trying to get there? Who wins and who loses?

Dr. Abir-Am's own ideological orientation can be inferred from her description of Weaver's tour through Europe, in the company of Lauder Jones, in 1933: "In every country, the two were received with the aristocratic honours befitting representatives of a powerful plutocracy." She goes on to assert that the RF's director of natural sciences was actually the mastermind of a far-reaching power play within the scientific community. Weaver, the failed physicist, promoted the "colonization" of the "underdeveloped area" of biology "by physicists who lacked any real interest in biological problems."

In her view, not only are Weaver's motives suspect, but his emphasis on funding the "transfer" of tools from physics and chemistry may actually have *impeded* the development of molecular biology—because it led RF grantees to overemphasize the importance of technology while neglecting the "biological roles" of the molecules they studied. Why, Dr. Abir-Am asks, did such longtime recipients of Rockefeller aid as Linus Pauling (California Institute of Technology, a former NRC fellow) and William T. Astbury (University of Leeds, twice a fellow in natural sciences) fail to solve the structure of DNA before James Watson and Francis Crick's famous paper on the double helix in 1953? Her answer, which she attempts to document in two brief case histories, is that Weaver held them back.

She takes note of Pauling's 1951 discovery of the alpha-helix structure of proteins (which strongly influenced the thinking of Watson and Crick), and of the fact that Astbury took good X-ray diffraction pictures of DNA in the early fifties, using equipment bought with Rockefeller money. She concludes that both men, having come close to solving the DNA puzzle, were blinded by their long association with Weaver and his "complete reliance on technology transfer as the automatic guarantee of biological progress." Under the sway of this seductively simple concept, people like Pauling and Astbury could not appreciate the importance of "alternative research problems pertaining to the function (i.e., not to the structure only) of their biological specimens."

Implied in this criticism of Weaver and the Foundation is a faith in the power of philanthropoids that goes far beyond anything claimed by Weaver or his most enthusiastic supporters. There is, of course, no way to test Dr. Abir-Am's hypothesis; but to suggest that a man like Pauling, whose confidence in his own genius was legendary and whose accomplishments were prodigious, might have achieved more if Weaver had not subtly deflected him seems almost willfully naive.

In a similar vein Dr. Abir-Am faults Weaver for rejecting a 1937 grant request from the British researchers Joseph Needham and Conrad Waddington, who wanted to conduct systematic studies in the relatively new field of chemical embryology. The real reason Weaver shied away from this project, she concludes, was "a corporate ethos which bent funding strategies towards 'safe investments.' " By taking his cue from overcautious leaders of the British scientific establishment (who were never comfortable with the two Cambridge "radicals"), Weaver missed his best chance to midwife a truly revolutionary "new biology."

The assumption here is that if Weaver and his circuit riders had only dispensed their favors in a different direction, chemical embryology instead of molecular biology might have flowered over the next two decades. While such a hypothesis cannot be tested, it is true that the science of embryology has so far not lived up to its bright promise of the early 1930s. Did Weaver strangle this discipline in its cradle by withholding support at a crucial juncture? A close look at Weaver's actual decision-making process, and a brief glance at the subsequent history of the discipline, suggests otherwise.

On October 20, 1937, Weaver, who had been director of the Division of Natural Sciences at the Rockefeller Foundation for nearly six years, faced what was for him a particularly distasteful task. He had to admit to his boss that his division's elaborate screening system—based on a never-ending round of interviews and site visits conducted by his staff of circuit riders—had failed to produce an unequivocal recommendation on an important grant request.

Two researchers at Cambridge—Joseph Needham (a thirty-seven-year-old reader in biochemistry) and Conrad H. Waddington (a thirty-one-year-old former MRC fellow who was now a demonstrator in zoology)—had asked for some $6,000 for research in chemical embryology. Their goal was to pinpoint the factors that determine how a single-celled egg grows to a functioning adult organism. No one doubted the importance of such an investigation. But were Needham and Waddington the right men for the job?

After canvassing the applicants' colleagues in Cambridge, as well as scientists in related fields in the United States and on the Continent, Weaver and his European representative Wilbur E. Tisdale were distressed to find that opinion on Needham and Waddington was split virtually down the middle. For every warm endorsement from a usually reliable source, there was a strongly negative opinion from an equally reputable informant.

This was the kind of administrative dilemma that had prompted Weaver's predecessor, Herman Spoehr, to return to the life of a research scientist after less than a year as director of the Division of Natural Sciences. Spoehr could not get used to making important decisions—saying yes or no to applicants—on the basis of less than compelling data. The mathematically trained Weaver was more comfortable with the limitations of the Foundation's fact-gathering procedures. But the Needham-Waddington case had him rattled. When he marched into President Raymond Fosdick's office on October 20, he was carrying "two sheets of paper—one of which contained the boiled-down versions of the favorable comments on Needham and Waddington and the other of which contained the adverse comments."

As he admitted in a letter to Tisdale a few days later, the two lists presented "a striking and curious contrast":

> Ordinarily, I suppose, we would be the first to claim that this sort of thing does not happen,—or at least does not frequently happen,—in the NS. Objective criteria usually exist for judging the importance of a problem and the capacity of a man. In this particular instance I think it is reasonably clear why the divergence does exist. Since Needham is working in the "No Man's Land" between two more orthodox disciplines, he is viewed skeptically by each.

Weaver and his staff were certainly familiar with interdisciplinary rivalry in the sciences. Since 1933 they had been concentrating on problems that spilled across academic boundaries; and many of their most successful grants had gone to men like Linus Pauling and Harold Urey and Niels Bohr who paid little attention to administrative distinctions that got in the way of creative science.

Similarly, Needham had pioneered in the biochemical manipulation of embryonic tissues, so it was not surprising that some traditional embryologists (whose lives had been devoted to describing the embryo in anatomical terms) worried about his "scanty" grasp of anatomy and tissue structure, while some traditional biochemists (unused to working with animal tissues at all) found his biochemistry "sloppy." Both Tisdale and Weaver were inclined to discount this criticism as the natural reaction of scientific conservatives to an outspoken risk taker who ignored academic conventions in the search for new knowledge.

What gave Weaver pause was the vehemence of the negative opinions—and the possibility that these reflected not the objective judgments of disinterested professionals but the strong personal feelings that both Needham and Waddington had aroused among their colleagues. Needham was a polymath, a man of broad learning who did not hesitate to press his conclusions, about history and politics as well as science, on anyone who would listen. As for the more affable Waddington, he had recently been divorced; in the academic community of the day, this left him open to charges of personal mismanagement if not actual moral turpitude.

In addition, Weaver and Tisdale had recently had firsthand experience with Needham's high-handed personal style and penchant for bending facts to suit his own desires. Just a few months earlier Needham had submitted to the authorities at Cambridge a detailed proposal for a new biochemical laboratory—in effect, *his* new biochemical laboratory since the longtime professor of biochemistry at Cambridge, Sir Frederick Hopkins, was about to retire and Needham was in line to succeed him.

According to Needham's proposal the Rockefeller Foundation was ready to provide two-thirds of the estimated cost of £137,000 if the university could manage the rest. News of this "commitment" came as a surprise to Weaver and Tisdale, who remembered only a series of speculative conversations with Needham (and Hopkins) dating back to 1934. A check of the correspondence files and of Weaver's and Tisdale's diaries turned up no evidence of anything more than an expression of interest in biochemistry at Cambridge, and a willingness to consider any future proposals on their merits. Tisdale explained the Foundation's position to university officers, and the proposal for a new laboratory was quietly shelved.

To complicate matters further, the Foundation had recently been involved in an embarrassing incident that concerned Waddington, although he himself was evidently not to blame. Waddington's immediate superior at Cambridge had asked the RF for £1,200 as a supplement to the young man's salary, in lieu of an MRC stipend that was being canceled. The Foundation had approved a grant—only to have the university refuse to accept the money. The official explanation was that the request had not been routed through proper channels, but Tisdale suspected that personal animosity toward Waddington had motivated the unusual action.

After reviewing all these matters with Fosdick in the fall of 1937, Weaver decided to turn down Needham and Waddington's new proposal for $6,000. Weaver's reasoning, as he outlined it to Tisdale, says a great deal about the assumptions that underlay his long stewardship of the Division of Natural Sciences. His primary goal was to support scientific discovery. But he recognized that no philanthropist acts in a vacuum in choosing his beneficiaries. Patrons who want to have a long-term effect on the development of science must take into account the reaction of the larger scientific community whenever a controversial grant is being considered.

From the tone of his correspondence with Tisdale, it seems likely that if money alone had been at stake, Weaver would have gone ahead with the grant to Needham and Waddington. Considered solely on its scientific merits, the proposal was the kind of gamble that appealed to Weaver. (The Foundation had already given, and would continue to give, modest support to embryological research.) But among those people with serious reservations about the project were Edward Mellanby of the MRC, Henry Dale of the Royal Society, and several senior officers of Cambridge University. In Dr. Abir-Am's view Weaver abandoned Needham and Waddington at the behest of these establishment figures, who disapproved of the researchers' politics. But the facts do not warrant such an extreme interpretation.

In reporting the bad news to Needham and Waddington, Tisdale assured them that their future requests for aid would be considered without prejudice. The very next year Waddington was awarded a Rockefeller Fellowship for study in the United States. In 1947 he accepted a call to the chair of animal genetics at the University of Edinburgh, where he remained until his retirement in 1975. Needham stayed at Cambridge but gradually switched his interest to the history of science. His many-volume *Science and Civilization in China* is considered a classic in its field.

In accord with criteria that he articulated for the Board of Trustees in 1933, Weaver's Division of Natural Sciences set out to invest in scientific disciplines that were "sufficiently developed to merit support, but so imperfectly developed as to need it." In retrospect, it is clear that embryology in the 1930s failed to meet the first part of this test. Nor has the last half century seen much progress. A comprehensive, multi-author survey of the field published in 1985 characterized embryology in the latter half of the twentieth century as "primitive . . . compared to other divisions of science."

The continuing challenge of embryology is to explain how so many different adult structures can develop from an undifferentiated fertilized egg. The German zoologist Hans Spemann won a Nobel Prize in 1935 for showing that the process of differentiation proceeds under chemical control. Both Needham and Waddington had studied with Spemann, and like other embryologists of the time, they had high hopes that the specific chemical "organizer" responsible for inducing differentiation in embryonic tissue would soon be identified. These hopes "ended in disillusionment"; half a century later the "nature of embryological events [remains] mysterious and unsolved."

Dr. Abir-Am suggests that this failure of embryology to develop into a mature science can be blamed on Weaver's failure to support Needham and Waddington in 1937. But it is hard to see how any infusion of money could have remedied the intellectual confusion that has held back chemical embryology. In the words of today's embryologists, theirs is a discipline without "a coherent explanatory framework," a field whose progress has been hampered by "an apparent block to conceptual advance at the very centre."

Even in the 1930s some skeptics noted that there were *too many* candidates for the posited role of "organizer"—a plethora of substances capable of triggering differentiation of bits of embryological tissue in laboratory dishes but unlikely to play any part in the normal development of an embryo. One of the strangest findings came from Needham and Waddington themselves, who demonstrated in 1936 that even nonbiological substances, like the common laboratory dye methylene blue, could "organize" the growth of embryonic tissue.

Faced with too many unassimilated experimental facts, the foremost embryologist of the day, Hans Spemann, moved toward a less and less precise "holism" and began speaking of the "physical" properties of something he called the "embryologic field." The fact that Spemann, an early supporter of the National Socialist party, liked to refer to the "organizer" as a kind of "führer" did not help matters. By

1939 Needham himself was writing that the "chemical nature" of embryological events was "more complex than the first explorers thought."

It would appear that the RF's decision not to support Needham and Waddington in 1937 was based on a sound use of what might be called the administrative calculus. Weaver and Tisdale were tempted by the proposal. But to support it, they had to go directly counter to the advice of many prominent British scientists whose cooperation was essential to the RF's long-term goals in Great Britain. Weaver had to ask himself, Were the prospects for a significant advance in chemical embryology at Cambridge so bright that they outweighed the risk of alienating the leaders of the British scientific community and thereby closing off channels of communications on which Tisdale and the other circuit riders relied so heavily? Weaver's reluctant conclusion: no.

Hindsight confirms that the kind of chemical embryology that Needham and Waddington wanted to pursue in the thirties would almost certainly have proved fruitless; the tools and concepts required to understand the chemical control of embryonic development simply did not exist at the time. Indeed, even with the postwar breakthroughs in tissue culture and techniques of micromanipulation, the current state of embryology is best described, in the words of its own practitioners, as "a puzzlement."

# 43

# Tricks of the Trade

Looking back on his career after retiring from the RF, in 1959, Weaver wondered if he had not sometimes been overly concerned with precision. He knew that he occasionally offended people by his inability to tolerate "sloppy" English in speech or writing; and he recognized in himself what could only be called an obsession with quantification.

A summer job working for the state Geological Survey in the muskeg swamps of northern Wisconsin left him with the unconscious habit of counting his paces whenever he went for a walk or even climbed a flight of stairs. Recalling the impact that this summer of blazing trails and pacing off survey lines had on him, he confessed,

> I almost never go up a flight of stairs without knowing how many steps there are. Time after time I take a walk, quite a walk, and at the end of it I will find myself saying "1364," and will not have the slightest recollection of the fact that I have been counting my steps. But I have.

Despite his passion for precision, however, Weaver was too much the mathematical physicist to be taken in by the false impression of certainty that mere number crunching can give. He was fond of saying that good science helps us "think about uncertainty." He wrote a popular book on the mathematics of probability to show nonscientists

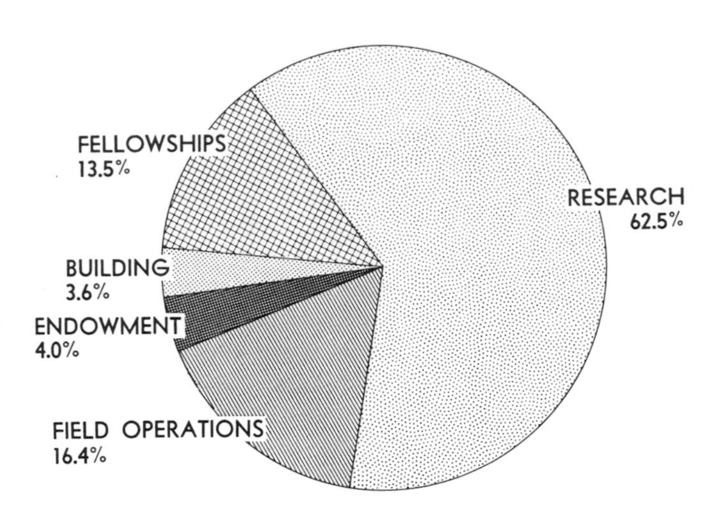


*How the Rockefeller Foundation divided its appropriations for the natural sciences (by type of support) in the Weaver years (1932–1959). Total appropriations: $88,348,093.*

how precisely one could figure the odds on recurring events and how practical such knowledge could be—as long as one was careful not to confuse a good gamble with a sure thing.

When Pnina Abir-Am suggests that Weaver somehow halted the development of chemical embryology in its tracks by withholding funds at a crucial moment, she surely overestimates the power of science patronage. But a well-heeled philanthropoid with a well-defined "shopping list" can wield considerable clout in the scientific community. As Weaver conceded; "It is unrealistic to assume that the announcement of a 'field of special interest' by a foundation does not involve any element of influence on potential recipients." Nor did he see this as something to apologize for. How else, he asked, could the staff of an organization like the Rockefeller Foundation be maintained on "a high level of competence and morale" unless the officers felt they were having a significant effect on the direction and pace of scientific research?

Weaver was frankly elitist in his dealings with the scientific community and in his assessment of his own profession. What he found "clearly intolerable" was the thought of "only moderately able persons using the lure of financial support to influence the activities of more able persons."

The ideal philanthropoid, Weaver came to believe, is "a rather strange individual. He has to be just good enough, but not too good." That is, foundation officers should be smart enough to form their own opinions about developments in a particular field yet "professionally unselfish" enough to avoid imposing those opinions on anyone else. Few philanthropoids meet this high standard. From personal experience, Weaver knew how hard it was for ambitious, hardworking circuit riders to submerge "their own ideas, their own enthusiasms, their own prejudices," in the endless rounds of correspondence and conversation with current and prospective grantees.

But philanthropoids can and should strive for a

> middle ground between pure passivity (in which the foundation operates only as a post-office box into which written requests are deposited) and improper influence. A "request" should be only the opening move in a mutually profitable discussion [between] the professionally competent officers of the foundation . . . and the scholars or scientists. . . . If these two groups cannot meet on equal intellectual terms, something is wrong with one or both of the groups.

For a man who took great pride in his ability to construct precise, even elegant English sentences, Weaver had surprisingly little faith in the written word when it came to evaluating scientific researchers. To judge the worth of a proposal submitted by a scientist, he concluded, it is necessary to look at

> the intellectual capacity, the record of imagination, of dedication and of interest of *the man who proposes to do the work.* The [written] specification of the problem is almost always unimportant. . . . A horribly precise and extensive statistical plan by somebody who really does not have the intellectual competence to do a good job is nothing but a snare and a delusion. . . . The application, even though it is in triplicate, even though it is in twelve duplicate copies, really never proves anything. . . . We get some very stupid ones from some good people, and we get some enormously elaborate and spectacular and polished ones from people who are not very good.

Even less trustworthy, in Weaver's estimation, is written testimony from the applicant's colleagues and superiors: "I have the highest disrespect for letters of recommendation." Such letters, he believed, are almost invariably exercises in hyperbole, entries in a tall-tale competition that induces otherwise honest people to praise candidates in the most extravagant terms on the assumption that the other candidates will be receiving similarly extravagant praise from *their* supporters.

Given the unreliability of such sources, the conscientious philanthropoid has no choice but to become a circuit rider. To do it right, a circuit rider must be more than a scientifically literate "tape recorder on legs." In order to win the confidence of their informants, circuit riders for Weaver's Division of Natural Sciences were called upon to offer a high level of "intellectual companionship"—without becoming "too chummy" with people whose work they had, ultimately, to judge.

While foundation officers have the right, even the duty, to offer advice when an applicant requests it, they have to resist the temptation to participate in the detailed planning of a project. Above all, circuit riders have to know when to back off from a relationship with a successful applicant:

> The grant once made should be [treated] like an adolescent son or daughter: the parent is deeply concerned and may even be worried, but he should have the good sense to realize that his direct personal contribution to the situation is essentially over, and the child must now stand on his own feet.

Striking the proper balance between involvement and detachment is always, in Weaver's words, "one of the most difficult as well as one of the most subtle problems" facing a conscientious foundation officer.

A less subtle but no less difficult task for most philanthropoids is saying no. The Foundation's technical term for rejection in Weaver's day was "declination." While circuit riders often tried to talk a researcher out of submitting a proposal that had little chance of success, formal applications that failed to make the grade always received written declinations. This was partly to spare embarrassment all around.

> But the main point [Weaver noted] is that it is almost impossible to give an oral declination without being maneuvered into explanations, and explained declinations are not only

> embarrassing, they almost always bounce back. Almost as sure as taxes, the petitioner will return within a year and say, in effect, "Doctor, you did us a great service in declining our request last year. Your reasons for the declination were absolutely sound, and in fact you put your finger on precisely the weak points in our set-up. You will be very happy to know that we have now completely remedied these difficulties and so, with eager confidence, we are now back to you with a new request, which has been drawn up in exact accordance with your own advice."
>
> It would obviously be cruel . . . to look anybody in the eye and say, "I'm sorry, brother, but you just don't rate." But this, of course, *is* the real reason. And since the real reason cannot be (in general) reported . . . the actual declination should practically always be made in writing. . . . Written declinations should be brief, firm, kindly, and should be equipped with no handles which the petitioner can grab hold of.

Weaver especially liked the formula supplied by Frederick P. Keppel, the longtime president of the Carnegie Corporation, who used to say, "The perfect declination should be just like a large and beautifully polished sphere. It should be heavy, it should have nothing you could take hold of, and it should be something which simply could not be picked up and hurled back at you."

Basic to Weaver's success as a science patron was his adherence to a few policy guidelines that he had distilled from the experience of his colleagues and predecessors in the Rockefeller philanthropies.

First came a predilection for basic research. The Division of Natural Sciences under Weaver typically supported scientists who sought answers to fundamental questions about nature or who were developing techniques to facilitate the search for such answers. Even Florey's penicillin project had originally appealed to Miller because it dealt with a biological phenomenon of possibly "universal" significance.

Second, Weaver and his circuit riders concentrated their energies on a particular "problem" in science. This made it easier for them to keep up with developments in the laboratories and to identify projects that promised substantial returns for relatively modest investments. Of course, as the experience of Edwin Embree showed, concentration is no virtue if one chooses the wrong problem to concentrate on.

Weaver's choice was inspired; as he had guessed, the times were ripe for a quantitative attack on some of the long-standing mysteries of biology.

Third, having defined a field of concentration, Weaver and his circuit riders insisted on flexibility in interpreting their own rules. When Florey's project veered away from fundamental biology toward the development and manufacture of a still-untested drug, the Division of Natural Sciences continued to back him, despite the obvious risks.

Fourth, by maintaining close contact with current and prospective grantees, senior scientists, academic administrators, and policymakers like Edward Mellanby, Weaver's circuit riders stayed abreast of nonscientific matters that might affect the RF's "investments"—from the purely personal (like the falling-out between Chain and Florey) to the broadly political (like the rise of nazism in German universities).

Although they assumed the "moral responsibility" of judgment, Weaver and his circuit riders did not set themselves up as independent judges of competence in the biological sciences. Rather, they relied on the collective judgment of the leading scientists of Europe and the United States. Yet the RF's relationship to the international scientific community in Weaver's day differed from the basically passive role that the Foundation had assumed toward the National Research Council a decade earlier. Under Weaver the Division of Natural Sciences had a definite policy of its own, which focused on quantitative biology. In pursuit of that policy the Rockefeller Foundation became an integral part of the scientific community's own news-gathering and decision-making apparatus—an apparatus no less influential for being largely informal and "unofficial."

# 44

# More Tricks

In Weaver's view, nothing justified the "tedious and expensive" procedures of circuit riding so much as the fellowship program.

The circuit riders spent from a third to a half of their time on the road; wherever they went, they worked both sides of the fellowship business. Interviewing senior scientists, they made it a point to learn which noted researchers were also good teachers who regularly attracted bright young assistants to their laboratories. At the same time, they kept their eyes open for bright young people who would benefit from studying for a year or two with a good teacher. With the results of their various surveys in hand, the circuit riders could then serve as matchmakers.

In the words of Harry Miller (who died in 1979, at the age of eighty-six), the challenge was to identify the "future leaders of science" and smooth their way to success. Rather than sit back and wait for applicants, the circuit riders went out looking for them:

> We in the Paris Office knew, collectively, all the great scientists. We went around asking each of them who were the most promising young men in their field. We made up these long lists, and then we compared them. If one name kept showing up on several lists, we went to visit him. We made no fuss; essentially we sneaked in unannounced, and met the candidate in his lab.

If Miller made it sound easy, that was partly because he was contrasting his prewar experiences in Europe with his later adventures in South America, where he searched for fellowship candidates in countries that had virtually no academic (much less scientific) traditions.

Even in Europe the process was far from mechanical. Miller quickly discovered that the recommendations of some of the most famous scientists could not be trusted: "Both Einstein and Madame Curie were too softhearted. Young people from all over Europe came to work with them. Young Poles especially flocked to Madame Curie." Unfortunately, most of them had no secure academic post or prospects, and no chance of winning a fellowship. "I had the nasty job of turning down *all* of Einstein's and Madame Curie's requests," Miller recalled. "I never met Einstein; I just corresponded with him. But I used to jump into a taxi and hop over to see Madame Curie, on the other side of Paris." To ease the embarrassment all around, Miller might tell Madame Curie that the "Committee" in New York had rejected a candidate because the application violated one of their rules. In fact, there was no such committee: "The New York office accepted virtually all our recommendations. And there were no 'rules.' While I was the Fellowship officer, I was the 'rules.' "

Even when the circuit riders sought information on a specific candidate, they preferred indirect to direct queries. Asked to assess all the "promising youngsters" who had recently come to his attention, a senior scientist might go on for hours comparing the potential of the solid but somewhat unimaginative Dr. So-and-So with that of the more brilliant but also more unstable Dr. Such-and-Such. In the course of these conversations, new names might come up, which the circuit riders would then add to their lists for future investigation.

Weaver was not a man given to hyperbole, but he thought that circuit riding, as practiced in the Division of Natural Sciences, was a near-perfect method for evaluating fellowship candidates—especially for identifying early signs of excellence among the quieter, less aggressive members of the scientific community.

Yet Weaver always warned newcomers to philanthropy that the RF's fellowship program was a model they should not be in a hurry to imitate. The creation of "a rather elite type of fellowship" with high standards and generous stipends is a relatively safe form of science patronage. In the words of one advocate, "As a rule, young scientists are more creative, more flexible, and less expensive than older scientists. They're a better investment."

But there is a catch. All such programs (the Rockefeller, the Guggenheim, the Fulbright, and so on) are basically looking for the same thing: the best and brightest of the next generation. Given the inevitable competition between these programs, it is reasonable to assume that all of the highly qualified candidates will win some award each year. So the best that the sponsors of a *new* fellowship program can hope to do is skim some of the cream off the top of existing programs. To fill these vacated slots, the sponsors of existing programs have no choice but to say yes to candidates who would have fallen below their cutoff in previous years. In Weaver's words,

> What you have really done, if you create one hundred of these new "elite" fellowships, is to have forced the level of some other organization down so that now one hundred fellowships at the bottom of the list are granted which would not otherwise have been granted. It is inescapably true that when you buy additional fellows, you buy them at the bottom of the heap. You may appear to buy them at the top, but this is always an illusion. You buy them at the bottom of the heap because that's the only place where they're for sale.

Depending on the size and quality of the heap, of course, the talent "at the bottom" may be quite good; but, as Weaver took a mordant pleasure in pointing out, how many philanthropists would care to announce to the world that the main impact of their new benefaction had been to give a second chance to other people's rejects?

In the course of his career as a philanthropoid, Weaver learned to be wary of *any* attempt to predict the social benefits that might result from a particular program or award. More and more, he came to appreciate the importance of tactical flexibility in the pursuit of an unchanging strategic goal, which, in his case, was the Foundation's mandate to contribute to the welfare of mankind through the advancement of knowledge. The Florey grant showed how enormous dividends could accrue from an initially modest investment. But there were all too many examples in the 1930s of ambitious plans that went awry through circumstances beyond the control of the planners.

Kasimir Fajans, the Polish-born chemist who had felt so secure in

Munich in 1933 despite his Jewish ancestry, fled to England three years later with the help of the Foundation. From Cambridge he went on to the United States, where he joined the faculty of the University of Michigan. As for the Institute of Physical Chemistry that Fajans had worked so hard to create, in conjunction first with the International Education Board and then with the Foundation, the building itself fell into the hands of his persecutors. This was only one of a number of Rockefeller-funded buildings, in Munich, in Berlin, in Göttingen, in Madrid, that were either destroyed during the Second World War or put to what Weaver called "very unworthy purposes."

The irony was that both the Foundation and the International Education Board had devoted a large part of their resources in the 1920s to so-called bricks-and-mortar philanthropy because it seemed the safest form of science patronage. It was clear that supporting specific research projects by independent investigators could lead to a high rate of failures or other embarrassing outcomes; but, so the reasoning went, no one could get in trouble for funding an up-to-date laboratory building.

The flaw in this reasoning was made abundantly clear to Weaver and his colleagues in the late thirties when they wrestled with the question of whether to honor an RF commitment to build an institute of physics in Hitler's Berlin. The commitment dated back to 1930 when Max Mason had agreed to finance the institute in partnership with the Kaiser-Wilhelm-Gesellschaft. The project had been "temporarily" shelved the following year because the KWG was unable to raise its share of the funds.

So Foundation officials were caught by surprise when Max Planck wrote to Mason in the summer of 1934 with the "happy news" that sufficient funds for the upkeep and general administration of the institute, as stipulated in the original plans, would be guaranteed through an annual subsidy from the Reich Ministry of Finance. Furthermore, the distinguished Dutch-born physicist Peter Debye had agreed to become director. Since the KWG had now met all the Foundation's requirements, Planck, in his capacity as president of the KWG, asked the Foundation to release the 1.5 million marks that had been promised four years earlier.

This request set in motion a flurry of communications between Weaver, Mason, Tisdale, and other officers and trustees of the Foundation. At issue was whether it was possible, legally and ethically, to break the commitment, or at least delay action until there was a change for the better in the political situation in Germany.

When Planck wrote to Tisdale of his eagerness to create "a modern physical institute of the first rank, the absence of which in Germany we have for years painfully deplored," Tisdale passed along the letter to Weaver with this comment: "The appeal leaves me quite cold when I realize that because of race prejudices they have exiled some of the very men who could have given them the physics which they now claim they so much need." At one point Mason tried to back out of his pledge on the grounds that an unfavorable dollar-exchange rate recently imposed by the Reichsbank would reduce the Foundation's contribution to much less than the sum needed to build an acceptable institute. But Planck used his influence to get the Reichsbank's ruling reversed; as proof that the Foundation's dollars would be converted to marks at a more favorable rate, he sent Mason a copy of a letter from a bank official that ended with the now-standard closing, "Heil Hitler."

Instructed by the RF's Executive Committee to release the funds once the Germans met all the requirements, Mason had no choice but to authorize payment of $360,436.75 to the KWG on March 15, 1935. Construction of the institute got under way in 1936, closely supervised by Debye, who was awarded a Nobel Prize that year. On November 24 a story revealing the Rockefeller connection to the new institute appeared on page one of the *New York Times* under the headline "ROCKEFELLER GIFT AIDS REICH SCIENCE." Several paragraphs were devoted to Raymond Fosdick's explanation that the Foundation was simply carrying out a promise made to the KWG in pre-Hitler days. In a statement drafted with the aid of the RF vice-president Thomas B. Appleget and Warren Weaver, Fosdick went on to say, "The world of science is a world without frontiers. It is quite possible, however, that the foundation would not have made the grant if it could have foreseen present conditions in Germany."

In the last paragraph of the story, the *Times* noted that "several distinguished scientists had been dropped from the Kaiser-Wilhelm-Institute since the ascendancy of the Nazi government," among them the "late Fritz Haber." (Debye, who was treated with surprising courtesy in the early years of the Hitler regime, stayed on as director of the Institute of Physics until 1939, when he was ordered to give up his Dutch passport and become a German citizen. Rather than obey, he returned to the Netherlands; a year later he settled in the United States, where he joined the faculty of Cornell University.)

The *Times* article prompted a letter to Fosdick from Felix Frankfurter, at that time a professor at the Harvard Law School. (He would be named an associate justice of the U.S. Supreme Court in 1939.) Without second-guessing the RF's decision to honor its pledge to the KWG, Frankfurter objected to Fosdick's inclusion of Hitler's Germany in his phrase "the world of science." Failure to draw a clear distinction between Nazi-style thought control and the spirit of free inquiry, Frankfurter said, was to "adulterate the spiritual coinage of the world."

It was not until long after the Second World War that the full truth about the debasement of the scientific community in Nazi Germany became known. The KWG's Institutes of Anthropology and Psychology were, according to one historian, the "spearhead of scholarly advance" in the prewar campaigns of sterilization and euthanasia that prefigured the mass exterminations of the war years. Eugen Fischer, director of the Institute of Psychology, publicly boasted that the "eugenic movement [had] been going far longer" than the Nazi party. Leading members of the Institutes of Anthropology and Psychology supported—as "historically justifiable on biological grounds"—plans for exterminating millions of Russians and Poles. Although these plans were later deemed logistically unsound, the scientists suggested a way to deal with the more manageable population of Gypsies in eastern Europe: they were to be deported to Poland and allowed to freeze to death.

Dr. Fischer, a member in good standing of the German scientific community during the 1930s, was not ungrateful for the generous support he received from the Nazi government. He wrote, "It is a rare and special good fortune for a theoretical science to flourish at a time when the reigning ideology welcomes it, and its findings can serve the policy of the state."

# 45

# New Directions

It has been said that the international influence that Warren Weaver wielded as a "manager of science" was a historical anomaly, never to be repeated in our age of big science and big government. This may seem no great loss to many in the scientific community, since effective control of the purse strings at both the National Science Foundation and the National Institutes of Health (the country's principal patrons of the basic sciences) is now in the hands of the scientists themselves—through a system known as peer review.

The system works essentially like this: Congress determines the total funds available to the NSF and the NIH each year. For fiscal 1987 the NSF budget was just over $1.6 billion; the NIH budget was just under $6 billion. Once Congress sets the budgetary limits, researchers apply for grants to carry out specific projects. Applications in a given field are rated by panels of scientists who are expert in that field. The applicants with the highest ratings get a piece of the pie. In the late seventies about half of all applicants to the NIH received awards; in 1986, when more people applied and budgets were held to minuscule increases under congressional deficit-control procedures, barely a third of the more than eighteen thousand applicants were funded.

Peer review is designed to insulate the grant-giving process from political pressures that might divert funds to a particular recipient, locality, or facility on some basis other than the scientific value of the research proposed. Similar mechanisms are used by most scientific journals to determine which papers submitted for publication will be printed. The assumption is that when scientists evaluate their peers, they will base their judgments on objective criteria.

Just how valid is this assumption? Critics of the NSF and the NIH have long contended that "grant proposals that are truly innovative and outside the mainstream often go unfunded," while special congressional appropriations for popular causes (like the "war on cancer" and "problems of the disabled") periodically distort the agendas and strain the budgets of the science-support agencies.

Partly in response to these criticisms, the NSF reviewed its own peer-review procedures a few years ago—and discovered a much greater degree of subjectivity than even the critics had suspected. Contrary to expectations, there was "substantial disagreement" among equally qualified reviewers in the same field concerning the "scientific value" of a particular proposal; and this held true in the natural sciences as well as the social sciences. In fact, the study showed that "about 25% of NSF decisions [to fund or not to fund] would have been reversed by a different panel." Stated another way, "the fate of a particular grant application is roughly half determined . . . by apparently random elements which might be characterized as the 'luck of the reviewer draw.' "

If scientists cannot be objective in evaluating other scientists, one might ask, who can? The important question, however, is not whether peer review should be replaced by something more "objective" but whether the current system needs to be *supplemented* with other systems of management and review.

No one has suggested doing away with peer review, because it is as correct today as it was in 1950, when the National Science Foundation was established, to say,

> Only a chemist [can] appraise the merits of a proposal . . . to explore the *Reaction of Alpha, Beta-Unsaturated Ethers with Grignard Reagents.* Only a mathematician [is] in a position to pass judgment on the desire of an . . . investigator to look into *Asymptotic Behavior and Stability Problems.*

Of course, it does not take an expert to draw up a list of social problems that need solving. When the Nixon administration and its congressional allies launched the so-called war on cancer in the 1970s, it was not the choice of target that was flawed. The flaw lay in assuming that solutions to recalcitrant scientific problems can be purchased—to order, as it were—through goal-oriented contracts with researchers. The lesson of the war on cancer was that no amount of money can produce a "quick fix" when fundamental knowledge is lacking.

As we have seen, Weaver's immediate predecessor at the Rockefeller Foundation, Herman Spoehr, placed high on his list of problems that needed urgent attention from the scientific community the mystery of photosynthesis: How do plants trap the energy of sunlight and store it in a form we can use for food and fuel? A better understanding of this process, Spoehr argued, would lead to more efficient farming methods and, therefore, to more abundant food crops.

Since Spoehr himself was an authority on the subject, it is possible that, had he remained at the Foundation, he might have persuaded the trustees to spend millions of dollars in support of research on photosynthesis in the 1930s. Whether this would have led to a quick breakthrough in understanding is doubtful. Although the "basic equation of photosynthesis" was known before the beginning of the nineteenth century, deeper insights into the process had to wait until more powerful tools of inquiry became available after the Second World War. And even with these tools, investigators are still a long way from bringing the key events of photosynthesis under control.

But there is another irony here. Having "missed" its chance to become a patron of photosynthesis research in the thirties, the Rockefeller Foundation, almost by accident, became involved a decade later in a worldwide "war on hunger" based on plant-breeding techniques. The seed from which the so-called green revolution grew was planted on February 3, 1941, during a lunchtime conversation between Raymond Fosdick, then president of the Foundation, and Henry Wallace, who had been Franklin D. Roosevelt's secretary of agriculture and was now his vice-president elect.

Wallace had just returned from Mexico, where he had been appalled by the sickly appearance of cornfields by the side of the road. Speaking not as a politician but as a successful breeder of hybrid corn, he told Fosdick that nothing would contribute more to the welfare of the Mexican people than a substantial increase in their per-acre yields of corn and beans.

Fosdick mentioned this to Weaver, who mentioned it to Dusty Miller. Following his involuntary exile from the Foundation's Paris office in

1940, Miller had been asked to put together some kind of natural-sciences program in Latin America; he had already been to Mexico to get a look at conditions there. Now he and Weaver recruited three distinguished American agricultural scientists to explore the options for aiding Mexican farmers. On the basis of the unanimous recommendations of this three-man survey team, the Foundation decided, in 1942, to organize an agricultural program in Mexico as an *operating* agency. The model was the RF's International Health Division, which had been active in Latin America for over two decades. All research and training would be conducted under the Foundation's auspices (in close cooperation with the Mexican government) rather than through grants to other institutions. The obvious reason for this decision was that there were no institutions in Mexico of sufficient stature to carry on the work independently.

The story of the success of this program under the leadership of J.George Harrar, a plant pathologist who later served as president of the Foundation, has been told elsewhere. In brief, the green revolution has enormously increased the production of food in a number of countries, helping to keep supplies in step with a world population that nearly doubled between 1950 and 1980 and that will reach six billion by 1990.

In Mexico food production doubled during the first twenty years of the RF program, with a threefold increase in per-acre yields of wheat and a twofold increase in corn and beans. From Mexico green-revolution programs in plant breeding spread to other countries in Latin America and Asia. The results are succinctly summarized by Michael Lipton of the University of Sussex, who wrote in 1985, "If the farmers of the Third World today used the same cereal varieties as in 1963–64, and everything else were unchanged, then tens of millions of people would this year die of hunger." Instead, there is now "a substantial world surplus of food."

With the establishment of the program in Mexico, the center of gravity of the Rockefeller Foundation's Division of Natural Sciences began to shift from experimental biology to agricultural science. This shift was accelerated after the Second World War as changing conditions in this country and abroad removed much of the rationale for the science-support programs that Weaver had initiated in the thirties.*

*In addition to the burgeoning agricultural projects under Harrar's direction, Miller conducted an opportunistic program in the natural sciences

In the United States the federal government assumed the lion's share of responsibility for the support of science through the vast postwar expansion of the National Institutes of Health and the creation of the National Science Foundation in 1950. These public agencies, and others with more specific mandates like the Atomic Energy Commission and the National Aeronautics and Space Administration, had far greater resources to invest in research grants and fellowships than private institutions like the Rockefeller Foundation could ever marshal.

The governments of postwar Europe also placed a high priority on public support of science, although the Foundation played an important, if temporary, role in helping rebuild shattered laboratories and educate young researchers whose training had been interrupted by the war.

We have seen how even so eminent a figure as Howard Florey relied on the Foundation for hard-to-get supplies through 1955. Gerard A. Pomerat, a biologist whom Weaver hired as a circuit rider in 1946, made a significant contribution, as unofficial consultant and disburser of transitional funds, to the restructuring of French science in the late 1940s; this was the period when the Centre National de la Recherche Scientifique (CNRS), a government agency outside the control of the universities, became firmly established as "the great guarantee of research quality in France."*

throughout Latin America from 1941 to 1960; all told, he traveled some half million miles, seeking out fellowship candidates, funding worthwhile research wherever he could find it, and trying to stimulate the development of self-sufficient scientific communities in countries like Brazil, Colombia, Uruguay, Peru, and Chile.

*Gerard Pomerat died in 1980, at the age of seventy-eight.

# 46

# The Art of Science Support

It is a curiosity (some might say an embarrassment) of our Age of Science that no one, including scientists, really understands the nature of the scientific enterprise. Although modern science is unsurpassed as a means of producing objectively verifiable information, the few attempts that have been made to define the "scientific method" in objective terms (from those of Francis Bacon to those of Karl Popper and Thomas Kuhn) satisfy neither philosophers nor historians of science nor working researchers. A well-known gibe in scientific circles holds that "the philosophy of science is just about as useful to scientists as ornithology is to birds."

Most working researchers simply ignore the problem, preferring to define science as "what scientists do when they do science." This, of course, begs the question of who qualifies as a scientist. One irrefutable answer is that a scientist is anyone in good standing as a member of the scientific community. From an organizational point of view, this community is characterized by the following:

- a consensus about the kinds of problems that are suitable at a given time for investigation (for example, the beginning of the universe and the origin of life on Earth, but *not* reincarnation or time travel)
- a commitment to certain restraints on the generating, interpreting, and reporting of data (for example, holding a press conference only

*after* an account of your results has been accepted by or at least submitted to a refereed journal)*
• a set of rules for allocating credit for discoveries (for example, senior versus junior authorship) and for punishing deviant behavior (for example, a collegial cold shoulder for poaching on someone else's problem; professional ostracism for outright plagiarism or proven fraud)

In the absence of any agreement about the nature of the scientific method, it is not surprising that the skills necessary to "do science" are taught by perhaps the oldest pedagogical system known, the apprentice system. Promising young researchers learn by assisting senior scientists at work; in the words of Julius Axelrod of Johns Hopkins University, who won a Nobel Prize in 1970 for his research on neurotransmitters, "You can't separate it, the science comes out of the relationship."

This is presumably why fellowship programs, which are designed to nurture this relationship, have been among the most successful forms of science patronage. Warren Weaver put it unequivocally: "If you do not have a laboratory in which young people are being inspired, in which their curiosity and their energy is being developed . . . then you do not have a situation that deserves large and enthusiastic support." He also liked to quote his mentor at the University of Wisconsin, Charles Slichter, who said that "every laboratory ought to have around it at least one damn fool who does not know what you can't do."

Weaver saw science as "a way of solving . . . a large class of important and practical problems; namely those in which the predominant factors are subject to the basic laws of logic and are usually quantitative in nature." But as a man who never lost his religious faith, he was also fascinated by the limits of science. He understood the shaky philosophical ground on which the impressive achievements of science rest. "Confirmation in science . . . has the attraction of being widespread," he said. "But it is by no means universal; the agreement is widespread only in the sense that the vote is preponderantly favorable within a small elite." In his autobiography Weaver testified to his belief

*A refereed journal is one that submits every article to a jury of the author's professional peers to determine whether the article is worthy of publication (with or without revisions).

> that the scientific enterprise is basically an artistic endeavor, that it has all the freedom of any other imaginative and creative activity, this activity being characterized by very special traditions of disinterested and unprejudiced open-mindedness together with a built-in protection against serious or prolonged error. Science is one of the most mature of the arts, combining a maximum of both freedom and discipline.

Since the scientific community cannot at present explain its methodology in objective terms or guarantee objectivity in its internal operations, how can the larger society that foots the bills ensure that it is getting full value for its science-support dollars?

The researchers who conducted the National Science Foundation's study of peer review came out in favor of a more adventurous approach by science patrons: "If it is difficult to determine which project will lead to a major breakthrough, then granting agencies should fund a wide range of research so as to reduce the probability that development of important ideas will be delayed because of lack of support." To put it another way, in order to encourage creative risk taking in science, we must be willing to take risks in the support of science. This is easier said than done, but it seems clear that the greater the number of channels of support that are available, the greater the chances that "risky" projects will find funding.

As it happens, there has been a trend in recent years toward more flexibility in the organization of science patronage. Major universities like MIT, Harvard, Yale, and Stanford have joined with biotechnology companies in cooperative ventures whose purpose is to fund basic research and, in the long run, to reap profits through the commercial exploitation of licenses and patents. In 1985, industrial firms provided more money to support biomedical research in the United States than the NIH did. It has been estimated that nearly one-fourth of all funds for biotechnology research in academic settings come from "university-industry research relationships."

Critics of these partnerships, which already number more than five hundred, worry about what will happen when conflicts arise between the protection of proprietary rights and the tradition of prompt public access to the fruits of academic research. Similar fears were voiced when the government got into the business of supporting science on a large scale after the Second World War. It will be several years before the new joint ventures between independent universities and profit-seeking entities can be judged in terms of profitability, let alone scientific and social utility.

Warren Weaver often cited statistics to show how the relative importance of private support of scholarship had dwindled since the early days of the great foundations. In the second decade of the century, he wrote,

> the annual grants of the General Education Board and the Carnegie Corporation were roughly one-fifth of the total of the annual budgets of all colleges and universities in the country; indeed, the Carnegie Corporation in 1913 spent $5.6 million, whereas the federal government's total expenditure for education in that same year amounted to $5 million.

By the mid-1960s the federal government was spending some $15 billion annually on research and development activities; and federal grants and contracts accounted for more than two-thirds of the total research expenditures at American colleges and universities. The two largest private philanthropies in the world as of January 1987, the Howard Hughes Medical Institute (HHMI) and the Ford Foundation, each had total assets in excess of $5 billion, but the hundreds of millions of dollars they spend each year to support research is only a small proportion of the public funds available for the same purpose.

For truly vast undertakings innovative ways have been found to pool public and private money. Responsibility for funding green-revolution programs around the world was assumed in 1972 by the Consultative Group on International Agricultural Research (CGIAR), a consortium of public and private donors whose members include the Rockefeller and Ford foundations. In 1985, thirty-eight governments, international agencies, developmental organizations, and private philanthropies contributed $181 million to the thirteen agricultural research stations in the CGIAR system.

The Howard Hughes Medical Institute became the world's largest private philanthropic entity in December 1985 with the sale of its principal asset, the Hughes Aircraft Company, to General Motors for over $5 billion. Created in 1953 as a tax dodge by the eccentric billionaire, the HHMI is now the most important nongovernmental patron of science in the world. Its annual expenditures are more than $230 million and will exceed $300 million by the 1990s; most of this clout is concentrated in the fields of genetics, cell biology and regulation, immunology, neuroscience, and structural biology.

Technically, the Hughes Institute is not a "foundation" but a "medical research organization," which means that instead of awarding grants, it must spend a certain percentage of endowment income each year on the actual conduct of medical research. Accordingly, it has entered into a series of "collaborations" with academic medical centers, building and equipping new laboratories and putting distinguished researchers on its payroll. By the fall of 1987 the number of these collaborations stood at twenty-seven, including a major unit at the Rockefeller University in New York.*

At the Rockefeller unit, which is devoted to fundamental and disease-related biology, two members of the University faculty, Gunter Blobel and Joseph R. Nevins, were appointed Hughes investigators. These appointments are considered extraordinary plums in the scientific community. Support is guaranteed for up to seven years before formal renewal. In addition, the HHMI provides very generous equipment allowances. No applications are accepted; appointments are made on the basis of track record only, and the choice of specific research projects is left to the investigators themselves. Progress is monitored annually by HHMI staff and through site visits by professional peers every three or four years.

In addition to its direct support of medical research, the HHMI agreed, under the terms of its settlement with the Internal Revenue Service, to spend $500 million over a ten-year period on foundation-like grants; major awards have been announced to bolster science education on the precollege, undergraduate, and graduate levels and to support "national scientific resources" like the Cold Spring Harbor Biological Laboratory.

The Rockefeller Foundation, with assets of $1.8 billion, has in recent years years devoted more than half of its annual appropriations to problems plaguing the so-called Third World. In 1984 a trustee task force was named to reevaluate the RF's programs in developing countries. The report of this task force, made public in April 1986, committed the Foundation to doubling its expenditures on international programs to between $250 and $300 million over the next five years. The long-range goal is to promote advances in agriculture, health,

*The Rockefeller University is the successor to the Rockefeller Institute for Medical Research, founded in 1901 by John D. Rockefeller, Sr.The Rockefeller University is a corporate entity totally separate from the Rockefeller Foundation.

and the population sciences (including contraception) that will fit the needs of specific developing countries. For example, recombinant DNA techniques and other sophisticated research approaches will be brought to bear on the improvement of crops like rice, sorghum, and millet that have the "greatest promise of alleviating hunger and improving nutrition."

Another program, known as INCLEN (International Clinical Epidemiology Network), trains physicians from developing countries in Africa, Asia, and Latin America to look at medical problems from a larger perspective. A major goal is to stimulate critical thinking about the behavioral aspects of disease and disease prevention. In addition to studies in epidemiology, biostatistics, and research design, INCLEN fellows learn to weigh the economic and social implications of medical interventions, with an eye toward framing cost-effective local and national health policies.

In a sense, such programs are a return to the Foundation's "missionary" roots—only instead of bringing proven solutions to the unenlightened, the RF today traffics in intellectual tools that will enable people in developing countries to define and solve their own problems.

# 47

# Experiments

Summing up his quarter century of experience as a philanthropoid, Weaver said,

> I think a great many people suppose that being an officer of a philanthropic foundation must be just about the simplest thing in the world, because what could be simpler than getting rid of money? Well, if all you have to do is to throw hundred dollar bills out of the window, then indeed one could get his day's work done in few seconds and go home and rest. But if you take the position that these dollars must be used for society in ways which promise to help the maximum number of individuals, both in this country and in other parts of the world, relative to their really important long range problems, then you face the almost impossibly difficult questions as to where you put any one dollar, what kind of leverage you can assign to it, the individuals to whom you unhappily have to say no, and how you can possibly learn enough about anybody so that you can feel comfortable about saying yes.

These difficulties are not unique to patrons of science. But while patrons of the arts may worry about saying yes or no to the "wrong" people, their errors of judgment do little harm outside their own field. By contrast, a patron of science must face the fact that his benefactions can cause suffering on a global scale. No amount of philanthropic support for researchers of the Davenport school could make eugenics a science. But large amounts of money in the wrong hands could make bad ideas more dangerous—with terrible consequences in this country and abroad.

In the search for "safe" innovations in science patronage, a contributor to the scientific journal *Nature* has suggested taking a leaf from the past—the eighteenth century, to be exact, when governments and private citizens offered large prizes as inducements for inventors to work on specific projects.

The model proposed by David Horrobin, managing director of a biomedical company, is the British government's offer in 1714 of £20,000 for the invention of a really reliable timepiece. To win the prize, an inventor had to provide a device that would allow sailors anywhere in the world to tell "Greenwich time" with enough accuracy to calculate their longitude to within a degree or less. Competition for the prize money was keen. Eventually, the British navy got a reliable "chronometer," and the inventor, a self-taught mechanic named John Harrison, got his prize money—although not without a long struggle against the judges, who kept asking him to improve the accuracy of his device even after he had satisfied the original requirements.

According to Horrobin, if governments today offered large enough prizes for solutions to such recalcitrant problems as "schizophrenia, eczema, multiple sclerosis and Alzheimer's disease," they would stimulate fresh thinking by experts and nonexperts alike, and foster creative "cross-pollination" between disciplines. Initial investment would be trivial; prize money would be awarded only on verification of a workable solution—the social benefits of which would presumably pay for the prize many times over. Furthermore, this kind of retrospective philanthropy would require no management of science at all.

Skeptics have argued that all such attempts to stimulate breakthroughs-to-order can at best produce technological, as opposed to scientific, advances. But Horrobin contends that advances in "practical research" are as likely to lead to "rapid progress in basic science" as the other way around. He cites the career of Louis Pasteur as evidence that "brilliant people devoting themselves to practical problems, such as the spoiling of wine or dying silkworms, may end up by discovering new fundamental principles."

Warren Weaver would have been amused, and I think stimulated, by Horrobin's suggestion—although, being an amateur historian of science, he might have pointed out that the king of Spain offered a generous prize for an accurate chronometer at the end of the sixteenth century, to no avail.

Historians confirm that "the idea of prizes for the solution of important practical problems was quite common in eighteenth century Europe." They also document the gradual disenchantment with this idea in the nineteenth century. The experience in France may be typical. In 1820 a wealthy French amateur of science named Baron Montyon died at the age of eighty-seven; his will provided 20,000 francs for the establishment of scientific breakthroughs. But within a few years the executors of the Montyon legacy, all members of the Académie des Sciences, were having difficulty finding worthy recipients; with the approval of the courts, they began disbursing funds in the form of grants to established researchers, among them a chemist named Louis Pasteur.

Of course, no analogies from the past can tell us whether a carefully thought-out system of prizes for specific achievements would be a useful supplement to current programs of science patronage. The only way to find out would be to conduct an experiment in philanthropy comparable to those initiated by Rockefeller and Carnegie and their advisers in the early decades of this century.

If Weaver learned one thing at the Rockefeller Foundation, it was that there is no "right" way to go about supporting scientific research:

> I have come to the conclusion that the only way in which to approach any intellectual enterprise, whether it is research in science or whatever, is to determine what things handicap the scholar from the complete and effective release of his potential capacity. And then when you find out what really handicaps him, the thing which support should do is to remove those handicaps, one at a time, in the order of their urgency.
>
> If the most urgent thing in the whole list is the building of a building or the revamping of an old laboratory or the putting on of a new wing or the putting on of an intermediate floor—if this is the top item in urgency, then it seems to me that this is the thing toward which philanthropic money should be directed. If release of time is the next most urgent thing, then this should be supported, etc., etc.
>
> There are cases in which some very unglamorous things

> stand high in this list. It is even conceivable that what a scholar needs most more than anything else is a little desk computer or a typewriter capable of handing the strange sort of type which is important in his work, or it may even be the salary of a secretary, or it may be some books, or it may be the freedom to travel, or it may even be the chance to go off to some quiet spot to sit down and think for a while.
>
> I am of the conviction that any philanthropic entity that wants to help scholarship should be willing to look at any and all of these things without prejudice.

Implicit in Weaver's formulation is the belief that a philanthropic entity, as a friendly outsider, may play a useful role not just in supplying the cash but also in helping creative people decide what their needs are in the first place.

Even if all they need is some wooden kitchen tables.

# Selected Bibliography

Abraham, Edward P. "Howard W. Florey." *Biographical Memoirs of Fellows of the Royal Society* 17 (1971): 255–302.

Andrade, Edward N. da C. *A Brief History of the Royal Society*. London, 1960.

Bates, Ralph. *Scientific Societies in the United States*. Cambridge, Mass., 1965.

Beyerchen, Alan D. *Scientists under Hitler: Politics and the Physics Community in the Third Reich*. New Haven, 1977.

Bickel, Lennard. *Rise Up to Life: A Biography of Howard Walter Florey, Who Gave Penicillin to the World*. New York, 1973.

Birch, Thomas. *The History of the Royal Society of London*. Introduction by A. Rupert Hall. Hildesheim, 1968.

Bronowski, Jacob, and Bruce Mazlish. *The Western Intellectual Tradition*. London, 1960.

Carnegie, Andrew. *The Gospel of Wealth and Other Timely Essays*. Cambridge, Mass., 1962.

Chain, Ernst. "Thirty Years of Penicillin Therapy." *Proceedings of the Royal Society B* 179 (1971): 293–319.

Chamberlain, John. *The Enterprising Americans: A Business History of the United States*. New York, 1963.

Christianson, Gale E. *This Wild Abyss: The Story of the Men Who Made Modern Astronomy*. New York, 1978.

Coben, Stanley. "Foundation Officials and Fellowships: Innovation in the Patronage of Science." *Minerva* 14 (Summer 1976): 225–40.

Colebrook, Leonard. "Alexander Fleming." *Biographical Memoirs of Fellows of the Royal Society* 2 (1956): 117–27.

Collier, Peter, and David Horowitz. *The Rockefellers: An American Dynasty*. New York, 1976.

Corner, George W. *A History of the Rockefeller Institute, 1901–1953: Origins and Growth*. New York, 1964.

Daniels, George H. "The Process of Professionalization in American Science: The Emergent Period, 1820–1860." *Isis* 58 (1967): 151–66.

———, ed. *Nineteenth-Century American Science*. Evanston, Ill., 1972.

Dunn, L. C. "Cross Currents in the History of Human Genetics." In *Heredity and Society*. New York, 1977.

Fisher, Donald. "The Rockefeller Foundation and the Development of Scientific Medicine in Great Britain." *Minerva* 16 (Spring 1978): 20–41.

Florey, H. W., E. Chain, N. G. Heatley, M. A. Jennings, A. G. Sanders, E. P. Abraham, and M. E. Florey. *Antibiotics*. Oxford, 1946.

Fosdick, Raymond B. *The Story of the Rockefeller Foundation: 1913–1950*. New York, 1952.

———. *Adventure in Giving: The Story of the General Education Board*. New York, 1962.

Gates, Frederick Taylor. *Chapters in My Life*. New York, 1977.

Goodspeed, Thomas Wakefield. *A History of the University of Chicago: The First Quarter Century*. Chicago, 1972.

Gray, George W. *Education on an International Scale*. New York, 1941.

Hall, Marie Boas. *All Scientists Now: The Royal Society in the Nineteenth Century*. Cambridge, 1984.

Hare, Ronald. *The Birth of Penicillin*. London, 1970.

Harrar, J. George. *Strategy toward the Conquest of Hunger: Selected Papers of J. George Harrar*. New York, 1967.

Hendrick, Burton J. *The Life of Andrew Carnegie*. London, 1933.

———. *The Benefactions of Andrew Carnegie.* New York, 1935.

Hollis, Ernest Victor. *Philanthropic Foundations and Higher Education.* New York, 1938.

Judson, Horace Freeland. *The Eighth Day of Creation.* New York, 1979.

Kevles, Daniel J. *The Physicists: The History of a Scientific Community in Modern America.* New York, 1978.

———. *In the Name of Eugenics.* New York, 1985.

Kohler, Robert E. "The Management of Science: The Experience of Warren Weaver and the Rockefeller Foundation Programme in Molecular Biology." *Minerva* 14 (Autumn 1976): 279–306.

———. "A Policy for the Advancement of Science: The Rockefeller Foundation, 1924–29." *Minerva* 16 (Winter 1978): 480–515.

Kohlstedt, Sally G. *The Formation of the American Scientific Community.* Urbana, Ill., 1976.

Kuper, Adam. *Anthropologists and Anthropology: The British School, 1922–1972.* New York, 1973.

Lomask, Milton. *A Minor Miracle: An Informal History of the National Science Foundation.* Washington, D.C., 1976.

Ludmerer, Kenneth M. *Genetics and American Society.* Baltimore, 1962.

MacFarlane, Robert Gwyn. *Alexander Fleming: The Man and the Myth.* Cambridge, 1984.

———. *Howard W. Florey: The Making of a Great Scientist.* New York, 1979.

MacLeod, Roy M. "The Support of Victorian Science: The Endowment of Research Movement in Great Britain, 1868–1900." *Minerva* 9 (April 1971): 197–230.

McKie, Douglas. *The Royal Society: Its Origins and Founders.* London, 1960.

Miller, Harry M. Recorded interview. Columbia Oral History Project, 1967.

Miller, Howard S. *The Legal Foundations of American Philanthropy, 1776–1844.* Madison, Wis., 1961.

———. *Dollars for Research: Science and Its Patrons in Nineteenth-Century America.* Seattle, 1970.

Morrell, Jack, and Arnold Thackray. *Gentlemen of Science: Early Years of the British Association for the Advancement of Science.* Oxford, 1981.

Moscow, Alvin. *The Rockefeller Inheritance*. New York, 1977.

Nevins, Alan. *John D. Rockefeller: A Study in Power*. New York, 1953.

Plotkin, Howard. "Edward C. Pickering and the Endowment of Scientific Research in America, 1877–1918." *Isis* 69 (1978): 44–57.

Purver, Margery. *The Royal Society: Concept and Creation*. London, 1967.

Reeves, Thomas C., ed. *Foundations under Fire*. Ithaca, N.Y., 1970.

Reingold, Nathan. "The Case of the Disappearing Laboratory." *American Quarterly* 29 (Spring 1977): 79–101.

Robinson, Robert. "The Early Stages of the Chemical Study of Penicillin." *Memoirs and Proceedings of the Manchester Literary and Philosophical Society* 92 (1950–51): 124–43.

Rockefeller, John D. *Random Reminiscences of Men and Events*. London, 1909.

Rockefeller Foundation. *Directory of Fellowships and Scholarships, 1917–1970*. New York, 1972.

Rosenberg, Charles. *No Other Gods: On Science and American Social Thought*. Baltimore, 1976.

Seligman, Ben B. *The Potentates: Business and Businessmen in American History*. New York, 1971.

Sprat, Thomas. *History of the Royal Society*. Edited by Jackson I. Cope and H. Whittemore Jones. St. Louis, 1958.

Stimson, Dorothy. *Scientists and Amateurs: A History of the Royal Society*. New York, 1968.

Weart, Spencer R. *Scientists in Power*. Cambridge, 1979.

Weaver, Warren. Recorded interview. Columbia Oral History Project, 1961.

———. *Lady Luck: The Theory of Probability*. New York, 1963.

———. *Science and Imagination: Selected Papers*. New York, 1967.

———. *U.S. Philanthropic Foundations: Their History, Structure, Management and Record*. New York, 1967.

———. *Scene of Change: A Lifetime in American Science*. New York, 1970.

Whewell, William. *Selected Writings in the History of Science* Chicago, 1984.

Williams, Trevor I. *Howard Florey: Penicillin and After*. Oxford, 1984.

Wilson, David. *In Search of Penicillin*. New York, 1976.

Zuckerman, Harriet. *Scientific Elite: Nobel Laureates in the United States.* New York, 1977.

Zuckerman, Harriet, and Robert K. Merton. "Patterns of Evaluation in Science: Institutionalisation, Structure and Functions of the Referee System," *Minerva* 9 (January 1971): 66–100.

# Notes

All notes are by chapter and paragraph. Each paragraph is identified by its order in the chapter and by its opening words.
The following abbreviations identify collections of manuscripts, letters, and other material in the Rockefeller Archive Center (RAC), Hillcrest, Pocantico Hills, North Tarrytown, NY 10591:

| | |
|---|---|
| RF | Rockefeller Foundation |
| RFA | Rockefeller Family Archives |
| IEB | International Education Board |
| GEB | General Education Board |
| BSH | Bureau of Social Hygiene |
| LSRM | Laura Spelman Rockefeller Memorial |

## Chapter 1: From Charity to Philanthropy

Paragraph 3 ("Historians of modern philanthropy . . ."): Barry D. Karl and Stanley N. Katz, "The American Private Philanthropic Foundation and the Public Sphere," *Minerva* 19 (Summer 1981): 243–44.

Paragraph 4 ("This was something new . . ."): Ibid.

Paragraph 7 ("The goal of competition . . ."): Peter Collier and David Horowitz, *The Rockefellers: An American Dynasty* (New York, 1976), p. 29; Ben

B. Seligman, *The Potentates: Business and Businessmen in American History* (New York, 1971), pp. 195–96.

Paragraph 8 ("Perhaps Morgan's greatest achievement . . ."): Seligman, *Potentates,* p. 198.

Paragraph 9 ("On September 6, 1901, . . ."): Alvin Moscow, *The Rockefeller Inheritance* (New York, 1977), pp. 74–75.

Paragraph 11 ("If the nineteenth century . . ."): Andrew Carnegie, *The Gospel of Wealth and Other Timely Essays* (Cambridge, Mass., 1962), p. 20.

Paragraph 13 ("The cynical may view . . ."): Howard S. Miller, *The Legal Foundations of American Philanthropy, 1776–1844* (Madison, Wis., 1961).

Paragraph 14 ("But Andrew Carnegie had . . ."): Carnegie, *Gospel,* p. 27.

Paragraph 15 ("In the June 1889 . . ."): Ibid.; Seligman, *Potentates,* pp. 203–5.

Paragraph 16 ("But Carnegie combined his . . ."): Carnegie, *Gospel.*

Paragraph 17 ("To help his fellow millionaires . . ."): Ibid.

Paragraph 18 ("In carrying out this . . ."): Seligman, *Potentates,* p. 140.

Paragraph 20 ("As an avowed Social Darwinist . . ."): Warren Weaver, *U.S. Philanthropic Foundations: Their History, Structure, Management and Record* (New York, 1967), p. 28.

Paragraph 21 ("Carnegie's gifts to his . . ."): Burton J. Hendrick, *The Benefactions of Andrew Carnegie* (New York, 1935); Seligman, *Potentates,* p. 184.

Paragraph 22 ("There had been few . . ."): Hendrick, *Benefactions.*

Paragraph 23 ("In both its scale . . ."): Ibid., pp. 13–14.

Paragraph 24 ("By the time of . . ."): Ibid.; Carnegie, *Gospel.*

Paragraph 25 ("One of the most . . ."): Raymond B. Fosdick, *The Story of The Rockefeller Foundation, 1913–1950* (New York, 1952), p. 6.

Paragraph 26 ("The grandeur of Carnegie's . . ."): Genesis, Numbers, and Leviticus.

Paragraph 27 ("Ledger A, a daily . . ."): Fosdick, *Story,* pp. 1–9; Collier and Horowitz, *Rockefellers,* p. 48.

## Chapter 2: Scientific Giving

Paragraph 1 ("Toward the end of . . ."): Frederick Taylor Gates, *Chapters in My Life* (New York, 1977), pp. 91–113; Thomas Wakefield Goodspeed, *A History of the University of Chicago: The First Quarter Century* (Chicago, 1972), p. 1–68.

Paragraph 2 ("The scope of the . . ."): Goodspeed, *History,* p. 498; Weaver, *Philanthropic Foundations,* p. 36.

Paragraph 3 (" 'The best philanthropy,' Rockefeller . . ."): John D. Rockefeller, *Random Reminiscences of Men and Events* (London, 1909), p. 142.

Paragraph 5 ("Throughout his career he . . ."): Alan Nevins, *John D. Rockefeller: A Study in Power* (New York, 1953), p. 281.

Paragraph 6 ("As a philanthropist Rockefeller . . ."): Gates, *Chapters,* p. 159; Rockefeller, *Reminiscences,* p. 181.

Paragraph 7 ("In hiring Gates, Rockefeller . . ."): Collier and Horowitz, *Rockefellers,* p. 38.

Paragraph 9 ("The phrase was a . . ."): Gates, *Chapters,* pp. 161–62.

Paragraph 10 ("Gates set out to . . ."): Ibid., p. 161.

Paragraph 11 ("Once Rockefeller's benefactions were . . ."): Ibid., p. 188.

Paragraph 12 ("By this time . . ."): Ibid., pp. 129–31.

Paragraph 13 ("On coming east to . . ."): Ibid.; RF 900.23.174, Alan Gregg's comments on personalities instrumental in developing original program and policies of the RF.

Paragraph 14 ("Rarely has the aspect . . ."): Gates, *Chapters,* pp. 131, 187–88.

Paragraph 15 ("This was the vision . . ."): Ibid., p. 242.

Paragraph 17 ("Long after he had . . ."): George W. Corner, *A History of the Rockefeller Institute, 1901–1953: Origins and Growth* (New York, 1964), p. 41.

Paragraph 18 ("In living to the . . ."): Collier and Horowitz, *Rockefellers,* p. 61 n.

Paragraph 21 ("Whether or not he . . ."): Corner, *History,* p. 16; Martin Bulmer and Joan Bulmer, "Philanthropy and Social Science in the Nineteen Twenties: Beardsley Ruml and the Laura Spelman Rockefeller Memorial, 1922–1929," *Minerva* 19 (Autumn 1981): 401; Weaver, *Philanthropic Foundations,* p. 37.

Paragraph 22 ("That Rockefeller could act . . ."): Gates, *Chapters,* pp. 197–98.

Paragraph 24 ("The resulting shift in . . ."): Ibid., pp. 198–205.

Paragraph 25 ("Between 1910 and 1913 . . ."): Daniel J. Kevles, *The Physicists: The History of a Scientific Community in Modern America* (New York, 1978), pp. 68–69; Fosdick, *Story,* pp. 14–23.

Paragraph 26 ("Even if circumstances had . . ."): Seligman, *Potentates,* p. 192.

Paragraph 27 ("In 1909 he published . . ."): Collier and Horowtiz, *Rockefellers,* p. 67.

Paragraph 28 ("In an era when . . ."): Seligman, *Potentates,* p. 211; Rockefeller, *Reminiscences,* p. 140.

## Chapter 3: Prudent Gamblers

Paragraph 8 ("In placing aid to . . ."): Weaver, *Philanthropic Foundations,* p. 34; Goodspeed, *History,* p. 28.

Paragraph 9 ("Between them Carnegie and . . ."): Collier and Horowitz, *Rockefellers,* p. 50.

Paragraph 10 ("One alternative was to . . ."): Ernest Victor Hollis, *Philanthropic Foundations and Higher Education* (New York, 1938), p. 78.

Paragraph 13 ("As it happened, this . . ."): Collier and Horowitz, *Rockefellers,* p. 35.

Paragraph 14 ("When the new technology . . ."): John Chamberlain, *The Enterprising Americans: A Business History of the United States* (New York, 1963); Rockefeller, *Reminiscences,* pp. 80–81; Collier and Horowitz, *Rockefellers,* p. 22; Seligman, *Potentates,* p. 188.

Paragraph 16 ("When his optimistic forecasts . . ."): Rockefeller, *Reminiscences,* pp. 9–10.

Paragraph 18 ("In 1901 the Rockefeller . . ."): Kevles, *Physicists,* p. 69; Hendrick, *Benefactions,* pp. 7–30; Raymond B. Fosdick, *Adventure in Giving: The Story of the General Education Board* (New York, 1962), p. 327; Fosdick, *Story,* pp. 10, 21; Corner, *History,* p. 37.

Paragraph 19 ("There were some observers . . ."): Collier and Horowitz, *Rockefellers,* p. 49.

## Chapter 4: Putting the Science into Scientific Giving

Paragraph 2 ("In 1901 President Charles . . ."): Kevles, *Physicists,* pp. 19–29, 70.

Paragraph 3 ("The genesis of the . . ."): Corner, *History,* pp. 575–84; Gates, *Chapters,* p. 179.

Paragraph 4 ("Although the European models . . ."): Corner, *History,* pp. 25–26, 39–41; Goodspeed, *History,* pp. 330–33.

Paragraph 5 ("The first order of . . ."): Corner, *History,* pp. 44–45.

Paragraph 6 ("It is fair to . . ."): Gates, *Chapters,* p. 183; Abraham Flexner quoted in Nathan Reingold, "The Case of the Disappearing Laboratory," *American Quarterly,* 29 (Spring 1977): 89; Corner, *History,* p. 45.

Paragraph 7 ("In April 1906 the . . ."): Corner, *History,* p. 45.

Paragraph 8 ("Once the brief fling . . ."): Ibid., pp. 65–69.

Paragraph 9 ("But final approval of . . ."): Ibid.

Paragraph 11 ("Significantly, the Carnegie Institution . . ."): Ibid., p. 52; *Carnegie Institution of Washington* (Washington, D.C., 1976).

Paragraph 12 ("But this was no . . ."): Corner, *History,* p. 51; Kevles, *Physicists,* pp. 69, 82–83.

Paragraph 14 ("The real pioneer in . . ."): Roy M. MacLeod, "The Support of Victorian Science: The Endowment of Research Movement in Great Britain, 1868–1900," *Minerva* 9 (April 1971): 197–230.

Paragraph 15 ("But the British example . . ."): Kevles, *Physicists,* p. 42.

Paragraph 17 ("But industrial laboratories . . ."): Ibid., pp. 9–11, 26, 63–67.

Paragraph 18 ("John D. Rockefeller was . . ."): Rockefeller *Reminiscences,* p. 177; Stanley Coben, "Foundation Officials and Fellowships: Innovation in the Patronage of Science," *Minerva* 14 (Summer 1976): 225–40.

Paragraph 20 ("The first Nobel Prizes, . . ."): Harriet Zuckerman, *Scientific Elite: Nobel Laureates in the United States* (New York, 1977).

Paragraph 23 ("In many cases . . ."): Ibid.

## Chapter 5: What's in a Name?

Paragraph 1 ("Most people think of . . ."): Mary Wollstonecraft Shelley, *Frankenstein; or, A Modern Prometheus,* ed. James Rieger (Indianapolis, 1974).

Paragraph 2 ("Mrs. Shelley's novel was . . ."): Ibid.

Paragraph 3 ("Certainly, Dr. Frankenstein's creator . . ."): Jack Morrell and Arnold Thackray, *Gentlemen of Science: Early Years of the British Association for the Advancement of Science* (Oxford, 1981).

Paragraph 4 ("From the beginning, according . . ."): *A Supplement to the Oxford English Dictionary,* ed. R. W. Burchfield, vol. 3 (Oxford, 1982); Sally G. Kohlstedt, *The Formation of the American Scientific Community* (Urbana, Ill., 1976), p. 45; Morrell and Thackray, *Gentlemen,* p. 186.

Paragraph 5 ("How important was this . . ."): Joseph Ben-David, "The Scientific Role: The Conditions of Its Establishment in Europe," *Minerva* 4 (Autumn 1965): 15.

Paragraph 8 ("The 'seven liberal arts' . . ."): *The Oxford English Dictionary*, ed. James Murray (Oxford, 1928); Morrell and Thackray, *Gentlemen*, p. 19.

Paragraph 9 ("Late medieval European philosophy . . ."): Ralph Bates, *Scientific Societies in the United States* (Cambridge, Mass., 1965), p. 2.

Paragraph 10 ("As the importance of . . ."): *Oxford English Dictionary.*

Paragraph 12 (" 'In a direct and . . ."): Morrell and Thackray, *Gentlemen*, p. 96.

Paragraph 15 ("At the 1834 meeting . . ."): *Supplement to the OED.*

Paragraph 16 ("Another coinage of Whewell's . . ."): *Oxford English Dictionary.*

Paragraph 17 ("It was not until . . ."): Ibid.; MacLeod, "Support," pp. 197–230.

Paragraph 18 ("At first the British . . ."): Morrell and Thackray, *Gentlemen*, p. 45; Kohlstedt, *Formation*, p. 42; Marie Boas Hall, *All Scientists Now: The Royal Society in the Nineteenth Century* (Cambridge, 1984), p. 48; Dorothy Stimson, *Scientists and Amateurs: A History of the Royal Society* (New York, 1968), pp. 190–91.

Paragraph 19 ("The gathering in Berlin . . ."): Ibid.

Paragraph 20 ("To rally the forces . . ."): Edward N. da C. Andrade, *A Brief History of the Royal Society* (London, 1960).

Paragraph 21 ("Damning as this statistic . . ."): Ibid.; Douglas McKie, *The Royal Society: Its Origins and Founders*, (London 1960); Margery Purver, *The Royal Society: Concept and Creation* (London, 1967); Thomas Sprat, *History of the Royal Society*, ed. Jackson I. Cope and H. Whittemore Jones (St. Louis, 1958).

Paragraph 22 ("Nevertheless, Babbage, Whewell, Dalton . . ."): Maurice Crosland, "From Prizes to Grants in the Support of Scientific Research in France in the Nineteenth Century: The Montyon Legacy," *Minerva* 17 (Autumn 1979): 356–61; Russell Moseley, "The Origins and Early Years of the National

Physical Laboratory: A Chapter in the Pre-History of British Scientific Policy," *Minerva* 16 (Summer 1978): 222–32; Babbage quoted in Everett Mendelsohn, "The Emergence of Science as a Profession in Nineteenth Century Europe," in *The Management of Scientists,* ed. Karl Hill (Boston, 1964), pp. 22, 31.

Paragraph 23 ("The reformers' decision to . . ."): Hall, *All Scientists,* pp. 58, 111; Stimson, *Scientists,* p. 119.

Paragraph 25 ("Despite the ridicule of . . ."): Morrell and Thackray, *Gentlemen,* pp. 12, 44, 88, 245–46.

Paragraph 26 ("Brisk sales of 'annual' . . ."): Ibid., pp. 150–55, 309–20.

Paragraph 27 ("These awards were the . . ."): Morrell and Thackray, *Gentlemen,* 270–71; Hall, *All Scientists,* pp. 18, 40–46.

Paragraph 28 ("The founders of the . . ."): *Nature* 313 (14 February 1985): 610.

## Chapter 6: Who Pays the Piper?

Paragraph 1 ("By midcentury the scientific . . ."): Hall, *All Scientists,* pp. 81–82; Andrade, *History.*

Paragraph 2 ("In the closely linked . . ."): Morrell and Thackray, *Gentlemen,* p. 241; George H. Daniels, "The Process of Professionalization in American Science: The Emergent Period, 1820–1860," *Isis* 58 (Summer 1967): 151–66.

Paragraph 3 ("More than anything else, . . ."): William Whewell, *Selected Writings in the History of Science* (Chicago, 1984).

Paragraph 4 ("Whewell knew exactly what . . ."): Ibid., pp. 8, 124.

Paragraph 5 ("Not all of Whewell's . . ."): Ibid.; Morrell and Thackray, *Gentlemen,* p. 271.

Paragraph 6 ("For Whewell, it was . . ."): Whewell, *Writings,* pp. 15, 124–28, 198, 298.

Paragraph 7 ("Whewell had no qualms . . ."): Ibid., 124–25, 298; Morrell and Thackray, *Gentlemen,* p. 276.

Paragraph 8 ("Whewell's linguistic strategy was . . ."): Whewell, *Writings,* (preface to "The Philosophy of the Inductive Sciences," 1840).

Paragraph 10 ("In 1849, thanks to . . ."): Hall, *All Scientists,* pp. 147–50, 163; MacLeod, "Support," pp. 198.

Paragraph 12 ("Francis Bacon, whose writings . . ."): Ben-David, "Role," p. 24.

Paragraph 13 ("To bring about what . . ."): Andrade, *History;* Purver, *Royal Society; The Seventeenth Century,* ed. Andrew Lossky (New York, 1967), p. 57; Gale E. Christianson, *This Wild Abyss: The Story of the Men Who Made Modern Astronomy* (New York, 1978); Benjamin Farrington, *The Philosophy of Francis Bacon* (Chicago, 1966), p. 47.

Paragraph 14 ("But the Accademia dei . . ."): Lossky, *Century;* Purver, *Royal Society;* Farrington, *Philosophy.*

Paragraph 15 ("Like their intellectual godfather, . . ."): Purver, *Royal Society;* Sprat, *History.*

Paragraph 17 ("When the coronation of . . ."): Ibid.

Paragraph 18 ("At least half of . . ."): McKie, *Royal Society,* p. 35; Purver, *Royal Society,* p. 72; Sprat, *History,* p. xiii.

Paragraph 19 ("Apologists for the new . . ."): *Oxford English Dictionary;* Purver, *Royal Society,* pp. 95, 129–30.

Paragraph 20 ("The original plan was . . ."): Andrade, *History,* p. 5.

Paragraph 21 ("In 1684 the astronomer . . ."): Christianson, *Abyss,* pp. 395–96.

Paragraph 22 ("A pattern had been . . ."): MacLeod, "Support."

Paragraph 23 ("Academic interest in the . . ."): Mendelsohn, "Emergence"; MacLeod, "Support," p. 199; Stimson, *Scientists,* p. 202.

Paragraph 24 ("Once the universities had . . ."): *Science* 236 (May 22, 1987): 983.

Paragraph 25 ("The poor showing of . . ."): Moseley, "Origins," pp. 222–23; MacLeod, "Support," p. 200.

Paragraph 27 ("A few respondents declared . . ."): MacLeod, "Support," pp. 208–11.

Paragraph 28–30 ("Early the next year, . . ." "Strange's instrument of choice, . . ." and "But Appleton's appeal to . . ."): Ibid.

Paragraph 31 ("But by 1875, when . . ."): Ibid., pp. 212–18.

Paragraph 32 ("The very success of . . ."): Ibid., pp. 215–30.

Paragraph 33–37 ("There were bureaucrats who, . . ." "Some critics argued that, . . ." "Meanwhile, the royal astronomer, . . ." "In December 1880 the, . . ." and "There is little doubt . . ."): Ibid.

Paragraph 38 ("And yet in 1919, . . ."): Medical Research Committee, *Report: 1918–1919,* cited in Donald Fisher, "The Rockefeller Foundation and the Development of Scientific Medicine in Great Britain," *Minerva* 16 (Spring 1978): 20–41.

## Chapter 7: Beggars and Choosers

Paragraph 2 ("In 1877 Edward C. Pickering . . ."): Howard Plotkin, "Edward C. Pickering and the Endowment of Scientific Research in America, 1877–1918," *Isis* 69 (1978): 44–57.

Paragraph 3 ("After a decade of . . ."): Ibid.

Paragraph 4 ("Two years later Pickering's . . ."): Kevles, *Physicists,* pp. 109–10.

Paragraph 5 ("Meanwhile, Pickering, who moved . . ."): Ibid.

Paragraph 6 ("Obviously, this kind of . . ."): Plotkin, "Pickering"; Daniel J. Kevles, "George Ellery Hale—The First World War and the Advancement of Science in America," *Isis* 59 (1968): 421–36.

Paragraph 7 ("The tide, however, was . . ."): Plotkin, "Pickering."

Paragraph 10 ("Behind the enthusiasm of . . ."): MacLeod, "Support."

Paragraph 11 ("The importance of what . . ."): Daniels, "Process," pp. 156, 160.

Paragraph 14 ("Huxley's efforts on behalf . . ."): MacLeod, "Support," pp. 200–1.

Paragraph 18 ("It is not mere . . ."): Daniel J. Kevles, *In the Name of Eugenics* (New York, 1985), p. 208.

## Chapter 8: Missionaries and Professionals

Paragraph 1 ("On the night of . . ."): George W. Gray, *Education on an International Scale* (New York, 1941), pp. 11–17; IEB 1.3.49, Rose's log.

Paragraph 2 ("To get from Denmark . . ."): Coben, *Foundation,* p. 230; IEB 1.2, Rose to Buttrick, 20 March 1924.

Paragraph 3 ("The image is pure . . ."): Gray, *Education,* p. 10; RF 900.23.174, Gregg's comments.

Paragraph 4 ("Rose was aware that . . ."): Gray, *Education;* Fosdick, *Adventure,* p. 229.

Paragraph 5 ("Just as Rose was . . ."): RF 900.22.165, Minutes of conference of members and officers at Princeton, N.J., 23–24 February 1925.

Paragraph 11 ("Of course, the GEB . . ."): Fosdick, *Adventure,* p. 28.

Paragraph 12 ("In taking on this . . ."): Ibid., p. 3; Weaver, *Philanthropic Foundations,* p. 24.

Paragraph 13 ("To coordinate an all-out . . ."): Fosdick, *Adventure,* pp. 4–5, 189.

Paragraph 15 ("The General Education Board . . ."): Ibid., pp. 9, 13–24.

Paragraph 16 ("For example, to help . . ."): Ibid., pp. 25–33.

Paragraph 17 ("Circuit riding for the . . ."): Ibid., p. 18.

Paragraph 18 ("The insights into local . . ."): Ibid., pp. 21, 39–50.

Paragraph 19 ("The innovations of the . . ."): Ibid., pp. 63–66.

Paragraph 20 ("Like Frederick Gates, Rose . . ."): Fosdick, *Story,* p. 12; Gray, *Education,* p. 5 n.

Paragraph 21 ("Aware that the success . . ."): Fosdick, *Adventure,* pp. 66–76.

Paragraph 22 ("As for Rose himself, . . ."): Ibid.

Paragraph 24 ("The facts seemed clear. . . ."): Fosdick, *Story* p. 31; Collier and Horowitz, *Rockefellers,* p. 63; Gates, *Chapters,* p. 226; Moscow, *Inheritance,* pp. 91–92.

Paragraph 25 ("In 1909 all this . . ."): Ibid.

Paragraph 26 ("Stiles had the answer . . ."): Fosdick, *Story,* p. 10; Gates, *Chapters,* p. 226.

Paragraph 28 ("Eventually, the campaign against . . ."): Fosdick, *Story,* p. 32.

Paragraph 29 ("Not surprisingly, both Gates . . ."): Ibid.; Gates, *Chapters,* p. 229.

## Chapter 9: A Missionary for Science

Paragraph 1 ("A tug-of-war between the . . ."): Gates, *Chapters,* pp. 207–10; Fosdick, *Story,* p. 3.

Paragraph 2 ("Gates not only defined . . ."): Gates, *Chapters,* pp. 208–9.

Paragraph 3 ("Gates had already introduced . . ."): RFA Record Group 2, Box 21, JDR Jr. Personal, Letter Book No. 1, A–M (General Education Board), John D. Rockefeller, Jr., to John D. Rockefeller, Sr., 21 June 1905; RFA Record Group 2, Box 21, JDR Jr. Personal, JDR Sr.–JDR Jr. Correspondence 1885–1914, John D. Rockefeller, Jr., to John D. Rockefeller, Sr., 31 December 1906; Gates, *Chapters,* pp. 207–10.

Paragraph 4 ("As for the specific . . ."): Gates, *Chapters.*

Paragraph 7 ("At the same time, . . ."): Rockefeller, *Reminiscences,* p. 177.

Paragraph 8 ("His very predilection for . . ."): RFA John D. Rockefeller, Jr., to John D. Rockefeller, Sr., 31 December 1906.

Paragraph 9 ("His son still had . . ."): Ibid.

Paragraph 10 ("What happened between the . . ."): Fosdick, *Story,* p. 15.

Paragraph 11 ("If the Rockefellers and . . ."): Ibid., pp. 14–20; Karl and Katz, "Foundation," p. 255.

Paragraph 13 ("The first Board of . . ."): Fosdick, *Story,* pp. 22–25.

Paragraph 14 ("By the end of . . ."): Ibid., p. 11; Rockefeller, *Reminiscences,* p. 148.

Paragraph 16 ("Beyond that, John D. . . ."): Fosdick, *Story,* p. 21.

Paragraph 17 ("In 1915, for example, . . ."): RF 900.5.42, Letter of Gift, John D. Rockefeller, Sr., 6 March 1914; *RF Annual Report for 1915;* RF 200, Eugenics Investigators, Charles B. Davenport to Jerome D. Greene, 12 May 1914.

Paragraph 19 ("While all these gifts . . ."): Fosdick, *Story,* pp. 22–23.

Paragraph 20 ("The list of no-nos . . ."): RF 900.21.163, Greene memo, "Principles and Policies of Giving," 22 October 1913.

Paragraph 21 ("Many of the Founder's . . ."): *RF Annual Report for 1917.*

## Chapter 10: The Greening of American Science

Paragraph 1 ("Nineteen thirteen, the year . . ."): Coben, *Foundation; Nature* 319 (30 January 1986): 344.

Paragraph 2 ("On October 24, 1913, . . ."): Reingold, "Case."

Paragraph 3 ("But this rejection was . . ."): Ibid.; Kevles, *Physicists,* p. 111.

Paragraph 4 ("Not surprisingly, Hale was . . ."): Ibid.

Paragraph 5 ("Hale's plan was to . . ."): Ibid.; Kevles, "Hale."

Paragraph 6 ("But Pickering was not . . ."): RF 915.1.6, 24 May 1916; Reingold, "Case"; Kevles, "Hale."

Paragraph 7 ("Speaking for the new . . ."): RF 915.1.6, 24 April 1916; Reingold, "Case."

Paragraph 8 ("Greene's reactions, as expressed . . ."): RF 900.23.174, Gregg's comments.

Paragraph 9 ("Greene recognized that 'the . . ."): Fosdick, *Adventure,* pp. 140–41; RF 915.1.1, Greene to Flexner, 13 March 1916; Greene to Pickering, 27 April 1916; RF 915.1.6, Greene to Board of Trustees, 24 May 1916; Reingold, "Case."

Paragraph 11 ("In addition, Greene was . . ."): Ibid.

Paragraph 12 ("Preoccupied as they were . . ."): RF 900.22.166, Report of the Committee on Appraisal and Plan, 11 December 1934, p. 6.

Paragraph 13 ("The frustrated Greene confided . . ."): Ibid.

Paragraph 14 ("This is not what . . ."): RF 915.1.6, May 24, 1916.

Paragraph 16 ("The major event was . . ."): Bates, *Societies,* pp. 134–35; Plotkin, "Pickering"; Kevles, *Physicists,* p. 112.

## Chapter 11: Battle Scars

Paragraph 1 ("Vincent was one of . . ."): Ibid.; RF 900.22.165, Minutes; 900.23.174, Gregg's comments.

Paragraph 3 ("At this time the . . ."): Fosdick, *Story,* pp. 80–92.

Paragraph 4 ("Construction of a vast . . ."): Robert E. Kohler, "A Policy for the Advancement of Science: The Rockefeller Foundation, 1924–29," *Minerva* 16 (Winter 1978): 491–92; RFA Record Group 2, Box 21, JDR Jr. Letter Book No. 3, R–Z (Rockefeller Foundation): John D. Rockefeller, Jr., to John D. Rockefeller, Sr., 17 April 1917.

Paragraph 6 ("In December 1917 Vincent . . ."): Reingold, "Case," 79–81, 91–96; Fosdick, *Story,* pp. 146–46; Kevles, *Physicists,* pp. 149–50.

Paragraph 7 ("However, opposition to the . . ."): Ibid.

Paragraph 9 ("During its first two . . ."): RF 900.22.166, Report, pp. 8–11; Fosdick, *Story,* pp. 25–28.

Paragraph 10 ("To say that the . . ."): *RF Annual Report for 1915,* p. 19.

Paragraph 11 ("Public outrage at . . ."): Collier and Horowitz, *Rockefellers,* p. 123.

Paragraph 13 ("Meanwhile, the RF trustees . . ."): Fosdick, *Story,* p. 26.

Paragraph 14 ("But an even more . . ."): Ibid., pp. 27–28.

Paragraph 16 ("A long round of . . ."): Reingold, "Case," pp. 94–96.

Paragraph 17 ("Having just been freed . . ."): *RF Annual Report for 1922,* p. 11.

Paragraph 19 ("Following another round of . . ."): Reingold, "Case"; Fosdick, *Story,* pp. 146; Kevles, *Physicists,* pp. 149–50.

Paragraph 20 ("The funding of fellowships . . ."): Fosdick, *Adventure,* pp. 298–99; RF annual reports.

## Chapter 12: Triumph—And Disappointment

Paragraph 7 ("In the best Rockefeller . . ."): Fosdick, *Story,* pp. 33–35; Gates, *Chapters,* p. 228.

Paragraph 9 ("With Rose as the . . ."): Fosdick, *Story,* pp. 41–42.

Paragraph 10 ("But the creating of . . ."): *RF Quarterly Bulletin,* July 1929; October 1932.

Paragraph 11 ("There remained other problems . . ."): Fosdick, *Story,* p. 45.

Paragraph 12 ("Not even the hookworm . . ."): Ibid., p. 35.

Paragraph 13 ("On his first swing . . ."): IEB 1.2.39, Wickliffe Rose to Mrs. Gorgas, 4 August 1921; Fosdick, *Story,* p. 59.

Paragraph 14 ("When Rose returned from . . ."): Ibid.

Paragraph 17 ("This recognition kindled Rose's . . ."): IEB 1.2.38, Rose to Col. Arthur Woods, 4 May 1923.

Paragraph 18 (" 'All important fields of . . ."): Ibid.; also IEB 1.5.72, "Formulation of IEB Program," April 1923.

Paragraph 19 ("Characteristically, Rose moved quickly . . ."): Ibid.

Paragraph 20 ("To illustrate what he . . ."): Ibid.

## Chapter 13: Trickle Down

Paragraph 1 ("If one had to . . ."): IEB 1.5.72, Rose to John D. Rockefeller, Jr., 29 November 1922; John D. Rockefeller, Jr., to IEB Board, 19 January 1923; Rose to John D. Rockefeller, Jr., 8 February 1923; Gray, *Education,* pp. x, 4–6.

Paragraph 2 ("The IEB began . . ."): IEB 1.5.72, John D. Rockefeller, Jr., to Rose, 27 January 1923.

Paragraph 3 ("Whatever Junior may have . . ."): IEB 1.5.72, John D. Rockefeller, Jr., to Rose, 1 November, 22 November 1923; Rose to John D. Rockefeller, Jr., 30 October, 7 November 1923.

Paragraph 4 ("After a similar experience . . ."): Gray, *Education,* p. x; Fosdick, *Adventure,* p. 227.

Paragraph 5 ("Rose himself had no . . ."): Ibid.; IEB 1.2.36, Rose to Buttrick, London, 5 December 1923.

Paragraph 6 ("During December 1923 he . . ."): IEB 1.2.36, Rose to Buttrick, London, 23 December 1923.

Paragraph 7 ("On a typical day . . ."): Ibid.; *Grafton Elliot Smith: The Man and His Work,* ed. A. P. Elkin and N. W. G. Macintosh (Sydney, 1974), p. 13.

Paragraph 8 ("He found them to . . ."): IEB 1.2.36, Rose to Buttrick, London, 5 December 1923.

Paragraphs 9 and 10 ("In his campaign against . . ." and "After long days of . . ."): IEB 1.2.36, Rose to Buttrick, London, 23 December 1923.

Paragraph 11 ("Rose the missionary was . . ."): IEB 1.2, Rose to Buttrick, Berlin, 9 March 1924; Rose to Buttrick, Praha, undated 1924.

Paragraph 12 ("But most important to . . ."): IEB 1.2, Rose to Buttrick, Berlin, 9 March 1924.

Paragraph 13 ("Rose could not wait . . ."): IEB 1.2, Rose to Buttrick, Aix-les-Bains, 3 February 1924.

Paragraph 14 ("A steady stream of . . ."): IEB 1.3.41, Lillian Adams to Rose, 4 March 1924.

Paragraph 15 ("Rose, who repeatedly stressed . . ."): IEB 1.3.49, Rose's log, first European trip, December 1923–April 1924, p. 121; Gray, *Education,* p. 102.

Paragraph 16 ("All told, Rose's trip . . ."): IEB 1.3.42, Rose to C. E. Bechofer Roberts, 13 June 1924; Gray, *Education,* pp. 25, 46–47; IEB 1.2, Rose to Buttrick, Stockholm, 20 March 1924.

Paragraph 18 ("In all, during its . . ."): RF 915.22.168, Agenda for Special Meeting, 11 April 1933.

Paragraph 19 ("Not surprisingly, most of . . ."): Gray, *Education,* pp. 45, 65–66.

Paragraph 20 ("The most expensive project . . ."): Ibid., pp. 37–44; Fosdick, *Story,* pp. 176–80.

Paragraph 21 ("Rose also served as . . ."): Fosdick, *Adventure,* p. 230.

Paragraph 22 ("Before Rose's tenure . . ."): RF 900.22.166, Report; Fosdick, *Adventure,* pp. 228–30.

Paragraph 24 ("The smallest of all . . ."): IEB 1.3.48, Wallace Buttrick to Einstein, 2 February 1924; Gray, *Education,* pp. 31–32.

Paragraph 25 ("Rose cabled his approval; . . ."): Ibid.; IEB 1.3.48, Dr. Heinrich Poll to Abraham Flexner (RF translation), 22 February 1924; Flexner to Poll, 10 March 1924; Einstein to IEB, 27 March 1924; Poll to Flexner (RF translation), 28 March 1924; W. W. Brierley to Einstein, 1 August 1924.

Paragraph 26 ("A search of his . . ."): Ibid.; IEB 1.2.38, Lillian Adams to Col. Arthur Woods, 31 January 1924.

Paragraph 27 ("In Rose's IEB, flexibility . . ."): Gray, *Education,* pp. 29, 33, 96–97.

Paragraph 28 ("The IEB came to . . ."): Ibid.; Fosdick, *Story,* pp. 151, 169.

## Chapter 14: A Scientific Education

Paragraph 1 ("Of course, Rose did . . ."): IEB 1.3.42, Rose to Bechofer Roberts, 13 June 1924.

Paragraph 2 ("On June 13, 1924, . . ."): Gray, *Education,* pp. 19–20, 55–56; Kevles, *Physicists,* pp. 126–38.

Paragraph 3 ("These early circuit riders . . ."): Gray, *Education,* p. 56, 110.

Paragraph 4 ("Some scientists found . . ."): Ibid., pp. 19–20.

Paragraph 5 ("This in no way . . ."): Personal communication, Paris, 8 November 1981.

Paragraph 6 ("Mandelbrojt's professor asked him . . ."): Ibid.; IEB, Trowbridge to Rose, 28 May 1925; Tisdale's log, 23 August 1926.

Paragraph 8 ("Between 1924 and 1929 . . ."): Coben, *Foundation,* pp. 226–27; Gray, *Education,* p. 22.

Paragraph 9 ("Once Wickliffe Rose became . . ."): Gray, *Education,* p. vii.

Paragraph 10 ("Especially memorable was the . . ."): Ibid.

Paragraph 11 ("It should come as . . ."): Ibid., pp. x–xi.

Paragraph 12 ("He pursued his self-education . . ."): Coben, *Foundation,* p. 237; IEB 1.4.55, Rose to Flexner, 8 July 1924.

Paragraph 13 ("A year and a . . ."): *Scientific American,* December 1925, pp. 661–64; IEB 1.1.11, Rose to Millikan, 13 January 1926.

Paragraph 14 ("But Millikan did reply, . . ."): IEB 1.1.11, Millikan to Rose, 21 January 1926.

## Chapter 15: A Cautionary Tale

Paragraph 1 ("The Rockefeller philanthropies were . . ."): Coben, *Foundation,* pp. 229, 238.

Paragraph 2 ("One rationale for these . . ."): Ibid., p. 239.

Paragraph 3 ("In daring to select . . ."): Ibid., pp. 235–36.

Paragraph 4 ("There is no doubt . . ."): Ibid., p. 225, 229.

Paragraph 10 ("Once science patronage entered . . ."): *Carnegie Institution of Washington, 1976,* p. 2.

Paragraph 11 ("The Carnegie Institution's support . . ."): Ibid. p. 1; *Carnegie Institution of Washington Year Book, 1904,* p. 33; *Carnegie Institution Report of the President, 1976–1977,* p. 25.

Paragraph 13 ("A graduate of Harvard . . ."): RF 200, Eugenics Investigators, Davenport to Greene, 12 May 1914.

Paragraph 14 ("Initial funding for the . . ."): Kevles, *Eugenics,* pp. 54–55.

Paragraph 15 ("That same year (1920) . . ."): Kenneth M. Ludmerer, *Genetics and American Society* (Baltimore, 1962), pp. 97–104; Kevles, *Eugenics,* pp. 97, 103; Madison Grant and Charles Stewart Davison, eds., *The Alien in Our Midst* (New York, 1930), pp. 158–79.

Paragraph 16 ("Many prominent geneticists . . ."): Charles Rosenberg, *No Other Gods: On Science and American Social Thought* (Baltimore, 1976), pp. 89–97; Kevles, *Eugenics,* pp. 45–48, 110–16; Ludmerer, *Genetics,* pp. 85–95; L. C. Dunn, "Cross Currents in the History of Human Genetics," in *Heredity and Society* New York, 1977), p. 23.

Paragraph 17 ("Davenport directed the Carnegie . . ."): Kevles, *Eugenics,* p. 199.

Paragraph 18 ("The Carnegie's belated change . . ."): *Carnegie Institution of Washington Year Book 1934.*

Paragraph 21 ("Davenport had his public . . ."): Dunn, "Cross Currents," pp. 23–24; Charles Davenport, *Naval Officers, Their Heredity and Development* (Washington, D.C., 1919), cited in Rosenberg, *Gods,* p. 92.

Paragraph 22 ("But American geneticists who . . ."): Ludmerer, *Genetics,* p. 97, 124, Rosenberg, *Gods,* p. 96; Kevles, *Eugenics,* pp. 122–32.

Paragraph 25 ("Warren Weaver, director of . . ."): Weaver, *Philanthropic Foundations,* pp. 119–20.

## Chapter 16: Choosing Up Sides

Paragraph 1 ("Edwin Embree was his . . ."): RAC Edwin Embree Collection, Box 1, Folder 3 (Rockefeller Programs—Early History), untitled manuscript, ca. 1930; Kohler, "Policy," pp. 500–501.

Paragraph 3 ("To begin with, Embree's . . ."): Ibid.; Kohler, "Policy," pp. 491–92.

Paragraph 4 ("Vincent's argument was seconded . . ."): IEB 1.5.72, John D. Rockefeller, Jr., to Rose, 18 October 1923.

Paragraph 6 ("The powerful directors of . . ."): RF annual reports; Fosdick, *Story,* p. 21.

Paragraph 7 ("But in that same . . ."): Fosdick, *Story,* p. 105; Fisher, "Foundation."

Paragraph 8 ("It was not Vincent's . . ."): Kohler, "Policy," p. 493; RF 900.22.165, Conference, pp. 3–4; RF 900.23,174, Gregg's comments; RAC Embree Collection, untitled manuscript, p. 2.

Paragraph 9 ("Embree was yet another . . ."): *Current Biography,* December 1948, p. 17.

Paragraph 11 ("In truth, the Foundation . . ."): RF 900.22.165, Conference, p. 9.

Paragraph 13 ("From the beginning Embree's . . ."): RF Embree's log, 9 February 1924; RF 900.22.165, Conference, p. 11.

Paragraph 14 ("This proposal was certainly . . ."): RF 900.22.165, Conference.

Paragraph 15 ("Embree tried to offer . . ."): Ibid.

Paragraph 16 ("From those trustees who . . ."): Ibid.

Paragraph 18 ("In cobbling together his . . ."): RF Board minutes, 23 February 1925.

## Chapter 17: The Vanishing Eugenicists

Paragraph 4 ("The first impetus for . . ."): Kevles, *Eugenics,* pp. 59–64.

Paragraph 6 ("It was on such . . ."): RF 410D, Anthropology, Australian Aborigines, Gregory to Grant, 24 December 1923.

Paragraph 7 ("Gregory argued that the . . ."): Ibid.

Paragraph 8 ("Just in case Fosdick . . ."): RF 410D, Grant to Fosdick, 29 December 1923.

Paragraph 9 ("Addressing this appeal to . . ."): RFA Record Group 2, Economics, Box 1, 20 February 1911–26 June 1914; RF annual reports; Kevles, *Eugenics,* pp. 54–55.

Paragraph 11 ("The chairman of the . . ."): BSH, Eugenics, Fisher to Katherine B. Davis, 24 November 1924.

Paragraph 13 ("In 1920 and again . . ."): BSH, Eugenics, Fosdick to Davis, 31 January 1921; Davis to Fosdick, 21 February 1921.

Paragraph 15 ("There was irony in . . ."): Dunn, "Cross Currents," p. 22; Kevles, *Eugenics,* p. ix.

Paragraph 16 ("Since American eugenicists . . ."): RF 410D, Fosdick to Ruml, 3 January 1924.

Paragraph 18 ("The original budget . . ."): RF 410D, Gregory to Grant, 24 December 1923; Wilkins to Wissler, 29 December 1924.

Paragraph 19 ("Far from being put . . ."): RF Board minutes, 27 February 1924; RF 410D, Davenport memo on meeting of 13 February.

Paragraph 20 ("Before submitting a . . ."): RF 410D, Anthropology, Australian Aborigines.

Paragraph 21 ("The two unwritten rules . . ."): Ludmerer, *Genetics,* p. 76, 85.

Paragraph 22 ("Without informing the eugenicists, . . ."): RF Board minutes, 27 February 1924; RF 410D, Anthropology, Australian Aborigines.

Paragraph 23 ("Meanwhile, Davenport, who was . . ."): RF Embree's log, 29 February 1924; 410D, Davenport to Embree, 3 March 1924 and 1 April 1924; Dunn, "Cross Currents," p. 24; Stephen Gould, in *Natural History,* December 1980.

Paragraph 24 ("On May 7, 1924, . . ."): RF Board minutes.

Paragraph 25 ("When Davenport learned that . . ."): RF Embree's log, 30 April, 2 May 1924; RF 410D, Gregory to Embree, 8 May 1924; Embree to Gregory, 8 May 1924; Gregory to Embree, 12 May 1924.

## Chapter 18: Anthropology in the Antipodes

Paragraph 1 ("The power of . . ."): RF 410D, Annual Report of the Australian National Research Council, 1929–1930, enclosed with Smith to Embree, 30 September 1924.

Paragraph 2 ("But there was a . . ."): RF 410D, Embree to Gregory, 8 May 1924.

Paragraph 3 ("But on his . . ."): RF 410D, Smith to Embree, 30 September 1924.

Paragraph 4 ("Impressed by this frank . . ."): Ibid.; also Smith to Embree, 5 November 1924; Embree to Smith, 13 November 1924.

Paragraph 5 ("On November 7 . . ."): RF Board minutes.

Paragraph 6 ("In a letter . . ."): RF 410D, Embree to Smith, 13 November 1924.

Paragraph 7 ("When Elliot Smith responded . . ."): RF 410D, Embree to Smith, 11 December 1924; RF Board minutes, 23 February 1925.

Paragraph 8 ("Disappointed as Embree . . ."): RF Board minutes, 7 July 1924.

Paragraph 9 ("One of the key . . ."): RF 410D, Smith to Embree, 20 January 1925; LSRM, Ruml to Arthur Woods, 22 May 1925.

Paragraph 10 ("Wissler was a fellow . . ."): *Dictionary of American Biography;* RF 410D, Embree's log, 27 February 1924.

Paragraph 11 ("All told, Embree . . ."): RF 410D, Embree to Vincent, 17 October 1925.

Paragraph 12 ("In a letter that . . ."): RF 410D, Embree to Vincent, 20 December 1925.

Paragraph 13 ("What Embree did not . . ."): RF 410D, ANRC, Evaluator's Report, Sydnor Walker, 7 July 1938.

Paragraph 14 ("On May 26, 1926, . . ."): RF Board minutes.

Paragraph 15 ("In his own detailed . . ."): RF 410D, ANRC, Wissler report, p. 53, 10 March 1926.

Paragraph 16 ("As if to justify . . ."): RF 410D, ANRC, Radcliffe-Brown to Embree, 6 January 1927.

Paragraph 17 ("But the smooth start . . ."): RF 410D, ANRC, Gibson to Embree, 20 December 1926.

Paragraph 18 ("The two researchers, Professor . . ."): Ibid.

Paragraph 19 ("After consulting with his . . ."): RF 410D, ANRC, Embree to Chapman, 18 February 1927; Chapman to Embree, 30 June 1927.

Paragraph 20 ("By the time this . . ."): RF Board minutes; 410D, ANRC, Embree to Chapman, 30 March 1927.

Paragraph 21 ("When the Australians finally . . ."): RF 410D, ANRC, Pearce to Chapman, 1 August 1927.

## Chapter 19: Double Downfall

Paragraph 4 ("As we noted earlier, . . ."): Kohler, "Policy," pp. 497–500.

Paragraph 5 ("When it became apparent . . ."): Ibid.

Paragraph 6 ("A three-man 'committee on . . ."): IEB 1.5.73, John D. Rockefeller, Jr., to Fosdick, 28 December 1925 (attached to letter of 30 December).

Paragraph 7 ("A few months later . . ."): Quoted in RF 900.22.166, Report, p. 21.

Paragraph 8 ("With the founder, the . . ."): IEB 1.5.73, John D. Rockefeller, Jr., to Rose, 18 November 1926 (attached to letter of 22 November).

Paragraph 9 ("As for the International . . ."): Ibid.

Paragraph 12 ("According to this plan . . ."): RF 900.22.166, Report; see also 900.17.123, Fosdick to Thomas Debevoise, 21 December 1927.

Paragraph 15 ("When he realized that . . ."): IEB 1.2.37, Rose to Fosdick, 10 February, 1 March 1928.

Paragraph 16 ("In the first place, . . ."): Ibid.

Paragraph 17 ("Rather than rush the . . ."): IEB 1.2.37, Rose to Fosdick, 10 February 1928; RF 900.17.123, John D. Rockefeller, Jr., to Fosdick, 13 October 1927; RF 900.17.124, John D. Rockefeller, Jr., to Fosdick, 14 March 1928; John D. Rockefeller, Jr., to Rose, 14 March 1928; Fosdick to John D. Rockefeller, Jr., 19 March 1928; IEB 1.3.48, Rose to W. W. Brierley, 22 May 1928.

Paragraph 19 ("The legacy that Edwin . . ."): RF 900.17.122, Fosdick to Simon Flexner, 28 October 1926; Kohler, "Policy" pp. 497–500.

Paragraph 21 ("Richard Pearce of the . . ."): RF 410D, Pearce to Radcliffe-

Brown, 27 May 1927; Pearce to Gibson, 15 June 1927; Pearce to Radcliffe-Brown, 21 October 1927; RF Executive Committee minutes, 25 May 1927.

Paragraph 22 ("Pearce's letters to the . . ."): Ibid.

## Chapter 20: Scandals

Paragraph 2 ("The foundation was responsible . . ."): ANRC Annual Report 1929, filed in RF 410D.

Paragraph 3 ("This was the very . . ."): Stanley D. Porteus, *A Psychologist of Sorts: The Autobiography and Publications of the Inventor of the Porteus Maze Tests* (1969), p. 274.

Paragraph 4 ("He also saw right . . ."): RF 410D, ANRC Annual Report 1929, p. 29; Porteus, *Psychologist,* pp. 321–23.

Paragraph 5 ("Porteus's change in attitude . . ."): Porteus, *Psychologist,* p. 321.

Paragraph 6 ("It is not clear . . ."): RF 410D, ANRC, Evaluator's Report, Sydnor Walker, 7 July 1938.

Paragraph 7 ("For Fosdick, whose principal . . ."): Ibid.

Paragraph 8 ("Despite Pearce's decision to . . ."): RF 410D, ANRC, Memo to Edmund E. Day, 13 May 1930; Radcliffe-Brown to Vincent, 23 June 1939; Adam Kuper, *Anthropologists and Anthropology: The British School, 1922–1972* (New York, 1973), pp. 63–64.

Paragraph 9 ("While pleading for an . . ."): Ibid.

Paragraph 11 ("So uncertain were . . ."): RF 410D, ANRC, Radcliffe-Brown to Mason, 17 November 1930; A. J. Gibson to Day, 3 March 1931.

Paragraph 12 ("Although the Foundation's new . . ."): RF Board minutes.

Paragraph 13 ("The Foundation's gesture of . . ."): RF 410D, ANRC, Elliot Smith to "The Council of the Rockefeller Foundation," 14 October 1932; Dr. Raymond Firth, quoted in Sydnor H. Walker's log, 1 January 1933; RF Board minutes.

Paragraph 14 ("By the time all . . ."): Kuper, *Anthropologists,* p. 52.

Paragraph 16 ("Radcliffe-Brown was a self-made . . ."): Ibid., p. 56.

Paragraph 17 ("He believed that by . . ."): Ibid.

Paragraph 18 ("It is not surprising . . ."): Ibid., p. 52; RF 410D, ANRC, Elliot Smith to the "Council of the Rockefeller Foundation."

Paragraph 19 ("Even after the crisis . . ."): RF Alan Gregg's log, 21 September 1932.

Paragraph 20 ("A few days later . . ."): RF 410D, ANRC, Radcliffe-Brown to Day, 3 October 1932; Kuper, *Anthropologists,* p. 64.

Paragraph 21 ("In 1935 the Foundation . . ."): RF Board minutes; RF 410D, ANRC annual reports; Sydnor H. Walker's log, 1 January 1933.

Paragraph 22 ("On the morning of . . ."): RF 410D, ANRC, Sir George A. Julius to Day, 26 June 1934; ANRC Annual Report 1934, January 1935; *London Observer,* 25 May 1934; E. S. Hornung to Dr. William S. Carter, 7 June 1934.

Paragraph 23 ("Chapman's family and acquaintances . . ."): Ibid.

Paragraph 24 ("In the course of . . ."): Ibid.

Paragraph 25 ("These revelations did not . . ."): Ibid.

Paragraph 26 ("Right after Chapman's death, . . ."): Ibid.; RF 410D, ANRC, A. P. Elkin to Day, 8 November 1933.

Paragraph 28 ("As auditors unraveled the . . ."): RF 410D, ANRC, Julius to Day.

Paragraph 31 ("The problem was not . . ."): Kuper, *Anthropologists,* p. 90.

## Chapter 21: But Is It Science?

Paragraph 1 ("Max Mason was the . . ."): Warren Weaver, *Scene of Change: A Lifetime in American Science* (New York, 1970), pp. 28–29; Kevles, *Physicists,* pp. 122–26, 247; Warren Weaver, "Oral History," RF Record Group 13 (Oral History), pp. 114–18, 121–30; *Science* 117 (20 February 1953): 174–76.

Paragraph 2 ("But his promise as . . ."): Ibid.

Paragraph 3 ("Perhaps because they sensed . . ."): Fosdick, *Story,* p. 157.

Paragraph 4 ("On balance, Mason was . . ."): Weaver, "Oral History," pp. 124–25; 295.

Paragraph 5 ("His closest friends, Weaver . . ."): Ibid., pp. 115, 127–28.

Paragraph 6 ("Despite his sympathy for . . ."): Ibid., pp. 122, 130.

Paragraph 7 ("Nevertheless, the appointment of . . ."): *RF Annual Report for 1929.*

Paragraph 8 ("Unable to pry any . . ."): RFA Record Group 2, Economics, Box 1 (20 February 1911–26 June 1914); RFA Record Group 2, Cultural Interests Series, Box 60, Folder 598, Advisory Committee Minutes, 9 July 1925, 3 February 1927.

Paragraph 9 ("During these years Raymond . . ."): RFA Record Group 2, Rockefeller Boards Series, Box 1, Folder 1, Memo of John D. Rockefeller, Jr., 27 February 1926.

Paragraph 10 ("When the matter of . . ."): RFA Record Group 2, Cultural Interests Series, Box 3, Folder 13, Advisory Committee Minutes, 7 February 1929.

Paragraph 11 ("Just a month earlier . . ."): RFA Record Group 2, Cultural Interests Series, Box 3, Folder 13, Advisory Committee Minutes, 11 January 1929.

Paragraph 12 ("Appleget's investigation of this . . ."): Ibid.

Paragraph 13 ("To Appleget's surprise, Dr. . . ."): Ibid.

Paragraph 14 ("From this tortuous circumlocution, . . ."): Ibid.

Paragraph 15 ("The just completed reorganization . . ."): RF 200, Eugenics Research Association, Mason to Vincent, 3 June 1929.

Paragraph 16 ("The next time the . . ."): RF Mason's log, 11 December 1930.

Paragraph 17 ("The severing of connections . . ."): Bulmer and Bulmer, "Philanthropy,"; RAC Embree Collection, untitled manuscript; Kohler, "Management," p. 488.

Paragraph 18 ("One of the commitments . . ."): LSRM Minutes, June 1928.

Paragraph 19 ("Among other objectives, the . . ."): RF Staff Conference Minutes, 26 January 1931.

Paragraph 21 ("The bureau had been . . ."): Collier and Horowitz, *Rockefellers,* pp. 105–7; BSH File Memo, 18 February 1932, with enclosures from Laughlin.

Paragraph 22 ("Mason's unequivocal rejection of . . ."): Rosenberg, *Gods,* p. 96.

Paragraph 23 ("In the absence of . . ."): BSH Memos, R. Topping to Lawrence Dunham, 24 and 27 September 1932; and File Memo, 4 October 1932.

Paragraph 24 ("Both the Carnegie Institution . . ."): BSH, Rudin to Dunham, 22 November 1932.

Paragraph 25 ("Two months later Hitler . . ."): Alan D. Beyerchen, *Scientists under Hitler: Politics and the Physics Community in the Third Reich* (New Haven, 1977), pp. 126–32; Ludmerer, *Genetics,* pp. 100–120; Dunn, "Cross Currents," p. 25; Kevles, *Eugenics,* pp. 116–17.

Paragraph 26 ("The leaders of the . . ."): Ibid.

Paragraph 27 ("Three years later *Rassenhygiene* . . ."): Nora Levin, *The Holocaust: The Destruction of European Jewry, 1933–1945* (New York, 1973), pp. 301–3.

Paragraph 28 ("Apparently, even in wartime . . ."): Raul Hilberg, *The Destruction of the European Jews* (Chicago, 1967), pp. 304, 561–62.

Paragraph 29 ("American eugenicists were not . . ."): Ludmerer, *Genetics,* p. 169; Rosenberg, *Gods,* pp. 95–96.

## Chapter 22: The Search for Relevance

Paragraph 2 ("In the early 1930s, . . ."): *RF Annual Report for 1917,* p. 20; RF 900.22.166, Report, p. 33.

Paragraph 3 ("Of course, compared . . ."): Kevles, *Physicists,* pp. 236–37.

Paragraph 4 ("Employment opportunities for young . . ."): Ibid., pp. 250, 274–76; Weaver, "Oral History," pp. 283–84.

Paragraph 8 ("The Rockefeller Foundation's rejection . . ."): RF 900.22.167, Verbatim Notes on Princeton Conference of Trustees and Officers, 29 October 1930.

Paragraph 9 ("The discussion coincided with . . ."): Kevles, *Physicists,* pp. 180–84, 236–40.

Paragraph 10 ("In the same month . . ."): *New York Times,* 5 October 1930, cited in Kevles, *Physicists,* p. 239.

Paragraph 11 ("The RF Trustees did . . ."): RF 900.22.167, Verbatim Notes.

Paragraph 12 ("In the reorganization of . . ."): RF 910.3.25, Edmund Day Memo, 30 July 1931, with enclosures.

Paragraph 13 ("To bolster this argument . . ."): Ibid.

Paragraph 15 ("Max Mason thought that . . ."): RF 910.3.25, Selskar M. Gunn to Day, 31 December 1931; Staff Conference, 21 January 1932.

Paragraph 16 ("Early in 1932 Leonard . . ."): RF 100S, Anthropology Survey, Condensed Report, pp. 2–3; Day to Mason, 17 February 1932.

Paragraph 19 ("Day was preparing a . . ."): RF 100S, Outhwaite to Day, 17 March 1933; RF 900.22.168, Agenda for Special Meeting, 11 April 1933, pp. 38–52, 88–104.

## Chapter 23: Betting on Uncertainty

Paragraph 1 ("In the last decade . . ."): Weaver, *Scene of Change,* p. 1; Weaver, "Oral History," p. 206.

Paragraph 2 ("In describing his . . ."): Weaver, *Scene of Change*, pp. 1–7.

Paragraph 3 ("In many ways Reedsburg . . ."): Ibid., p. 2.

Paragraph 4 ("Weaver's own memories of . . ."): Ibid., pp. 4–12.

Paragraph 5 ("His father ran a . . ."): Ibid., pp. 12–13.

Paragraph 6 ("The shy, sickly, . . ."): Ibid., p. 13.

Paragraph 7 ("It wasn't until . . ."): Ibid., pp. 24–25.

Paragraph 8 ("Under Slichter's tutelage, Weaver . . ."): Ibid., pp. 28–32; Weaver, "Oral History," p. 29.

Paragraph 9 ("Weaver would not . . ."): *RF Annual Report for 1938*, p. 39.

Paragraph 10 ("According to the historian . . ."): Kohler, "Management."

Paragraph 11 ("Ironically, Weaver was not . . ."): Weaver, "Oral History," pp. 121–22.

Paragraph 12 ("Weaver's remark is significant . . ."): Ibid.

Paragraph 16 ("Not at all . . ."): *RF Annual Report for 1958*, p. 12.

Paragraph 17 ("Weaver was not only . . ."): Warren Weaver, *Lady Luck: The Theory of Probability* (New York, 1963).

Paragraph 18 ("Weaver was fond of . . ."): Weaver, *Scene of Change*, pp. 144, 191; Weaver, "Oral History," pp. 212–15.

## Chapter 24: Taking the Plunge

Paragraph 1 ("If there is a . . ."): Weaver, *Scene of Change*, p. 50.

Paragraph 2 ("Weaver loved the challenge . . ."): Ibid., p. 51.

Paragraph 3 ("He and his wife . . ."): Ibid., pp. 45, 58–59; Weaver, "Oral History," pp. 58–59.

Paragraph 4 ("So in the fall . . ."): Weaver, *Scene of Change*, p. 59.

Paragraph 5 ("Nevertheless, he took a . . ."): Ibid., pp. 59–63; Weaver, "Oral History," pp. 241–46.

Paragraph 6 ("Asked if he had . . ."): Ibid.

Paragraph 7 ("Weaver felt free to . . ."): Ibid.

Paragraph 8 ("He returned to Madison . . ."): Weaver, *Scene of Change,* p. 62.

Paragraph 9 ("As much as he . . ."): Ibid.; Weaver, "Oral History," p. 31.

Paragraph 13 ("The dilemma facing this . . ."): Kevles, *Physicists,* p. 168.

Paragraph 15 ("Weaver reported for work . . ."): Weaver, "Oral History," pp. 246–47.

Paragraph 16 ("The sense of disorientation . . ."): Ibid., pp. 262, 295, 304.

Paragraph 17 ("Weaver learned something about . . ."): Ibid.

Paragraph 19 ("In many southern schools, . . ."): RF Weaver's log, 2–4 March 1932.

Paragraph 20 ("But Weaver did not . . ."): Weaver, "Oral History," p. 276.

Paragraph 21 ("In 1930 Director Reginald . . ."): RF 200D, Long Island Biological Station, Mason to A. W. Page, 18 February 1930; Mason's log, 15 October 1931.

Paragraph 22 ("Harris's request met with . . ."): RF 900.3.20, Staff Conference, 4 November 1931; RF 200D, Long Island Biological Station, Thompson to Harris, 13 November 1931; Weaver's log, 18 April 1932.

Paragraph 23 ("By his own testimony . . ."): Weaver, *Scene of Change,* p. 69.

## Chapter 25: Riding the Circuit

Paragraph 1 ("Although the Paris office . . ."): Weaver, "Oral History," pp. 295, 324; RF 700.1.4, Staff Memo, 10 June 1927.

Paragraph 2 ("Before long, as the . . ."): *RF Quarterly Bulletin,* George E.

Vincent, October 1927, pp. 85–89; RF 700.1.3, George W. Bakeman to Vincent, 11 March 1927.

Paragraph 3 ("A minor annoyance to . . ."): Weaver, *Scene of Change*, p. 64; *RF Quarterly Bulletin*, October 1927.

Paragraph 5 ("What he found waiting . . ."): Weaver, "Oral History," pp. 250, 304.

Paragraph 6 ("Gunn ran a lively . . ."): Ibid., pp. 261–62, 304–5.

Paragraph 7 ("The dean of the . . ."): Ibid., pp. 305–7.

Paragraph 8 ("Second in command for . . ."): Ibid.

Paragraph 9 ("Harry M. (Dusty) Miller . . ."): Harry M. Miller, "Oral History," RF Record Group 13 (Oral History), pp. 1–7; Harry M. Miller, personal communication, 30 April 1979; Weaver, "Oral History," pp. 320–21.

Paragraph 10 ("The Foundation's Paris staff . . ."): Weaver, "Oral History," pp. 306–9.

Paragraph 11 ("The Paris outpost of . . ."): RF Biography File, Harry M. Miller, Warren Weaver speech on Miller's birthday, 3 June 1957.

Paragraph 12 ("In those days the . . ."): Miller, personal communication; Weaver, *Scene of Change*, p. 69.

Paragraph 13 ("On his first tour . . ."): Weaver, *Scene of Change*, pp. 66–69; Weaver, "Oral History," pp. 309–15.

Paragraph 14 ("This was all pretty . . ."): Ibid.

Paragraph 16 ("Nowhere else did university . . ."): RF Robert Lambert's log, 5 January 1932.

## Chapter 26: Self-Education

Paragraph 1 ("When he returned from . . ."): Weaver, *Scene of Change*, p. 69.

Paragraph 2 ("Weaver, of course, had . . ."): Mason's log, 5 February 1931; *RF Annual Report for 1933*, pp. 204–6.

Paragraph 3 ("The first grant of . . ."): Weaver, "Oral History," p. 336; *RF Annual Report for 1932*, p. 247.

Paragraph 4 "But Pauling was so . . ."): Weaver, "Oral History," p. 335.

Paragraph 5 ("The Board no longer . . ."): Hollis, *Foundations*, p. 79.

Paragraph 6 ("But when considering issues . . ."): Weaver, *Oral History*, pp. 296–98.

Paragraph 8 ("Most knowledgeable observers, including . . ."): Ibid., p. 125; Hollis, *Foundations*, p. 80.

Paragraph 9 ("Part of the problem . . ."): Weaver, "Oral History," pp. 123–24.

Paragraph 10 ("The job of an . . ."): Ibid., pp. 296–97.

Paragraph 11 ("Most trying of all . . ."): Ibid., p. 298.

Paragraph 13 ("By October he was . . ."): RF 915.1.1., Staff Conference, Warren Weaver and Max Mason, 18 October 1932; Weaver to Jones, 26 January 1933.

Paragraph 14 ("After meeting with Alan . . "): RF 915.1.1., Weaver to Jones, 19 November 1932; Weaver, *Oral History*, p. 250; Fosdick, *Story*, p. 46.

Paragraph 15 ("In the fall of . . ."): RF 915.1.1., Weaver to Jones, 19 November 1932; Jones to Weaver, 14 December 1932.

Paragraph 16 ("Understandably, Jones objected to . . ."): Ibid.

Paragraph 17 ("While conceding that there . . ."): Kohler, "Management," p. 290.

## Chapter 27: To the Barricades

Paragraph 1 ("In January 1933 Weaver . . ."): RF 915.1.6, 27 January 1933.

Paragraph 2 ("In bringing this lofty . . ."): Ibid.

Paragraph 3 ("What it all added . . ."): RF 900.22.168, Agenda, p. 63.

Paragraph 7 ("Albert Einstein, who was . . ."): Beyerchen, *Scientists,* pp. 10–14.

Paragraph 13 ("Throughout this long deliberative . . ."): RF 900.22.168, Agenda.

Paragraph 14 ("To emphasize that his . . ."): Weaver, "Oral History," p. 333; Kohler, "Management," p. 288.

Paragraph 15 ("Along with concentration, a . . ."): RF 900.22.168, Agenda, p. 63.

Paragraph 18 ("To qualify for backing, . . ."): Ibid.; RF Weaver's log, 31 December 1936, for parapsychology; RF 200D, Long Island Biological Association, Reginald Harris to Weaver, 14 October 1932; Harris to A. W. Hill, 3 August 1932.

Paragraph 19 ("To indicate why he . . ."): RF 900.22.168, Agenda.

Paragraph 21 ("With RF income shrinking . . ."): Ibid., p. 76.

Paragraph 25 ("Because so much was . . ."): RF 915.1.6, Warren Weaver, "The Science of Man," 29 November 1933; RF 915.1.7, "The Natural and Medical Sciences Cooperative Program," 13 December 1933.

Paragraph 26 ("In an effort to . . ."): RF 915.1.7, "Cooperative Program," p. 5.

Paragraph 27 ("With the advantage of . . ."): RF 915.1.7; *RF Annual Report for 1933,* p. 199.

## Chapter 28: Reason and Unreason

Paragraph 2 ("As soon as possible . . ."): RF Weaver's log; Beyerchen, *Scientists,* p. 11.

Paragraph 3 ("The purge of German . . ."): Beyerchen, *Scientists,* pp. 13–16.

Paragraph 4 ("In addition to what . . ."): RF logs of Harry Miller and Wilbur Tisdale.

Paragraph 5 ("As early as March . . ."): Ibid.; for Weiss, Tisdale's log, 22 March 1933.

Paragraph 6 ("As news of such . . ."): RF Executive Committee Action 33055, 12 May 1933; see also RF *Trustees Bulletin,* Special Research Aid Fund for Deposed Scholars, 1 October 1939.

Paragraph 7 ("This 'temporary' program was . . ."): Ibid.

Paragraph 8 ("In the spring of . . ."): RF Weaver's log, 8 May 1933.

Paragraph 9 ("Weaver never tired of . . ."): Ibid.; Weaver, "Oral History," p. 358.

Paragraph 10 ("Even aside from the . . ."): RF Weaver's log, 8 May 1933.

Paragraph 12 ("Having just suffered through . . ."): Weaver, "Oral History," p. 356.

Paragraph 13 ("Other signs of the . . ."): Ibid., p. 354; Weaver's log, 18 May 1933.

Paragraph 14 ("There was no mistaking . . ."): Ibid.

Paragraph 17 ("Weaver, after making clear . . ."): Ibid.

Paragraph 18 ("By the time Weaver . . ."): Beyerchen, *Scientists,* pp. 18–19.

Paragraph 21 ("Because he had no . . ."): RF 200D, Long Island Biological Association, Bernstein: Harris to Mason, 28 October 1933.

Paragraph 22 ("Although he had to . . ."): Ibid.

## Chapter 29: Despair and Hope

Paragraph 2 ("In January 1933, when . . ."): *Nature* 319 (30 January 1986): 344; RF Weaver's log, May 1933, pp. 24–25; Beyerchen, *Scientists,* p. 42.

Paragraph 3 ("The president of the . . ."): Abraham Pais, *Inward Bound* (New York, 1986), p. 134; Max Born, quoted in Beyerchen, *Scientists,* p. 63.

Paragraph 4 ("The seventy-five-year-old Planck . . ."): Beyerchen, *Scientists,* p. 43; Kevles, *Physicists,* pp. 279–81.

Paragraph 5 ("Weaver and Jones spoke . . ."): RF Weaver's log, 24–25 May 1933, p. 44.

Paragraph 6 ("In fact, the representatives . . ."): Kevles, *Physicists,* p. 280; RF 717, KWI Physics and Cell Physiology, Jones's log, 16 July 1932.

Paragraph 7 ("The Institute of Cell . . ."): Ibid.

Paragraph 8 ("The general director of . . ."): Beyerchen, *Scientists,* p. 60; RF 717.

Paragraph 9 ("Inevitably, the talk turned . . ."): RF Weaver's log, 26 May 1933, p. 47.

Paragraph 10 ("If Weaver found nothing . . ."): Beyerchen, *Scientists.*

Paragraph 11 ("At this stage in . . ."): Ibid., pp. 60–61.

Paragraph 12 ("Although the Rockefeller circuit . . ."): RF Weaver's log, 24–25 May 1933; Alan Gregg's logs, 25 October 1933, 15 June 1934; Harry Miller's log, 13 December 1933, pp. 19–20.

Paragraph 13 ("At the end of . . ."): RF Weaver's log, 29 May 1933, p. 50.

Paragraph 14 ("In Munich, where Hitler . . ."): Ibid., p. 60.

Paragraph 15 ("It was with such . . ."): RF Weaver's log, 24–25 May 1933, p. 40.

Paragraph 16 ("In retrospect, it is . . ."): Ibid., p. 35.

Paragraph 18 ("Before the end of . . ."): RF *Trustees Bulletin,* 1 October 1939, pp. 1–2; Beyerchen, *Scientists,* p. 22; RF 401D, University of Cambridge, Max Born to Jones, 12 September 1933; RF Paris Action, Special Research Aid, 13 August 1933.

Paragraph 19 ("The luckiest of the . . ."): RF Weaver's log, 24–25 May 1933, p. 38; RF 401D, University of Sheffield, Pharmacology, Biochemistry.

Paragraph 20 ("Another young Freiburg émigré . . ."): RF *Trustees Bulletin,* 1 December 1942, pp. 4–8.

Paragraph 21 ("While the Foundation worked . . ."): RF Miller's log, 13 December 1933, pp. 16–19.

Paragraph 22 ("Haber himself had left . . ."): Ibid.; Beyerchen, *Scientists,* p. 62.

Paragraph 23 ("Although no mention whatsoever . . ."): Beyerchen, *Scientists,* p. 62; RF 717, Kaiser-Wilhelm Institutes, Physics, Thomas Appleget to Mason, 30 July 1934.

Paragraph 24 ("Weaver's detailed diary entries . . ."): RF 100.78.424, Marcovich to Hagemeister, 2 June 1933.

Paragraph 27 ("Max Mason had asked . . ."): RF 900.22.166, Report; Kohler, "Management," pp. 291–96.

Paragraph 28 ("When the full Board . . ."): RF 900.22.166, Report, pp.25, 57; Weaver, "Oral History," p. 333.

Paragraph 29 ("For his part, Weaver . . ."): Kohler, "Management," p. 295; RF 900.22.166, Report, p. 58.

## Chapter 30: Some Wooden Kitchen Tables

Paragraph 1 ("Nineteen thirty-five was the . . ."): RF Weaver, "Oral History," pp. 359, 379, 383–87; Weaver, *Philanthropic Foundations,* p. 246; Weaver, *Scene of Change,* pp. 69–75.

Paragraph 2 ("Although Weaver later characterized . . ."): Weaver, "Oral History" p. 283; *Science* 231 (3 January 1986): 63; *Science* 234 (14 November 1986): 821; *Nature* 323 (23 October 1986): 663.

Paragraph 4 ("By his own calculations, . . ."): Weaver, "Oral History," pp. 379, 383–86; Kohler, "Management," p. 297.

Paragraph 5 ("The more time Weaver . . ."): Weaver, "Oral History," p. 532.

Paragraph 6 ("He was equally proud . . ."): *RF Annual Report for 1938,* pp. 203–19; Warren Weaver, *Science* 170 (6 November 1970): 581–82.

Paragraph 7 ("As one measure of . . ."): Weaver, *Scene of Change,* p. 73.

Paragraph 8 ("The Nobelist James Watson, . . ."): Watson quoted in Horace Freeland Judson, *The Eighth Day of Creation* (New York, 1979), pp. 45, 72.

Paragraph 12 ("As Harry Miller remembered . . ."): Miller, "Oral History," pp. 22–23; Miller, personal communication.

Paragraph 15 ("Florey's flair for . . ."): Lennard Bickel, *Rise Up to Life: A Biography of Howard Walter Florey, Who Gave Penicillin to the World* (New York, 1973), p. 29; Gwyn MacFarlane, *Alexander Fleming: The Man and the Myth* (Cambridge, 1984), p. 79.

Paragraph 16 ("What Sherrington found for . . ."): Bickel, *Rise,* p. 21; MacFarlane, *Fleming,* p. 82.

Paragraph 17 ("Florey's promise as . . ."): RF, *Directory of Fellowships and Scholarships, 1917–1970* (New York, 1972), pp. ix, 98; MacFarlane, *Fleming,* p. 99.

Paragraph 18 ("Thee high point of his . . ."): MacFarlane, *Fleming,* pp. 107, 125.

Paragraph 19 ("On his return to . . ."): Ibid.

Paragraph 20 ("He spent the next . . ."): E. P. Abraham, "Memoir of Lord Florey," *Biographical Memoirs of Fellows of the Royal Society* 17 (1971): 258–59; Bickel, *Rise,* pp. 23, 40–42.

Paragraph 21 ("At Cambridge he also . . ."): Bickel, *Rise,* p. 39; Abraham, "Memoir," p. 264.

Paragraph 22 ("His own researches during . . ."): MacFarlane, *Fleming,* p. 107; Abraham, "Memoir," p. 258.

Paragraph 23 ("In his search of . . ."): Alexander Fleming, "On a Remarkable Bacteriolytic Element Found in Tissues and Secretions," *Proceedings of the Royal Society,* Series B, 93 (1922): 306; David Wilson, *In Search of Penicillin* (New York, 1976), pp. 50–52.

Paragraph 24 ("But Florey was intrigued . . ."): Bickel, *Rise,* p. 35; MacFarlane, *Fleming,* p. 200.

## Chapter 31: A Bushranger of Research

Paragraph 1 ("In the fall of . . ."): Bickel, *Rise,* p. 41; MacFarlane, *Fleming,* p. 151.

Paragraph 2 ("There were a few . . ."): Abraham, "Memoir," p. 260.

Paragraph 3 ("By the time Florey . . ."): Bickel, *Rise,* p. 41.

Paragraph 5 ("Any other newly appointed . . ."): Bickel, *Rise,* p. 24; MacFarlane, *Fleming,* p. 217.

Paragraph 6 ("Fletcher rejected this request, . . ."): Ibid.

Paragraph 7 ("The following year Fletcher . . ."): Bickel, *Rise,* p. 46.

Paragraph 8 ("Once again, it was . . ."): Ibid., p. 47; MacFarlane, *Fleming,* pp. 234–36; Trevor I. Williams, *Howard Florey: Penicillin and After* (Oxford, 1984), p. 45.

Paragraph 9 ("The professorship brought Florey . . ."): MacFarlane, *Fleming,* pp. 234, 244; Abraham, "Memoir," p. 261.

Paragraph 10 ("As an Oxford professor, . . ."): Abraham, "Memoir," p. 262; MacFarlane, *Fleming,* p. 255.

Paragraph 11 ("Chain had arrived in . . ."): Ernst Chain, "Thirty Years of Penicillin Therapy," *Proceedings of the Royal Society,* series B, 179 (1971): 295.

Paragraph 12 ("At Cambridge, Chain had . . ."): Ibid.; *Nature* 281 (25 October 1979): 715–17.

Paragraph 14 ("The investigation of . . ."): RF 401.36.457, Florey to Miller,

18 December 1939; MacFarlane, *Fleming,* p. 257; Abraham, "Memoir," p. 263; Chain, "Therapy," p. 302.

Paragraph 15 ("But for every . . ."): MacFarlane, *Fleming,* p. 244, 257.

Paragraph 18 ("The £2 million would . . ."): Ibid.; MacFarlane, *Fleming,* p. 257.

Paragraph 19 ("The disappointment must have . . ."): Bickel, *Rise,* p. 11; MacFarlane, *Fleming,* p. 257.

Paragraph 20 ("Meanwhile, Florey, having assembled . . ."): RF 401D, Oxford University, Biochemistry.

Paragraph 23 ("There was no way . . ."): *RF Annual Report for 1935,* p. 69; Fosdick, *Story,* p. 127.

Paragraph 24 ("Tizzy, as he was . . ."): *RF Quarterly Bulletin,* April 1929; RF annual reports.

Paragraph 25 ("In his memo to . . ."): RF 401D, Oxford University, Biochemistry.

Paragraph 26 ("Two weeks later Tisdale . . ."): RF Tisdale's log, 15 June 1936.

Paragraph 28 ("In recording his first . . ."): Ibid.

Paragraph 29 ("The laconic Tisdale was . . ."): RF 401.36.457, Paris Office Minutes, 26 June 1936.

Paragraph 31 ("But the distinction was . . ."): RF 401D, Oxford University, Biochemistry, Tisdale to Florey, 6 July 1936.

## Chapter 32: Keeping the Team Together

Paragraph 2 ("As part of their . . ."): RF O'Brien's log, 7 September 1936.

Paragraph 3 ("In all, the Foundation . . ."): *RF Annual Report for 1936,* pp. 3, 181, 348; Tisdale's log, 14 March 1936.

Paragraph 4 ("Unlike the large grant . . ."): *RF Annual Report for 1936,* p. 59, 221–23; RF 900.22.166, pp. 83–90.

Paragraph 5 ("The value of very . . ."): Ibid., p. 83, 89; *RF Annual Report for 1929,* p. 216; *RF Annual Report for 1935,* pp. 179.

Paragraph 6 ("There is no question . . ."): H. W. Florey, E. Chain, N. G. Heatley, M. A. Jennings, A. G. Sanders, E. P. Abraham, and M. E. Florey, *Antibiotics* (Oxford, 1946), p. 637; Chain, "Therapy," p. 296; RF 401.36.457, p. 3.

Paragraph 7 ("Still, it was a . . ."): Chain, "Therapy," pp. 296–97.

Paragraph 8 ("Unlikely as it sounds . . ."): Alexander Fleming, "On the Antibacterial Action of Cultures of a Penicillium with Special Reference to Their Use in the Isolation of *B Influenzae," British Journal of Experimental Pathology* 10 (June 1929): 226–36.

Paragraph 9 ("One might think that . . ."): "Alexander Fleming," *Biographical Memoirs of Fellows of the Royal Society* 2 (1956): 117–24.

Paragraph 10 ("No one including Fleming, . . ."): Ronald Hare, *The Birth of Penicillin* (London, 1970), pp. 94–102; Robert Robinson, "The Early Stages of the Chemical Study of Penicillin," *Memoirs and Proceedings of the Manchester Literary and Philosophical Society* 92 (1950–51): 125; Chain, "Therapy," 302–3.

Paragraph 11 ("By the time Chain . . ."): Wilson, *Search,* pp. 120–32.

Paragraph 12 ("Fleming himself had become . . ."): Ibid., p. 69, 124.

Paragraph 13 ("Chain was originally . . ."): Chain, "Therapy," pp. 298–302; Chain et al., "Penicillin as a Chemotherapeutic Agent," *Lancet,* 24 August 1940, pp. 226–28.

Paragraph 14 ("In the summer of . . ."): Ibid., Bickel, *Rise,* p. 56; E. Chain and H. W. Florey, "Penicillin," *Endeavour,* January 1944, pp. 3–14.

Paragraph 16 ("Florey's predecessor, Professor Georges . . ."): Bickel, *Rise,* p. 68; Wilson, *Search,* p. 156; Chain, "Therapy," p. 297.

Paragraph 17 ("In their later public . . ."): Abraham, *Fleming,* pp. 263, 288; Chain, "Therapy," p. 301.

Paragraph 18 ("In the fall of . . ."): Chain, "Therapy," p. 297; Bickel, *Rise,* p. 55; RF 401.36.457, Florey to Miller, 18 December 1939; Mellanby quoted in O'Brien's log, 6 May 1937.

Paragraph 19 ("Of all people, the . . ."): RF Miller's log, 14 June 1938.

Paragraph 20 ("He had repeatedly sounded . . ."): Ibid.; O'Brien's log, 2 November 1936, 6 May 1937, 14 March 1938, 17 May 1938.

Paragraph 21 ("Undaunted by Mellanby's admonition . . ."): RF Tisdale's log, 14–20 August 1938.

Paragraph 22 ("He explained that key . . ."): Ibid.

Paragraph 23 ("It was shortly after . . ."): Bickel, *Rise,* pp. 56–62; Chain, "Therapy," p. 297.

Paragraph 24 ("Florey has left no . . ."): Bickel, *Rise,* p. 66; MacFarlane, *Fleming,* p. 289.

## Chapter 33: Lebensraum

Paragraph 1 ("Florey's daughter remembers . . ."): Chain, "Therapy," p. 301; Bickel, *Rise,* p. 66.

Paragraph 2 ("This project, which ran . . ."): Florey et al., *Antibiotics,* p. 637; Bickel, *Rise,* p. 70; RF Tisdale's log, 28 November 1938; Miller's log, 24–25 December 1938.

Paragraph 3 ("If by this time . . ."): RF Miller's log, 24–25 November 1938.

Paragraph 4 ("Early in 1939 Florey . . ."): Bickel, *Rise,* p. 69.

Paragraph 8 ("In Oxford this escalation . . ."): Ibid., pp. 69–70; MacFarlane, *Fleming,* p. 298; RF Paris–New York war correspondence file, Paris, no. 35, Miller to Weaver, 6 November 1939.

Paragraph 9 ("Florey was pleased by . . ."): Bickel, *Rise,* p. 71; MacFarlane, *Fleming,* p. 298.

Paragraph 10 ("Through the summer . . ."): RF 401D, Oxford University,

Robinson: Weaver to Fosdick, 13 March 1939; Weaver's log, 4 May 1939; Board minutes, 5 April 1939.

Paragraph 11 ("What Florey certainly did . . ."): *RF Annual Report for 1940*, pp. 12–15.

Paragraph 12 ("Meanwhile, at the request . . ."): RF 900.26.216, "The Foundation and the War," Memo from Appleget to Fosdick, 9 October 1939; 700.183.1314, Gunn to Fosdick, war correspondence file, Paris, no. 8, February 1939.

Paragraph 13 ("All in all, the . . ."): RF 700.183.1314, Fosdick to Gunn, 6 March 1939; annual reports.

Paragraph 14 ("The final dismemberment of . . ."): RF 700.1.4, Gunn to Fosdick, 21 April 1939; Miller, personal communication; RF 700.15.113, O'Brien to Gregg, 11 October 1939.

Paragraph 15 ("Since no one could . . ."): RF 900.26.216, Memo.

Paragraph 16 ("There was only one . . ."): RF 700.183,1319, Miller to Tisdale, 27 April 1939; Miller, "Oral History," pp. 12–14, 21.

Paragraph 17 ("On April 28 Hitler . . ."): RF 700.14.104, Paris Office Staff Conference, 3 May 1939.

Paragraph 18 ("Having completed all . . ."): RF O'Brien's log, 3 June 1939.

Paragraph 19 ("Still, the mere existence . . ."): RF 700.183.1314.

Paragraph 21 ("Ernst Chain, who had . . ."): RF 700.14.104, Paris Office Staff Meeting, 25 August 1939; Wilson, *Search*, p. 156.

Paragraph 22 ("Gunn resisted the suggestion . . ."): RF 700.14.104, Paris Office Staff Meeting, 24 August 1939; Gunn to Fosdick, 26 August 1939.

Paragraph 23 ("By August 29 . . ."): RF Paris Office Staff Meetings, 28–29 August 1939; 700.15.113, O'Brien to Gregg, 11 October 1939.

Paragraph 24 ("On Thursday, August 31 . . ."): RF Miller, *Oral History*, pp. 13–17; RF 700.14.105, Melson to Marshall, 19 September 1939.

## Chapter 34: Embattled

Paragraph 1 ("In the first days . . ."): Eva Figes, *Little Eden: A Child at War* (New York, 1987), p. 51; RF 401D Oxford Biochemistry, Miller to Weaver, 6 and 13 November 1939.

Paragraph 3 ("In this application Florey . . ."): Bickel, *Rise,* pp. 72–73.

Paragraph 4 ("Mellanby replied within . . ."): Ibid.

Paragraph 5 ("This response seemed . . ."): Abraham, "Memoirs," pp. 263–64.

Paragraph 6 ("With Chain's three-year British . . ."): Ibid.; MacFarlane, *Fleming,* p. 298.

Paragraph 7 ("But no sooner had . . ."): RF 401D Oxford University, Biochemistry, Gordon to O'Brien, 29 September 1939.

Paragraph 8 ("Of more immediate concern, . . ."): RF 401D Oxford, Biochemistry, Florey to Miller, 6 October 1939.

Paragraph 9 ("Heatley's unfortunate situation . . ."): RF 700.15.114, Miller to Weaver, 8 September 1939; 700.14.106, Miller to Weaver, 2 October 1939.

Paragraph 10 ("Transatlantic telephone service . . ."): RF 700.14.106, Fosdick to Aldrich, 6 October 1939; 700.15.113, Gregg to O'Brien, 5 October 1939; *Trustees Bulletin,* 1 March 1946.

Paragraph 11 ("A skeleton staff of . . ."): RF 700.14.105, Gunn to Fosdick, 6 September 1939.

Paragraph 12 ("Once the La Baule . . ."): *Trustees Bulletin,* 1 November 1939.

Paragraph 13 ("Even before they received . . ."): RF 700.14.105, Gunn to Fosdick, 5 September 1939.

Paragraph 14 ("Other cases were not . . ."): *Trustees Bulletin,* 1 November 1939; Miller, personal communication.

Paragraph 15 ("In New York, meanwhile, . . ."): RF 700.14.106, Fosdick to Aldrich, 6 October 1939.

Paragraph 16 ("In a letter to . . ."): RF 700.14.105, Fosdick to Gunn, 7 September 1939.

Paragraph 17 ("But, Fosdick went on, . . ."): Ibid.

Paragraph 18 ("To the officers at . . ."): RF 700.15.114, Miller to Weaver, 8 September 1939.

Paragraph 20 ("Even in France the . . ."): RF 700.14.105, Paris Office Staff Conference, 5 and 7 September 1939; Gunn to Fosdick, 18 September 1939; RF 700.14.105, Gunn to Fosdick, 12 September 1939; Melson to Marshall, 19 September 1939.

Paragraph 21 ("The relative tranquillity of . . ."): RF 700.14.106, Fosdick to Gunn, 3 October 1939.

Paragraph 22 ("Gunn, however, had already . . ."): Ibid.; RF 700.14.105, Paris Office telephone log, 22 September 1939.

Paragraph 23 ("Gunn was understandably cautious . . ."): RF Miller's log, 13–15 September 1939; 700.183.1318, Miller to Weaver, 2 October 1939; 700.15.114, Miller to Weaver, 16 November 1939.

Paragraph 24 ("O'Brien, meanwhile, had gone . . ."): RF 700.15.114, O'Brien to Gunn, 4 October 1939.

Paragraph 25 ("Not surprisingly, both Miller . . ."): RF 700.15.113, O'Brien to Gregg, 11 October 1939.

Paragraph 26 ("During the last week . . ."): RF 700.14.105, O'Brien to Gunn, 16 September 1939.

Paragraph 27 ("O'Brien did not miss . . ."): Ibid.

## Chapter 35: Pure Science

Paragraph 1 ("After the fall of . . ."): RF 700.14.106, Paris Office telephone log, 6 October 1939, Gunn to Fosdick, 11 October 1939, Gunn to Appleget, 25 October 1939.

Paragraph 2 ("The French, like the . . ."): *RF Annual Report for 1919,* p. 11.

Paragraph 4 ("Miller assured Ephrussi that . . ."): RF 700.15.114, Miller to Weaver, 17 October 1939; 900.26.216, Board minutes 5–7 December 1939.

Paragraph 5 ("Meanwhile, at the rue . . ."): RF 700.14.105, Paris Office telephone log, 23 September 1939; 700.14.106, Gunn to Appleget, 24 October 1939.

Paragraph 6 ("In La Baule the . . ."): RF 700.14.105, Paris Office telephone log, 24 September 1939.

Paragraph 7 ("While all the circuit . . ."): RF 700.183.1314, Letort to Gunn, 18 October 1939; 700.15.114, O'Brien to Weaver, 21 October 1939.

Paragraph 8 ("Miller reminded Weaver that . . ."): RF 700.15.114, Miller to Weaver, 17 October 1939.

Paragraph 9 ("It was not until . . ."): RF 401D Oxford, Biochemistry, Miller to Weaver, 6, 16, and 24 November 1939; RF 700.15.114, 24 November 1939.

Paragraph 10 ("At the same time, . . ."): RF 401D Oxford, Biochemistry, Miller to Weaver, 6 November 1939.

Paragraph 11 ("On November 1 Miller . . ."): Ibid.

Paragraph 12 ("But Florey could hardly . . ."): Ibid.

Paragraph 13 ("In a detailed memo . . ."): Ibid.

Paragraph 15 ("It was standard procedure . . ."): RF 177.14.106, Miller to Weaver, 2 October 1939; 401D Oxford, Biochemistry, Miller to Weaver, 6 November, 1939; 700.15.114, Miller to Weaver, 24 November 1929.

Paragraph 16 ("It is significant that . . ."): RF 401D Oxford, Biochemistry, Miller to Weaver, 6 November 1939.

Paragraph 18 ("In fact, there is . . ."): Bickel, *Rise,* pp. 74–75; Chain, "Therapy," p. 298.

Paragraph 19 ("On November 29 Florey . . ."): RF 401.36.457, Florey to Miller, 20 November 1939.

Paragraph 20 ("But he was quick . . ."): Ibid.

## Chapter 36: Foundation at War

Paragraph 1 ("Having resolved to 'try . . .'"): Chain "Therapy," pp. 297–98; Bickel, *Rise,* p. 75.

Paragraph 2 ("The first section of . . ."): Wilson, *Search,* p. 152; RF 401.36.457, "Application to the Rockefeller Foundation for a Research Grant," 20 November 1939.

Paragraph 3 ("The second section, entitled . . ."): Chain et al., "Agent"; Chain and Florey, "Penicillin."

Paragraph 4 ("The goal of the . . ."): RF 401.36.457, "Application for a Research Grant."

Paragraph 5 ("In the third and . . ."): Ibid.

Paragraph 6 ("In addition, because of . . ."): Ibid.

Paragraph 7 ("By the time this . . ."): RF 700.15.114, Miller to Weaver, 24 and 28 November 1939; 700.15.107, Appleget to Gunn, 6 November 1939.

Paragraph 8 ("On December 8 Miller . . ."): RF 700.15.114, Weaver to Miller, 30 November 1939; Miller to Weaver, 9 December 1939.

Paragraph 9 ("From the context it . . ."): RF Board minutes, 6 December 1939.

Paragraph 10 ("In any case, despite . . ."): RF 401.36.457, Miller to Weaver, 9 December 1939.

Paragraph 11 ("Miller even tried to . . ."): RF 401.36.457, Miller to Florey, 11 December 1939.

Paragraph 12 ("Miller sent a copy . . ."): Ibid.

Paragraph 13 ("Then Miller added a . . ."): Ibid.

Paragraph 14 ("Finally, Miller reminded Weaver . . ."): Ibid.

Paragraph 15 ("The trustees had, in . . ."): *RF Annual Report for 1939;* RF 900.26.216, Board minutes, 5–6 December 1939; 700.14.106, New York Staff

Conference, 24 October 1939; 700.15.107, New York Staff Conference, 16 November 1939; 700.15.107, Gunn to Fosdick, 11 November 1939.

Paragraph 16 ("The Board went on . . ."): RF 900.26.216, Board minutes, 5–6 December 1939; RF 700.203.1435, Frank Hanson to Miller, 4 January 1940.

Paragraph 17 ("In truth, despite attempts . . ."): RF 700.14.106, Gunn to Fosdick, 13 October 1939.

Paragraph 18 ("How to draw this . . ."): RF 700.15.107, Fosdick to Gunn, 30 November 1939.

Paragraph 19 ("Two weeks later he . . ."): RF 700.15.107, Fosdick to Gunn, 13 December 1939.

Paragraph 20 ("When Fosdick spoke of . . ."): *RF Annual Report for 1939,* pp. 9–18.

Paragraph 21 ("Speaking directly to an . . ."): Ibid.

Paragraph 22 ("He noted that the . . ."): Ibid.

Paragraph 23 ("When he wrote that . . ."): RF 401.36.457, Florey to Miller, 18 December 1939.

Paragraph 24 ("By the time this . . ."): RF 401.36.457, Miller to Weaver, 31 December 1939.

Paragraph 25 ("In case this plan . . ."): Ibid.

Paragraph 26 ("Gunn's trip had come . . ."): RF 700.15.107, A. J. Warren to Wilbur A. Sawyer, 15 December 1939; 700.15.107, Fosdick to Gunn, 30 November 1939.

## Chapter 37: Legends and Luck

Paragraph 1 ("Gunn and his wife . . ."): RF 700.15.107, Gunn to Fosdick, 19 and 22 December 1939.

Paragraph 2 ("Miller's attempts to expedite . . ."): RF 401.36.457, Miller to Florey, 3 January 1939.

Paragraph 3 ("On his return to . . ."): RF 401.36.457, Miller to Weaver, 9 January 1940; 700.203, 1435, Hanson to Miller, 4 January 1940.

Paragraph 4 ("Informed of this turn . . ."): RF 401.36.457, Florey to Miller, 9 February 1939.

Paragraph 5 ("On the morning of . . ."): RF 700.15.108, Weaver cable to Miller, 5 February 1940; 700.203.1435, Miller to Weaver, 21 February 1940; 700.15.109, Staff memo, 29 June 1941.

Paragraph 6 ("Before he left, however . . ."): RF 401D, Oxford, Biochemistry, Miller to Florey, 19 February 1940; 401.36.457, La Baule Grant-in-Aid Action No. 3, 21 February 1940.

Paragraph 7 ("As might be expected . . ."): Chain, "Therapy," p. 298–99.

Paragraph 8 ("No one questions the . . ."): Ibid.

Paragraph 9 ("Even $5,000 for one . . ."): Ibid., p. 294.

Paragraph 10 ("In a biography of . . ."): Bickel, *Rise,* pp. 77, 88–89.

Paragraph 11 ("This exaggeration of the . . ."): *Nature* 281 (25 October 1979): 715–17.

Paragraph 14 ("Neither man was known . . ."): Bickel, *Rise,* p. 49; RF Gerard Pomerat's log, 19 May 1948.

Paragraph 15 ("This was, after all, . . ."): Bickel, *Rise,* p. 23.

Paragraph 16 ("In the summer of . . ."): RF Miller's log, 7 August 1945; 401.36.464, Miller to Florey, 17 September 1945.

Paragraph 17 ("Florey's reply was swift . . ."): RF 401D, Oxford, Biochemistry, Florey to Miller, 26 September 1945.

Paragraph 20 ("With Heatley growing the . . ."): Chain, "Therapy," p. 302; Abraham, "Memoir," p. 264.

Paragraph 21 ("Working with a new . . ."): Robinson, "Stages," p. 125; Chain, "Therapy," p. 302; Chain and Florey, "Penicillin."

Paragraph 22 ("There was a more . . ."): Chain, "Therapy," p. 301.

Paragraph 23 ("Over the next few . . ."): Ibid., p. 203; Chain et al., "Agent."

Paragraph 24 ("The production of this . . ."): Chain and Florey, "Penicillin."

Paragraph 25 ("Chain was so elated . . ."): Chain, "Therapy," p. 304; Williams, *Florey*, pp. 102–7.

Paragraph 26 ("Subsequent tests on rats . . ."): Ibid.

## Chapter 38: Evacuation

Paragraph 2 ("In La Baule the . . ."): RF 700.203.1435, Gunn to Fosdick, 20 February 21 and 26 March, 25 April 1940; 700.1.4, Gunn to Fosdick, 15 March 1940; Fosdick to Gunn, 29 March 1940; Gunn to Fosdick, 5 April 1940; 700.15.108, Bakeman to Appleget, 19 January and 15 April 1940; Fosdick to Gunn, 12 March 1940; O'Brien to Gregg, 23 January 1940; 700.15.109, Staff memo, 29 June 1940.

Paragraph 4 ("Under continuing pressure from . . ."): RF 700.15.108, O'Brien to Gregg, 23 January 1940.

Paragraph 5 ("The 'chasm' between the . . ."): RF Appleget's log, 15, 17, 22, and 26 January 1940; 700.15.108. Fosdick to Gunn, 12 March 1940.

Paragraph 6 ("Meanwhile, Fosdick's own doubts . . ."): RF 700.15.108, Gregg to Lambert, 5 February 1940.

Paragraph 7 ("Gunn's letters present an . . ."): RF 700.203.1435, Gunn to Fosdick, 21 and 26 March 1940; 700.1.4, Gunn to Fosdick, 5 April 1940; 700.15.108, Gunn to Fosdick, 6 May 1940.

Paragraph 10 ("On May 25, a . . ."): Bickel, *Rise*, p. 95.

Paragraph 11 ("Norman Heatley later recalled . . ."): Wilson, *Search*, p. 161; Abraham, "Memoir," p. 265.

Paragraph 13 ("From the moment the . . ."): RF 700.203.1435, Gunn to Fosdick, 10 May 1940; 700.15.109, Gunn to Fosdick, 14 May 1940; Fosdick cables to Gunn, 17 and 21 May 1940; 700.15.108, New York Staff Conference, 21 May 1940.

Paragraph 14 ("Yet even now Gunn . . ."): RF 700.203.1435, Gunn to Fosdick, 25 May 940.

Paragraph 15 ("Perhaps O'Brien, Miller, and . . ."): RF 700.15.109, Gunn to Fosdick, 25 May 1940.

Paragraph 16 ("By the end of . . ."): RF 700.15.109, Gunn to Fosdick, 29 May 1940.

Paragraph 17 ("O'Brien sailed on June . . ."): Miller, "Oral History," pp. 18–19.

Paragraph 18 ("The staff members who . . ."): RF Makinsky's log, 9–10 May 1940.

Paragraph 19 ("In the face of . . ."): Ibid., 13 June 1940; RF 700.203.1435, Letort to Gunn, 6 August 1940.

Paragraph 20 ("Paris fell on June . . ."): RF Makinsky's log, 10–20 June 1940.

Paragraph 21 ("Traveling without lights, Makinsky's . . ."): Ibid., 21–24 June 1940.

Paragraph 22 ("After being assured that . . ."): Ibid., 24–25 June 1940.

## Chapter 39: Almost Miraculous

Paragraph 1 ("In Oxford there was . . ."): Bickel, *Rise,* p. 103.

Paragraph 2 ("The large-scale animal trials . . ."): Chain et al., "Agent"; Bickel, *Rise,* p. 105; Wilson, *Search,* p. 164.

Paragraph 4 ("The paper announced the . . ."): Chain et al., "Agent"; Bickel, *Rise,* p. 105.

Paragraphs 5 and 6 ("In fact, his sense . . ." and "Mellanby objected to both . . ."): Bickel, *Rise,* p. 108.

Paragraphs 7 and 8 ("In reply, Florey noted . . ." and "Denying any intention of . . ."): Ibid., pp. 108–9.

Paragraph 10 ("At the beginning of . . ."): Abraham, "Memoirs," p. 265; Florey et al., *Antibiotics,* p. 639; Weaver, "Oral History," p. 591.

Paragraph 14 ("Oxford was spared direct . . ."): Bickel, *Rise,* p. 117.

Paragraph 15 ("Hoping to save space . . ."): Ibid., p. 101; Florey et al., *Antibiotics,* p. 603; Wilson, *Search,* p. 179.

Paragraph 16 ("“Fortunately, Florey knew a . . ."): Bickel, *Rise,* pp. 113–17; Wilson, *Search,* pp. 179–80.

Paragraph 17 ("Since it turned out . . ."): MacFarlane, *Fleming,* p. 325.

Paragraph 18 ("By the end of . . ."): Abraham, "Memoir," p. 265; Florey et al., *Antibiotics,* p. 645.

Paragraph 19 ("Another patient at the . . ."): Bickel, *Rise,* pp. 122–23; Wilson, *Search,* pp. 171–72.

Paragraph 20 ("Almost immediately his condition . . ."): Ibid.

Paragraph 21 ("The outcome was disheartening . . ."): Wilson, *Search,* p. 172.

Paragraph 22 ("Once again the collaborators . . ."): E. P. Abraham, E. Chain, C. M. Fletcher, H. W. Florey, A. D. Gardner, N. G. Heatley, and M.A. Jennings, "Further Observations on Penicillin," *Lancet,* 16 August 1941, pp. 177–88.

Paragraph 23 ("The British censors' lack . . ."): Bickel, *Rise,* pp. 132, 295.

Paragraph 24 ("A different kind of . . ."): Ibid., pp. 133–41; Abraham, "Memoir," p. 269.

## Chapter 40: Good as His Word

Paragraph 1 ("While Edward Mellanby seemed . . ."): Bickel, *Rise,* p. 154.

Paragraph 2 ("Even before the first . . ."): Ibid., p. 136.

Paragraph 3 ("The RF had already . . ."): RF 401.36.458, Florey to Miller, 20 December 1940.

Paragraph 4 ("These words must have . . ."): RF 700.15.109, New York staff memo, April 1940.

Paragraph 5 ("The Foundation also remained . . ."): *RF Annual Report for 1940,* pp. 8–12.

Paragraph 6 ("The Foundation's original program . . ."): Ibid., pp. 12–15.

Paragraph 7 ("The principal escape route . . ."): RF *Trustees Bulletin,* 1 March 1941; Makinsky's log, 19–22 May 1941.

Paragraph 8 ("He knew at least . . ."): RF Makinsky's log, 8 December 1940, 6 June 1941.

Paragraph 9 ("Clearly, Fosdick had been . . ."): RF Executive Committee minutes, 8 July 1940; 700.15.109, Fosdick to John D. Rockefeller, Jr., 24 July 1940.

Paragraph 10 ("When a three-man survey . . "): *RF Annual Report for 1940,* pp. 18–20; O'Brien's logs, August–October 1940.

Paragraph 11 ("Once the commission was . . ."): Ibid.; Weaver, "Oral History," pp. 562–92; Wilson, *Search,* pp. 187–89.

Paragraph 12 ("Weaver's boss at the . . ."): Weaver, "Oral History," pp. 559–61, 581–84.

Paragraph 13 ("During his stay in . . ."): Ibid., pp. 588–92.

Paragraph 14 ("Weaver had to spend . . ."): Ibid.; RF Weaver's log, 14 April 1941.

Paragraph 15 ("The Oxford pathologist briefly . . ."): Ibid.

Paragraph 16 ("When he got back . . ."): RF 401D, Oxford University, Florey, Grant-in-Aid Action, 19 June 1941; O'Brien to Weaver, 3 and 14 June 1941.

Paragraph 17 ("It took a coordinated . . ."): RF 401D, Oxford University, Florey: O'Brien to H. E. Whittingham, 11 June 1941; Whittingham to O'Brien, 12 June 1941; Florey to O'Brien, 14 June 1941; Frank B. Hanson's log, 17 June 1941; O'Brien to Makinsky, 14 June 1941.

Paragraph 18 ("On the morning of . . ."): Bickel, *Rise,* pp. 137–38; Wilson, *Search,* pp. 187–90; RF Makinsky's log, 27 June 1941.

Paragraph 19 ("After a year and . . ."): Ibid.; RF 401D, Oxford, Florey: Weaver to O'Brien, 8 July 1941; Heatley to Hanson, 20 September 1942.

Paragraph 20 ("Florey spent July 4 . . ."): RF 401.36.461, Florey to Weaver, 14 July and 9 September 1941; Hanson's log, 10 September 1941; *Trustees Bulletin,* 1 October 1943.

Paragraph 21 ("Florey's efforts to enlist . . ."): RF 401.36.461, Florey to Weaver, 9 September 1941; Florey et al., *Antibiotics,* pp. 650–54. Chain, "Therapy," p. 306; Wilson, *Search,* pp. 196–98.

Paragraph 22 ("By a coincidence that . . ."): RF 401.36.461, Hanson to Richards, 3 May 1943; A. N. Richards, "Penicillin," *Journal of the American Medical Association,* 22 May 1943, pp. 235–36.

Paragraph 23 ("This committee wielded great . . ."): Ibid.

Paragraph 24 ("But all of Florey's . . ."): Bickel, *Rise,* p. 153; Abraham, "Memoir," p. 266.

Paragraph 25 ("The results exceeded everyone's . . ."): Bickel, *Rise,* p. 186–89.

## Chapter 41: Miracles

Paragraph 1 ("As it happened, the . . ."): Bickel, *Rise,* p. 157.

Paragraph 2 ("Back in England, Florey . . ."): Ibid., p. 168; Wilson, *Search,* p. 212.

Paragraph 3 ("Four days later Florey . . ."): Bickel, *Rise,* p. 109; Wilson, *Search,* p. 234.

Paragraph 4 ("What rekindled Fleming's interest . . ."): Wilson, *Search,* pp. 213–16.

Paragraph 5 ("In response to Fleming's . . ."): Ibid.

Paragraph 6 ("One result of this . . ."): Bickel, *Rise,* p. 170; Wilson, *Search,* pp. 221–23.

Paragraph 7 ("But hearsay evidence about . . ."): Florey et al., *Antibiotics,* p. 659; Bickel, *Rise,* pp. 200–201.

Paragraph 8 ("This procedure ran directly . . ."): Ibid.

Paragraph 9 ("By the time of . . ."): Florey et al., *Antibiotics,* p. 661; Bickel, *Rise,* p. 187.

Paragraph 10 ("With the flow of . . ."): Chain, "Therapy," p. 306; Robinson, "Stages," pp. 127, 136.

Paragraph 11 ("As government and industry . . ."): RF 401.36.460, Board minutes, 25 September 1953.

Paragraph 12 ("The postwar files of . . ."): RF 401.36.464, Miller to Florey, 10 January, 14 March 1946; Florey to Miller, 12 and 27 February, 7 March 1946; Florey to Pomerat, 5 November 1946; Pomerat's log, 29 March 1949.

Paragraph 13 ("Of course, once Florey's . . ."): RF Pomerat's log, 20 May 1948, 19 May 1954; Bickel, *Rise,* pp. 248–51; Abraham, "Memoir," pp. 277–78, 290.

Paragraph 15 ("He had been a . . ."): Abraham, "Memoir," pp. 272, 278.

Paragraph 16 ("But most of his . . ."): Ibid., p. 283.

Paragraph 17 ("When the Royal Society . . ."): Bickel, *Rise,* pp. 291–92.

Paragraph 18 ("Three months later, on . . ."): Ibid., p. 292.

Paragraph 20 ("We have seen how . . ."): Wilson, *Search,* pp. 230–31.

Paragraph 21 ("The omission of Florey's . . ."): Ibid., p. 232; Abraham, "Memoir," p. 267.

Paragraph 22 ("Scenting a good story . . ."): MacFarlane, *Fleming,* p. 354; Abraham, "Memoir," pp. 268, 287; Bickel, *Rise,* p. 244; Wilson, *Search,* p. 234.

Paragraph 23 ("In failing to stake . . ."): *RF Annual Report for 1943.*

Paragraph 25 ("A columnist for the . . ."): Bickel, *Rise,* p. 237.

Paragraph 26 ("More to the point . . ."): Florey et al., *Antibiotics,* p. 670.

Paragraph 27 ("Confusion over the Rockefeller . . ."): Wilson, *Search,* pp. 193–94.

Paragraph 28 ("The propriety of these . . ."): Ibid.; RF 401.36.469, Dean Rusk to Weaver, 25 May 1955.

Paragraph 29 ("I do not know . . ."): RF 401.36.462, Fulton to Gregg, 4 October 1944.

## Chapter 42: A Puzzlement

Paragraph 1 ("Historians of science take . . ."): Robert Olby, *The Path to the Double Helix* (London, 1974); Judson, *Eighth Day,* pp. 45, 72; Kohler, "Management."

Paragraph 2 ("Scholarly opinion on Weaver . . ."): Pnina Abir-Am, "The Discourse of Physical Power and Biological Knowledge in the 1930s: A Reappraisal of the Rockefeller Foundation's 'Policy' in Molecular Biology," *Social Studies of Science* 12 (1982): 341–82.

Paragraph 3 ("Dr. Abir-Am's own ideological . . ."): Ibid., pp. 350–52.

Paragraph 4 ("In her view, not . . ."): Ibid., pp. 353–61.

Paragraph 5 ("She takes note of . . ."): Ibid.

Paragraph 7 ("In a similar vein . . ."): Ibid., pp. 361–67.

Paragraph 9 ("On October 20, 1937 . . ."): RF 401D, Strangeways Laboratory, Weaver to Tisdale, 22 October 1937.

Paragraph 10 ("Two researchers at Cambridge . . ."): Ibid., p. 2; RF 401D, Strangeways Laboratory, Tisdale to Weaver, 12 July 1937.

Paragraph 11 ("After canvassing the applicants' . . ."): Ibid.

Paragraph 12 ("This was the kind . . ."): RF 401D, Weaver to Tisdale, 22 October 1937.

Paragraph 13 ("As he admitted in . . ."): Ibid.

Paragraph 15 ("Similarly, Needham had pioneered . . ."): RF 401D,

Strangeways Laboratory, Streeter to Weaver, 14 September 1937; Tisdale to Weaver, 12 July 1937.

Paragraph 16 ("What gave Weaver pause . . ."): RF 401D, Strangeways Laboratory, Tisdale to Weaver, 8 January 1937.

Paragraph 17 ("In addition, Weaver and . . ."): RF 401D, Strangeways Laboratory, "Memorandum to the Vice-Chancellor and the Council of the Senate on the Question of the Extension of the Biochemical Laboratory, by Joseph Needham, Long Vacation 1936."

Paragraph 18 ("According to Needham's proposal . . ."): Ibid., p. 6; Weaver's log, 13 June 1936; Tisdale's log, 1 July 1934, 14 June 1936; RF 401D, Strangeways Laboratory, Weaver to Tisdale, 14 December 1936; Tisdale to Weaver, 8 January 1937.

Paragraph 19 ("To complicate matters further . . ."): RF 401D, Tisdale to Weaver, 8 January 1937.

Paragraph 21 ("From the tone of . . ."): RF 915.2.10 "Report of the Committee of Review, Appraisal and Advice, Division of Natural Sciences," May 1939.

Paragraph 23 ("In accord with criteria . . ."): RF 900.22.168, Agenda; *A History of Embryology: The Eighth Symposium of the British Society for Developmental Biology*, ed. T. J. Horder, J. A. Witkowski, and C. C. Wylie (Cambridge, 1985), pp. i–xvi.

Paragraph 24 ("The continuing challenge of . . ."): Ibid.

Paragraph 25 ("Dr. Abir-Am suggests that . . ."): Ibid.

Paragraph 26 ("Even in the 1930s . . ."): Ibid., pp. 181–226.

Paragraph 27 ("Faced with too many . . ."): Ibid.

Paragraph 29 ("Hindsight confirms that the . . ."): Ibid.

## Chapter 43: Tricks of the Trade

Paragraph 1 ("Looking back on his . . ."): Warren Weaver, "Oral History," pp. 92–95.

Paragraph 2 ("A summer job working . . ."): Ibid.

Paragraph 3 ("Despite his passion for . . ."): Weaver, *Lady Luck.*

Paragraph 4 ("When Pnina Abir-Am suggests . . ."): Weaver, *Philanthropic Foundations,* p. 128; Weaver "Oral History," pp. 451–52.

Paragraph 5 ("Weaver was frankly elitist . . ."): Ibid., p. 127.

Paragraph 6 ("The ideal philanthropoid, Weaver . . ."): Weaver, "Oral History," pp. 451–54.

Paragraph 7 ("But philanthropoids can and . . ."): Weaver, *Philanthropic Foundations,* p. 128.

Paragraph 8 ("For a man who . . ."): Weaver, "Oral History," pp. 430–35.

Paragraph 9 ("Even less trustworthy, in . . ."): Ibid.

Paragraph 10 ("Given the unreliability of . . ."): Ibid., pp. 444, 452–54.

Paragraph 11 ("While foundation officers had . . ."): Ibid., pp. 444, 486–87, 451–52.

Paragraph 12 ("A less subtle but . . ."): Ibid., pp. 481–83.

Paragraph 13 ("Weaver especially liked the . . ."): Ibid., p. 482.

## Chapter 44: More Tricks

Paragraph 1 ("In Weaver's view, nothing . . ."): Weaver, "Oral History," p. 438.

Paragraph 2 ("The circuit riders spent . . ."): Ibid., p. 444.

Paragraph 3 ("In the words of . . ."): Miller, personal communication.

Paragraph 5 ("Even in Europe, the . . ."): Ibid.

Paragraph 6 ("Even when the circuit . . ."): Weaver, "Oral History," pp. 430–48.

Paragraph 7 ("Weaver was not a . . ."): Ibid., p. 443.

Paragraph 8 ("Yet Weaver always warned . . ."): Ibid.; quote from Dr. Donald Brown in *Cell,* 14 February 1986, pp. 373–74.

Paragraph 9 ("But there is a . . ."): Weaver, "Oral History," p. 535.

Paragraph 11 ("Kasimir Fajans, the Polish-born . . ."): RF Trustees Action, 23 December 1935; Weaver, "Oral History," pp. 514–15.

Paragraph 13 ("The flaw in this . . ."): RF 717, KWI Physics and Cell Physiology.

Paragraph 14 ("So Foundation officials were . . ."): RF 717, KWI Physics and Cell Physiology, Planck to Mason (RF translation), 3 July 1934; Planck to Weaver (RF translation), 6 July 1934.

Paragraph 16 ("When Planck wrote to . . ."): RF 717, KWI Physics and Cell Physiology, Planck to Tisdale (RF translation), 29 August 1934; Tisdale to Weaver, 4 September 1934; Mason to Planck, 1 November 1934; Planck to Mason (RF translation), 25 February 1935.

Paragraph 17 ("Instructed by the RF's . . ."): RF Appleget's log, 4 January 1935; RF 717, KWI Physics and Cell Physiology, Mason to Planck, 15 March 1935.

Paragraph 19 ("The *Times* article prompted . . ."): RF 717, KWI Physics and Cell Physiology, Frankfurter to Fosdick, 24 November 1936.

Paragraph 20 ("It was not until . . ."): Review by Hugh Trevor-Roper of *Tödliche Wissenschaft: Die Aussonderung von Juden, Zigeunern und Geisteskranken, 1933–1945* (Deadly Science: The Elimination of Jews, Gypsies, and the Mentally Ill), by Benno Müller-Hill (Hamburg, 1984), in *Nature* 313 (31 January 1985): 407–8; see also Robert Jay Lifton, *The Nazi Doctors: Medical Killings and the Psychology of Genocide* (New York, 1986).

Paragraph 21 ("Dr. Fischer, a member . . ."): Ibid.

## Chapter 45: New Directions

Paragraph 2 ("The system works essentially . . ."): *Science* 238 (9 October 1987): 151; Philip Handler, "Basic Research in the United States," *Science* 204 (4 May 1979): 474–9; *Medical World News,* 25 August 1986, pp. 42–46.

Paragraph 4 ("Just how valid is . . ."): *Science* 204 (8 June 1979): 1064–65;

*Science* 222 (9 December 1983): editorial page; *Science* 233 (11 July 1986): 145–46.

Paragraph 5 ("Partly in response to . . ."): *Nature* 279 (14 June 1979): 575–76; *Science News,* 21 November 1981; p. 327; *Science* 214 (20 November 1981): 881; *Science* 215 (1 January 1982): editorial page; *Science* 215 (22 January 1982): 344–48.

Paragraph 7 ("No one has suggested . . ."): Milton Lomask, *A Minor Miracle: An Informal History of the National Science Foundation* (Washington, D.C., 1976), p. 74.

Paragraph 10 ("Since Spoehr himself was . . ."): Lubert Stryer, *Biochemistry* (New York and San Francisco, 1981), pp. 431–53.

Paragraph 11 ("But there is another . . ."): Fosdick, *Story,* pp. 184–91; E. C. Stakman, Richard Bradford, and Paul L. Mangelsdorf, *Campaigns against Hunger* (Cambridge, Mass., 1967), pp. 4–25.

Paragraph 12 ("Wallace had just returned . . ."): Ibid.

Paragraph 13 ("Fosdick mentioned this to . . ."): Ibid.; also Weaver, "Oral History," pp. 653–57; Harry Miller, personal communication; Miller, "Oral History," p. 85.

Paragraph 15 ("In Mexico food production . . ."): Ibid.; also see Stanley Johnson, *The Green Revolution* (New York, 1972); J. George Harrar, *Strategy to Conquer Hunger* (New York, 1967); *Science* 215 (26 February 1982): 1043; Stakman et al., *Revolution;* Harold M. Keele and Joseph L. Kiger, eds., *Foundations* (Westport, Conn., and London, 1984), p. 386; for Lipton quote see Edward C. Wolf, *Beyond the Green Revolution: New Approaches for Third World Agriculture,* WorldWatch Paper 73 (1986); Philip H. Abelson, *Science* 236 (3 April 1987): 9.

Paragraph 16 ("With the establishment of . . ."): Weaver, "Oral History," pp. 653–66; Weaver, *Scene of Change,* p. 108; Miller, "Oral History," pp. 43–44; Miller, personal communication; Keele and Kiger, *Foundations.*

Paragraph 17 ("In the United States . . ."): *New England Journal of Medicine* 304 (26 February 1981): 509–17; Lomask, *Miracle.*

Paragraph 19 ("We have seen how . . ."): RF Pomerat's logs, 1946–60; Pomerat, personal communication, June 1979; *Nature* 319 (27 February 1986):

729–32; *Nature* 320 (17 April 1986): 656; for early days of CNRS, see Spencer Weart, *Scientists in Power* (Cambridge, Mass., 1979).

## Chapter 46: The Art of Science Support

Paragraph 1 ("It is a curiosity . . ."): George Gale, "Science and the Philosophers," *Nature* 312 (6 December 1984): 491–95; *Science* 212 (22 May 1981): 873; *Science* 226 (7 December 1984): 1185–86; Steven Weinberg, "Newtonianism, Reductionism, and the Art of Congressional Testimony," *Nature* 330 (3 December 1987): 433.

Paragraph 2 ("Most working researchers simply . . ."): Harriet Zuckerman and Robert K. Merton, "Patterns of Evaluation in Science: Institutionalization, Structure, and Functions of the Referee System," *Minerva* 9 (January 1971): 66–100.

Paragraph 3 ("In the absence of . . ."): Robert Kanigel, "The Mentor Chain," *Johns Hopkins Magazine*, December 1981, i–viii.

Paragraph 4 ("This is presumably why . . ."): Weaver, "Oral History," pp. 433, 528–29.

Paragraph 5 ("Weaver saw science as . . ."): Weaver, *Scene of Change*, pp. 182, 190.

Paragraph 7 ("The researchers who conducted . . ."): *Science* 225 (22 January 1984): 347.

Paragraph 8 ("As it happens, there . . ."): *Science* 218 (24 December 1982): 1278–80; *Nature* 296 (4 March 1982): 1–2; *Science* 231 (17 January 1986): 242–46; *Science* 232 (6 June 1986): 1191; *Medical World News*, 25 August 1986, pp. 42–54.

Paragraph 10 ("Warren Weaver often cited . . ."): Weaver, *Philanthropic Foundations*, p. 168.

Paragraph 11 ("By the mid-1960s the . . ."): Philip Handler, "Research," *Science* 204 (4 May 1979); Keele and Kiger, *Foundations*, p. 372; for January 1987 Ford endowment figure, personal communication; for HHMI history, Ted Hearne, personal communication; *Medical World News*, 25 August 1986, pp. 42–54; *Medical Economics*, 6 October 1986, pp. 91–102; Joel Brinkley,

"The Richest Foundation," *New York Times Magazine,* 30 March 1986, pp. 32–39; *Science* 237 (18 September 1987): 1406–7.

Paragraph 12 ("For truly vast undertakings . . ."): *RF Annual Report for 1985,* pp. 16–17.

Paragraph 13 ("The Howard Hughes Medical . . ."): *Science* 237 (18 September 1987), 1406–7; *Medical World News,* 25 August 1986, 42–54; *Medical Economics,* 6 October 1986, 91–102; Brinkley, "Richest."

Paragraph 14 ("Technically, the Hughes Institute . . ."): Ibid.; *Nature* 323 (30 October 1986): 749; personal communication, Robert Potter, director of communications, HHMI.

Paragraph 15 ("At the Rockefeller unit, . . ."): Ibid.

Paragraph 16 ("In addition to its . . ."): *Science* 238 (9 October 1987): 150; Potter, personal communication.

Paragraph 17 ("The Rockefeller Foundation, with . . ."): *RF Annual Report for 1985;* "The Rockefeller Foundation in the Developing World" (Special report, 1986).

Paragraph 18 ("Another program, known as . . ."): Personal communication, Scott B. Halstead, acting director, Division of Health Sciences, RF.

## Chapter 47: Experiments

Paragraph 1 ("Summing up his . . ."): Weaver, "Oral History," p. 265 (quotation edited to make pronouns agree).

Paragraph 4 ("The model proposed by . . ."): *Nature* 316 (29 August 1986): 754; *Nature* 324 (20 November 1986): 221.

Paragraph 5 ("According to Horrobin, if . . ."): Ibid.

Paragraph 6 ("Skeptics have argued that . . ."): Ibid.

Paragraph 8 ("Historians confirm that . . ."): Crosland, "Prizes," pp. 356–61.

Paragraph 10 ("If Weaver learned one . . ."): Weaver, "Oral History," pp. 510–11.

# Index

Italicized page numbers indicate photographs